Forestry in a Global Context

For Elizabeth

Forestry in a Global Context

Roger Sands

Professor and Head
New Zealand School of Forestry
University of Canterbury
Christchurch
New Zealand

CABI Publishing

CABI is a trading name of CAB International

CABI Head Office
Nosworthy Way
Wallingford
Oxfordshire OX10 8DE
UK

CABI North American Office
875 Massachusetts Avenue
7th Floor
Cambridge, MA 02139
USA

Tel: +44 (0)1491 832111
Fax: +44 (0)1491 833508
Email: cabi@cabi.org
Web site: www.cabi.org

Tel: +1 617 395 4056
Fax: +1 617 354 6875
Email: cabi-nao@cabi.org

A catalogue record for this book is available from the British Library, London, UK.
A catalogue record for this book is available from the Library of Congress, Washington DC, USA.

Library of Congress Cataloging-in-Publication Data
Sands, Roger
    Forestry in a global context / Roger Sands.
        p.   cm
    Includes bibliographical references and index.
    ISBN 0-85199-089-4 (alk. paper)
    1. Forests and forestry. I. Title.
SD131.S26 2005
634.9—dc22

2005008078

ISBN-13: 978-0-85199-089-7
ISBN-10: 0-85199-089-4

First published 2005
Reprinted 2007, 2008

Printed and bound in the UK from copy supplied by the author by Cromwell Press, Trowbridge.

Contents

Chapter 3   The Environmental Value of Forests   67

Chapter 4   Forest Products   87

Chapter 5   Deforestation and Forest Degradation in the Tropics   121

Preface

This book was written to introduce forestry students and others in allied natural resource management disciplines to the world of forestry, from a historical and international perspective. The profession of forestry has its roots in the natural sciences. Only recently has there been general acceptance that social aspects are at least as important, if not more important. The management of the world's forests is a very controversial topic, capable of arousing great passion and creating considerable conflict. Major issues have been (and remain) deforestation in the tropical world and the logging of old growth forests in the temperate world. The current catchphrase is 'sustainable forest management' which is seen by many to be the hope for the future but is treated with suspicion by others who see it as an excuse to disguise poor practice. This book is not about techniques and tools, of which there are many excellent texts. Rather the book is about the relationship of people to forests.

In this book I have attempted to present and analyse conflicting arguments. While I have done my best to be objective, it is inevitable I will carry with me some of the preconceptions and values associated with 40 years exposure to a research and academic environment in traditional forestry. I have argued for a balanced approach. However, my point of balance may well differ from other readers of this book and I claim no particular advantage. The advice I give my students is: always consider and evaluate the evidence; do not be dogmatic; make up your own mind but be prepared to change it.

This book was based on, initially at least, lecture notes I inherited from Geoff Sweet and I thank him for this. I am indebted to Westoby (1989) for showing me that a proper understanding of forests and their relationship to people necessitates going back a very long way, to the evolution of the human species. The FAO on-line database and FAO publications were important sources of information for this book and I acknowledge this with gratitude. I thank Leon Bren, Mark Burgman, Mike Carson, Ben Cashore, Bob Fisher, John Turnbull, John Walker and Don Wijewardana for comment. I thank the authors of the Boxes in chapter 8. I am grateful to the library staff at the University of Canterbury, the Australian National University and CSIRO Division of Forestry and Forest Products. In particular I thank Jeanette Allen for preparing camera-ready copy and Helen Bavinton for editing.

Chapter 1

A History of Human Interaction with Forests

Forests and Humans before Agriculture

It is thought that the world was formed about 4500 million years ago and the first primitive life forms appeared about 4000 million years ago. Forests by contrast are relatively recent (about 350 million years ago) but are quite old compared to grasslands (about 25 million years ago) and positively ancient compared to modern human beings as we know them (less than 500,000 years ago). It is interesting to note that dinosaurs are often considered to be evolutionary failures because they became extinct. However they lasted for 160 million years (70–230 million years ago). Modern humans by contrast have not yet reached 500,000 years and have not stood the test of time as being a viable and enduring species on earth. Given the propensity of the human race for self destruction and environmental degradation, it is not at all impossible that humans could become quickly extinguished and be scarcely worth a mention over the time scales considered here.

Environmental Impacts on Plants and Animals before Humans

The time sequence for the development of forests, grasslands, animals and humans is shown in Table 1.1. For convenience the developments are boxed into geological periods but this does not mean that there were sharp and distinct differences across the boundaries. Neither should it be assumed that the process of evolution was smooth and continuous. Rather it was, and is, a process of origin, expansion, maximum diversity and then contraction (Boulter, 2002). There have been a series of mass extinctions followed by sudden accelerations in evolutionary activity. For example, there was a major extinction event about 250 million years ago that wiped out about 95% of all marine species. The development of forests and grasslands prior to human impact has been shaped by a range of disturbances including climate change, volcanic activity, changes in the disposition of the land masses, cosmic collisions, changes in sea level, and the movement of ice. The land masses have moved about considerably. Within the time period in Table 1.1, two great land masses, Laurasia and Gondwanaland moved together in the Triassic to form the single super-continent Pangea and subsequently parted company, moved around, fragmented and collided, often violently, in the passage to their configuration today. Over the same time period the climate changed from warm to cold to warm to cold, sea levels went up and down, land bridges between land masses came and went, ocean currents changed, large areas of the land masses were periodically covered with ice, meteors collided with the earth, and wind and water relocated soils. Consequently the geographic distribution of many plants and animals changed greatly over time.

Gondwanaland (Antarctica, Australia, India, South Africa and South America) started to break up about 130 million years ago at about the same time as the development of angiosperms. Consequently plants that evolved while the land masses were together could be represented in the evolutionary history of all Gondwanal countries while those that developed after separation could be quite distinctive. For example *Nothofagus*, *Dacrydium* and *Podocarpus* are present in New Zealand, South America, and Australia. *Dacrydium* and *Podocarpus* are conifers whose Triassic ancestors are generally considered to have established across Gondwanaland before it started to split up in the Cretaceous. *Nothofagus* is an angiosperm that developed later while Gondwanaland was splitting up and there is some argument about whether this genus has Gondwanal or Pacific origins. *Eucalyptus* is an angiosperm genus that is almost entirely confined to Australia and it developed relatively recently after New Zealand split from Australia 70 million years ago.

In the northern hemisphere forests contracted and expanded many times during the multiple recent ice ages of the Pleistocene. Soils and landforms also changed. The glacials were great natural deforestation events of greater magnitude than any anthropogenic (human-induced) deforestation of modern times. The recovery of the forests during the interglacials is testimony to their great resilience. There always remained during the glacials residual vegetation in the warmer areas near to the equator from which re-colonization and further evolution could occur. There was repeated fragmentation and coalescence of this residual vegetation and rates of speciation and extinction were high. The revegetation of bare ground during the interglacials was predominantly by forest rather than other vegetation types. The change in climate was the major factor determining the patterns of re-vegetation but soil type, competition, migration rate and source of propagules all contributed.

In northern Europe the pattern of re-vegetation was similar in many respects between interglacials. Iversen (1958) described the sequence of immigration of tree species from the south following the increase in temperature coming out of a glacial as, first, the boreal birch (*Betula*) and pine (*Pinus*) on unleached calcareous soils followed, with further increase in temperature, by a mixed deciduous woodland on soils that had progressed to brown forest soils. As the temperature started to decrease towards the next glacial the vegetation was replaced by a heath and conifer type and, with further decrease in temperature to the next glacial, by open tundra and steppe on arctic mineral soils. Much of North America was covered by ice in the last glacial. The vegetation patterns in North America today can be divided in very general terms into four based on current climate: a warm dry climate supporting grassland, a cold dry climate producing tundra, a warm wet climate favouring deciduous forest and evergreen rainforest, and a cold wet climate supporting boreal forest. However the pathway and rate of migration of individual species into these vegetation patterns following colonization of the land exposed from the receding ice was variable and complex as shown in pollen diagrams from sediment cores (Pielou, 1991). Mannion (1991) and Roberts (1989) reviewed the vegetation changes that occurred during the quaternary ice ages.

Currently there are about 4000 species of mammals, about 9000 species of birds, about 19,000 species of fish, about 250,000 species of vascular plants, about 1.6 million species of fungi and at least 10 million insect species most of which are beetles and most of which have not been described and never will be (Rosenzweig, 1995). Table 1.1 features the larger more conspicuous plants and animals. It should be remembered though that fungi and microorganisms have been an integral part of the biota right across this timescale.

Table 1.1. The development of plants and animals over the last 400 million years with particular reference to forests, grasslands and humans. Adapted from Calder (1983), Futuyama (1998), Osborne *et al.* (1996).

Geological period	Millions of years ago	Forests and other plants	Animals	Climate	Associated events
Devonian	408-360	Seed ferns	Boned fish Amphibians Life moves on to land	Greenhouse	Mass extinction event (tidal waves) Gondwanaland and Laurasia in collision
Carboniferous	360-286	Forest trees (early gymnosperms, including conifers) Giant lycopsid swamp forests Glossopteris fauna in Gondwanaland	Reptiles Insects	Ice sheets across southern continents Sea level lowers	Gondwanaland and Laurasia continue to collide Coal deposits
Permian	286-245	Gymosperms continue	Warm-blooded reptiles	Warm and dry	Pangea is in place Coal deposits Permian terminal catastrophe (245 million years ago) annihilated most marine species
Triassic	245-208	Conifers, cycads and ferns dominate the vegetation Bennettitales (precursor to Angiosperms)	Ancestral dinosaurs	Warm	Norian catastrophe (cosmic impact) extinguished many marine life and reptiles
Jurassic	208-144	Conifers, cycads and ferns dominate the vegetation	Dinosaurs Birds	Warm	Pangea starts to break up Oil formation
Cretaceous	144-66	Angiosperms develop and proliferate	Dinosaurs Marsupials Mammals Pre-primates	Warm Sea level at all time high at 93 million years ago	Oil formation Cretaceous terminal catastrophe (a cosmic impact) at 67 million years ago extinguishes dinosaurs and most marine animals and plants Gondwanaland breaks up Coal deposits
Tertiary	66-2	Angiosperms largely displace cycads and ginkgo and confine conifers mostly to cool temperate northern latitudes Grasslands develop	Mammals, birds, insects are supreme Grazing animals develop Human evolution through to *Homo habilis*	Warm but a cooling trend Ice in Antarctica Sea level lowers Oscillations in climate Volcanism	Magnetic pole reversals Eocene terminal turnover (37 million years ago)
Quaternary	2-present	Areas of grassland increase at the expense of forests Forest areas expand and contract with the ice-ages Human impacts on forest distribution	Human evolution progresses through to modern humans Modern cattle Neanderthals corne and go	Cool Successive ice ages Glaciation commences in the northern hemisphere Overall cooling North Africa desertification	Pleistocene overkill Agriculture, civilization, religion Industrial humans, green revolution Space travel and genetic modification Human-induced extinctions of species Human clearing of forests

In summary, environmental impacts before humans have had a major effect on the nature and location of forests. Recently though, and particularly in the last 10,000 years, humans have shown they also can radically affect the nature, distribution and pattern of forests (by forest clearing, land degradation, firing and species dispersal to mention just a few). Humans now have the technical capacity to create disturbances of similar magnitude and effect to the great mass extinctions of the past and to deforest the planet to the same extent as the Pleistocene ice ages. Hopefully they will not put this to the test.

The Development of the Forests

The first land plants were Bryophytes (liverworts, hornworts and mosses) and these appeared about 425 million years ago. These were followed by vascular plants that were either seedless (such as ferns) or containing seeds (such as conifers and angiosperms). Gymnosperm forests containing cycads, ginkgos and conifers developed quite early (about 350 million years ago) and conifer forests containing *Pinus* (pines), *Picea* (spruce) and *Abies* (fir) still dominate the cool temperate latitudes and the higher elevations of the northern hemisphere today. Conifers are also found in the southern hemisphere (e.g. *Agathis, Araucaria, Podocarpus,* and *Callitris*). However, they have smaller geographic ranges, are somewhat scattered and have been characterized as relicts or living fossils. Angiosperms (flowering plants) developed much later, about 120 million years ago. They have been particularly successful and are still expanding. Conifers are long-lived species but with a lower reproductive capacity than angiosperms. Consequently angiosperms have progressively displaced conifers except for the cooler parts (higher latitudes or altitudes) of the northern hemisphere where the superior adaptation of conifers to withstand seasonal cold has given them an advantage. Angiosperms now dominate the plant kingdom including forest trees.

The Development of Grasslands

The grasslands developed about 25 million years ago. Their distribution was largely determined by climate but soil-type and fire were also important. Grasslands developed and expanded prior to human development but human development certainly assisted in their spread. In fact the appearance of the grasslands is critical and perhaps more critical than any other factor in understanding the relationship between people and forests. Humans developed from arboreal apes in the forest and maybe there is some argument that humans have a deep-seated primordial attachment to the forest and yearn to go back. However, humans are creatures of the grasslands and not the forest. Human evolution accelerated only after, and probably because, the precursors of the human race came down from the trees, walked out of the forest, stood erect and looked into the distance.

The grasslands are particularly people friendly. They provide the bulk of the food used for human consumption. They spawned the development of animals that graze on grass. These animals were relatively easy to hunt and they supplied improved nourishment to produce hominids with progressively larger brains. Grasses dominate cultivated agriculture. The world's current consumption of edible plants is dominated, along with the potato, by the grasses wheat, maize and rice. Grasslands have a well-developed regenerative capacity and because of this they can recover from grazing and can persist in relatively dry environments. Consequently animals feeding from them also can persist in relatively dry environments. Also, grasslands regenerate after fire and fire promotes grassland at the expense of forests. Grasslands posed some threat for the welfare of forests prior to humans. However, with the advent of humans and their love affair with grasses, the odds became stacked further in favour of the grasses. The human race as we know it would not have existed without the advent of the grasslands and open country. Humans have flourished by clearing forests and deforestation is an inevitable consequence of the human condition. Most humans do not live in the forests but a small number of people have always lived in the forests and still do today. In past times these forest dwellers were prominent in riverine incursions into the forest where the riverine system was an important part of their food supply. This is still the case to some extent. However, in modern times deforestation of lowlands has often pushed forest dwellers into the hills. Even though forests are rich sources of biodiversity they are not a good food source for humans and they have a small carrying capacity for humans.

The Development of the Human Race

The human race evolved over the Tertiary and modern humans appeared in the Quaternary geological periods. The development of the human race alongside climate and associated events is shown in Tables 1.2 and 1.3. The hominid line is considered to come from the African apes, which split from the monkeys about 30 million years ago. About 10-15 million years ago bipedal apes (the hominid line) developed and separated from other ape lines, which developed into gorillas and chimpanzees. This coincided with a decrease in the amount of dense forest cover in Africa, an increase in the amount of scattered woodland and grassland, and an increase in the variety and availability of grazing animals. This in turn produced an increase in the variety of food available to these ape-men (or pre-hominids). Hominids as such first appeared about 5 to 7.5 million years ago. The first evidence of a stone axe emerged about 2.5 million years ago and this marked the beginning of the Palaeolithic archaeological age (old stone age). This was at about the same time as polar glaciation and global aridity accelerated the already existing trend from forest to woodland to grassland (savannah) in Africa.

Homo habilis, a scavenger, appeared just over 2 million years ago and lived at the forest fringes and in the woodlands and was predominantly a vegetarian. *Homo ergaster* appeared about 1.9 million years ago and *Homo erectus* about 1.7 million years ago. *Homo ergaster* (African *Homo erectus*) and *Homo erectus* were clearly anatomically more suited to the grassland than the forest and hunting commenced in a very rudimentary form. Consequently meat became more prominent in the diet and the trend away from scavenging towards hunting and gathering commenced. *Homo ergaster* also provided the first evidence of the controlled use of fire (about 1.4 million years ago) although evidence of shelters containing hearths was much later (about 400,000 years ago). The movement of the hominids out of the forests not only triggered an acceleration in the rate of human evolution but also marked the commencement of their achievements as long distance travellers. *Homo ergaster* and *Homo erectus* migrated out of Africa via land bridges that became exposed at a time of low sea levels (Tattersall, 1998). The migration of hominids from Africa throughout the world coincided with the Pleistocene ice ages which commenced about two million years ago and are still in progress. The marked changes in climate caused separation and rejoining of hominid populations and these conditions were ideal for human evolution and diversification.

Early *Homo sapiens* (archaic humans) developed about 1 million years ago and modern *Homo sapiens*, anatomically the same as we know them today, developed at least 100,000 years ago and possibly up to 500,000 years ago. There have been differences of opinion over whether modern humans developed from multiple sources or from one original source. Recent evidence supports that modern humans originated in Africa and that they migrated to other parts of the world, progressively out-competing all other archaic humans, including the Neanderthals (*Homo neanderthalensis*). Consequently, *Homo sapiens* is the only hominid surviving today. There is also some uncertainty about when hominids developed language. The development of language is considered to be a major factor in the evolution of humans and a key feature in the progressive increase in brain size that occurred from early hominids through *Homo habilis* to *Homo erectus* to *Homo sapiens*.

Table 1.2. Early evolution of humans and associated climate, forests and other plant communities, animals and other events. Compiled from Calder (1983), Tattersall (1993) and Vrba *et al.* (1995). (Mya = million years ago).

Period	Epoch	Millions of years ago	Humans	Forests and other plant communities	Animals	Climate	Tectonic and other events
Tertiary	Paleocene	65-53	Early primates	Angiosperms continue to expand at the expense of gymnosperms	Early horses	Equable, warm and wet	North America separates from Europe Rockies form
	Eocene	53-36	Higher primates			Equable, warm and wet	Australia separates from Antarctica India collides with Eurasia Eocene terminal turnover
	Oligocene	36-23	Extinction of primates in northern hemisphere	Grass develops and covers large areas	Cats, dogs, whales Apes and monkeys split	Equable, cool and dry	Sea level falls
	Miocene	23-5	Hominids diversify in Africa		Separation of hominid apes from other apes Antelopes, grazing animals Orangutans Modern cats and dogs Elephants	Warm, wet seasonal moving to cool, dry seasonal	Miocene disruption Antarctic deeps freeze Volcanic activity Northern glaciers Mediterranean dry-out and recovery Himalayas uplift
	Pliocene	5-2	*Homo habilis* (2 mya) Transition from forest dwelling to woodland dwelling Diet predominantly vegetarian	Cooling and drying in Africa about 2.5 mya causes some forests to change to woodland and grassland	Camels, bears, pigs, baboons, modern horses, early cattle Extra grassland favours development of grazing animals	Warmer seasonal, but cooling off about 2.5 mya	Andes uplift Current ice ages begin Oldowan stone axe (2.5 mya)
Quaternary	Pleistocene	2-0.01	*Homo ergaster, Homo erectus* and *Homo sapiens* (see Table 1.3)	(see Table 1.3)	(see Table 1.3)	Cool to cold	Ice ages (see Table 1.3)
	Holocene	0.01-present	*Homo sapiens* out-competes all other hominids (see Table 1.4)	Cultivation of plants (see Table 1.4)	Domestication of animals (see Table 1.4)	Warmer (interglacial)	Age of agriculture and development of civilization (see Table 1.4)

Dwellings or shelters, made from tree saplings, were evident 400,000 years ago and these contained hearths. Wood was also used widely from 400,000 years ago for fuel, charcoal, drying, ladders, artefacts and tools. Indeed the use of wood may have been much greater than we think because it readily decays and is not persistent in the archaeological record. In fact Tyldesley and Bahn (1983) considered that the use of wood in the Palaeolithic was so widespread that the age should more appropriately be called the Palaeoxylic (old wood age) rather than the Palaeolithic (old stone age). Most attention has been given in the literature to the use of stone axes for hunting and less attention to the tools used in felling trees and preparing wood for various end uses. The stone axes and tools of the day, however, would have met the need (Cole, 1970). The evidence for human presence, in Europe at least, in the relatively short-lived warm interglacials was scant compared with the longer colder glacial periods (Mannion, 1991). The interglacials were forested and the glacials were not. This reinforces that, by and large, humans were not forest dwellers and preferred open areas.

Table 1.3. Human development and associated climate, forests, other plant communities, animals and other events during the Pleistocene geological epoch (2 million to 10,000 years ago) and the Palaeolithic archaeological time scale (2.4 million to 10,600 years ago). Compiled from Calder (1983), Tattersall (1998) and Goudie (2000).

Period (years ago)	Humans	Forests, other plant communities, and animals	Weather	Associated events
2,400,000 – 2,000,000	*Homo habilis* (a scavenger that frequented the forest margins and woodlands in Africa)	Accelerated rate of change of forest to woodland to grassland in Africa. Further proliferation of grazing animals	Cold and dry. Current ice-ages begin	Beginning of the Palaeolithic age (stone age). First evidence of stone tools
2,000,000- 1,400,000	*Homo ergaster* (1.9–1.4 mya) and *Homo erectus*, both in Africa. Better suited to the grasslands than the forests. Omnivorous. Migration out of Africa	Grasslands continue to develop in Africa	Glaciation (2 mya)	First evidence of controlled use of fire. Acheulean hand axe
1,000,000	*Homo erectus* migrates to China and Java (and perhaps earlier)			
800,000-700,000	Early hominids in Spain (800,000 years ago)		Interglacial	
600,000			Glacial	
500,000	Perhaps the origin of modern *Homo sapiens* in Africa		Interglacial	
400,000-300,000	*Homo heidelbergensis* in Europe. Language	Wood used for dwellings, fuel, charcoal and tools	Glacial	Earliest shelters made from tree saplings. Convincing evidence of controlled use of fire
200,000	*Homo neanderthalensis* were foragers and scavengers. Early *Homo sapiens* were hunter gatherers		Interglacial	
150,000	Convincing evidence of modern *Homo sapiens*			
100,000	Population of *Homo sapiens* about 5 million		Glacial	
40,000 – 30,000	*Homo sapiens* widely spread globally (Cro Magnon man in Europe)		Glacial	Art, music, symbols, more advanced tools, elaborate burials
27,000	*Homo neanderthalensis* becomes extinct		Glacial	
20,000-10,000	*Homo sapiens* is the only surviving hominid and is widely distributed (but not in Ireland, Polynesia, Madagascar, New Zealand, the Caribbean or Antarctica)	Megafauna extinction throughout the world but not in Africa	Glacial	Early cultivation and domestication of animals. Use of fire (deliberate or accidental) to promote grassland

The impact of humans and their ancestors on the environment 50,000 and more years ago was small and inconsequential. This was because the populations were small and scattered and they had not yet developed the technologies and tools to have a significant effect on the environment. The expansion of the grasslands, although of great benefit to humans, was not at this stage assisted by humans. It would be misleading to suggest that grassland was overtaking forests globally on a grand scale. Grassland expansion occurred locally and humans chose to live in these areas. Global changes in vegetation were dominated by the ice ages and the forests were the prime re-colonizers during the interglacials.

Hunter-gatherers and Forests

Modern humans hunted to get their meat and gathered plant food where they could. This continued until the 'age of agriculture' when humans found they could grow their own food and domesticate their own livestock. It is difficult to put a start and end point to this period. Hunter-gathering replaced scavenging perhaps 500,000 years ago but the record of hunter-gatherers and how they lived is strongest from about 50,000 years ago. Human evolution appeared to take a quantum leap forward in technology and art about 40,000 to 50,000 years ago (Diamond, 1998) from which time also there was convincing evidence of widespread geographical dispersal of humans across the world and undisputed evidence of humans having a significant impact on the global environment. It is generally reckoned that the age of agriculture commenced about 10,000 years ago. There are, however, records of early agriculture as far back as 18,000 years ago. Also, some hunter-gatherers, such as the Australian aborigines, completely bypassed the age of agriculture and first confronted other humans from the industrial age less than 250 years ago. Murdock (1968) lists 27 modern hunter-gatherer groups that have existed recently enough to be studied and published by modern ethnographers. Currently about 0.001% of the world's population live by predominantly hunting and gathering.

Hunter-gatherers were by necessity close observers of nature. There was a strong selective pressure to ensure that they were good ecologists. They needed to be so in order to provide year round sustenance. They had considerable knowledge of the plants around them, their flowering and fruiting patterns and their value as a food source. They knew the migratory patterns of the animals and the pattern of the seasons. They were nomadic because they quickly depleted local food reserves and because their food supplies were seasonal (Bush, 1997). Modern hunter-gatherers show survival skills that contemporary humans have lost. The close association between hunter-gatherers and the natural world is still evident in modern hunter-gatherers and is an important component of their folklore and spirituality. Indeed it is their close association with the land that has caused most misunderstanding and conflict between modern hunter-gatherers and their industrial and agricultural co-inhabitants. Hunter-gatherers appear to have been well resourced and this is largely due to their intimate association with the land, plants and animals. Their life-spans were relatively short, but this was because of warfare and hunting accidents rather than starvation or infectious diseases. Consequently populations were kept low. The populations of more recent-day hunter-gatherers have fallen because of exposure to infectious diseases, habitat removal and competition with agriculturalists. Forest clear-cutting has completely destroyed the hunter-gatherer's environment in parts of Borneo and the Philippines (Southwick, 1996).

Hunter-gatherers, though good ecologists, were not especially good conservationists. They had no real concept of the finite nature of natural resources or of sustainable management. When they depleted the reserves around them, they just moved on. Their numbers were small and consequently, for the most part, their impacts on the environment were correspondingly small. Probably they were the first human agents of seed and plant dispersal because of their nomadic nature. The amounts of wood that they extracted from the forests for fuel and shelter would have been negligible. However, there were two areas where hunter-gatherers did have a significant effect on their environment: the extinction of large animals and the depletion of forest area. In both of these their controlled use of fire was an important factor.

Many large animals (megafauna) became extinct during the Pleistocene (Pleistocene overkill) ranging in time from 40,000 years ago in Australia to about 12,000 years ago in the Americas. There is some difference in opinion over the cause of these mass extinctions. Some consider that climate change was the major cause but others suggest that hunter-gatherers were a major contributing factor and probably the main cause. These megafauna had escaped many severe climatic cycles prior to the Palaeolithic. Their sudden disappearance in the Palaeolithic coincided with the time that modern humans were diversifying and developing superior hunting technologies and also competing with these large animals for habitat. Also it is a matter of record that when humans first entered oceanic islands in more modern times they were directly responsible for the extinction of the dodo in Mauritius, the moa in New Zealand, giant lemurs in Madagascar and the big flightless geese of Hawaii (Diamond, 1998). However, the matter remains controversial.

Fire was the first product of the natural world that humans learned to 'domesticate' (Pyne, 1995). Hunter-gatherers used fire to manipulate vegetation patterns and continue to do so to the present day. They used fire to increase the area of grassland, to inhibit woody regeneration on grassland, to provide green regeneration to attract grazing animals, to deprive game of cover, to drive game out of cover, and to promote and harvest insects and edible plants. Hunter-gatherers deliberately lit fires, but also took advantage of fires lit from lightning strikes that most often would be left to burn out. For thousands of years hunter-gatherers put the land to the torch. The Australian aborigines burned the land whenever they could and the frequent low intensity fires favoured open countryside rather than forest (Flannery, 1994). The Australian aborigines have been called 'fire-stick farmers'. It is possible that one of the reasons they did not develop cultivated agriculture is because of their success in using fire to optimize their hunting and gathering. The Indians in North America used fire to keep the prairies as open range and when the fires were withheld the trees came back. The hunter-gatherers of southern Africa used fire to maintain the grazing lands of the veldt. When Magellan rounded the southern tip of South America, he saw many fires lit by the indigenous peoples and he named this part of South America, Tierra del Fuego (land of fire).

The ecosystems produced by this continual firing not only tolerated fire but actually encouraged it and even required it for their maintenance. It is an over-simplification to say that fire promotes grassland over forest. Grasslands appear to be favoured by more frequent less intense burns and forests by less frequent more intense burns. However, fire sometimes promotes heath or shrub country rather than grass. Also some forest types are extremely well adapted to fire. Grasslands that are either under-grazed or over-grazed are susceptible to encroachment by woody perennials. However, on balance, the net effect of hunter-gatherers burning the vegetation whenever they could probably was an increase in the area of grassland and a consequent decrease in the area of forest. For example, the continual firing of the forests of New Zealand by Polynesians (Maori) in pre-European times in order to hunt the now extinct moa reduced the forest cover from 79% to 53% (Mark and McSweeney, 1990). Probably an equally important consequence of the burning by hunter-gatherers has been the profound effect that this has had on the composition and structure of both forested and non-forested ecosystems. This, in turn, has strongly influenced how they should be managed.

Modern hunter-gatherers tend to be forest or forest-fringe dwellers. This is consistent with the observation made earlier that the technological advancement of the human race depended on leaving the forest and occupying and promoting open areas.

Forests and the Age of Agriculture

The age of agriculture can be defined as the period in which humans cultivated plants for food production and domesticated animals for meat production, for milk and for the harnessing of animal power for ploughing and other activities. Agriculture commenced about 10,000 years ago. The earliest records come from New Guinea, Peru and Iran. The earliest crops cultivated were grasses: wheat and barley in the Middle East, rice in Southeast Asia, millet in Africa and maize in the Americas. Potatoes, peas, lentils, beans, capsicum and gourds were also important early crops. Table 1.4 shows the sequence of events during and past the age of agriculture.

Agriculture developed first in the grasslands and open woodlands. For example, the agricultural origins of the Incas of the Andes, the Pueblos of North America and the Bantus of Africa commenced in the open country where they had lived as hunter-gatherers (Winters, 1974). However, early agriculturalists would soon have learnt that crop productivity was best on fertile soils with good rainfall and these are the very soils that originally would have supported forests. As such, the incentive to clear forest for agriculture probably occurred quite early in the age of agriculture. There is evidence of widespread deforestation having occurred in temperate forests during the Mesolithic and the Neolithic (Goudie, 2000) and over a wider area and at an accelerated rate thereafter, particularly in northern Africa, Mesopotamia, the Mediterranean and temperate Europe (Dimbelby, 1976). This deforestation needs to be considered alongside the naturally occurring reforestation that was occurring in Europe, the Mediterranean and North America during the interglacial Mesolithic, approximately 8,000 to 10,000 years ago (see Roberts, 1989). Today there are extensive areas of agricultural and grazing land that arose from clearing forests, mostly on the best soils. Some of this clearing happened so long ago that it is forgotten as a deforestation event.

The early agriculturalists would have found that fire produced an initial burst of fertility but that their yields declined after cultivating an area for a few years (through nutrient depletion) and also that weeds progressively overtook the cultivated areas. They responded to this by abandoning the cultivation, moving on, burning a new area and starting again. Of course they did not understand the scientific basis of why they needed to do this. This marked the beginning of slash and burn agriculture (or shifting agriculture) that continues to this day. Slash and burn agriculture can be quite responsible and sustainable providing it is done at a low enough intensity to allow the vacated site to completely re-vegetate and

Table 1.4. Human development and associated events from 10,000–1,900 years before present (adapted from Mannion,1991 and others).

Period (years ago)	Humans	Plants and animals	Forests	Technology	Environment
10,000 – 8000 (Mesolithic)	Shifting (slash and burn) agriculture Population about 5 million	Cultivation of wheat, barley, peas, lentils, flax and vetch and domestication of cattle, goats, and sheep (all in the Near East)	Interglacial reforestation in the northern hemisphere Beginning of deforestation to support farming and grazing Fire used to promote grassland over forests	Axes used to fell forests	Incipient soil erosion
8000-5000 (Neolithic)	Trend from shifting to settled agriculture Agricultural communities in China	Reduction in number of cultivated plants Farming and grazing more productive Domestication of the pig, horse (Russia), llama and alpaca (South America) Cultivation of potato, capsicum, beans and maize in tropical America Rice in China	Continued deforestation Cultivation of the olive Decline of elms in Europe Commencement of the deforestation of southern Britain	Smelting of copper in Anatolia and Thailand Agricultural tools	More soil erosion and more wide-spread
5000 – 3000 (Bronze Age)	Great riverine civilizations commenced in Mesopotamia, Egypt and Pakistan Hierarchical societies and trade	Monocultures Transfer of plants and animals between regions Domestication of rice in India and Pakistan	Greatly accelerated deforestation to support farming, grazing and smelting Cultivation of peaches, apricots and citrus	Smelting of copper Irrigation Wheeled carts Plough Textiles Arts and crafts Saws	Accelerated soil erosion, siltation, salinization, desertification, overgrazing and abandonment of non-productive soils Infectious diseases Climate change towards drier conditions in Mesopotamia, Mediterranean and North Africa
3000 – 1900 (early Iron Age)	Civilizations in Greece, Rome, India and China Population reaches about 200 million	Further reduction in the number of food crops, and intensification of the production of food crops Continued transfer of plants and animals between regions.	Deforestation particularly to support smelting and shipbuilding Beginnings of silviculture (coppicing, thinning, pruning, tree breeding, plantation establishment)	Smelting of iron Glass Rudimentary agricultural technology (composting, terracing, fallowing, fertilizing, breeding)	No significant change in climate Soil erosion and land degradation continues

recover. The early agriculturalists probably never or rarely returned to a vacated site and consequently their slash and burn agriculture was relatively benign. (Slash and burn agriculture as practised in recent times, however, can be quite damaging and plays an important role in the current deforestation of tropical forests.)

Settled Agriculture

Some hunter-gatherer groups developed limited agriculture but maintained predominantly a hunting and gathering lifestyle (such as in New Guinea). Others embraced agriculture and grazing as their main form of subsistence and progressed from 'shifting' agriculture through to 'settled' agriculture where communities developed in settlements with larger numbers of people living at closer quarters. The higher densities of population brought health problems. Settled agriculturalists had worse teeth and bones, caught more communicable diseases and had a lower life expectancy than shifting agriculturalists (Cohen, 1989). It is likely that the plains and valleys supporting settled agriculture were initially deforested in part by

Neolithic agriculturalists. This trend continued in the early Bronze Age such that later accessible forests were mostly confined to the hills and mountains.

Smaller settlements grew into larger settlements and in some instances into sophisticated city-states. This heralded the beginning of hierarchical societies and 'civilization.' The key to the reason why some agriculturalists moved on to civilization and others did not was primarily access to good soil and plants suitable for domestication. The first great civilizations developed from about 5000 BC (7000 years ago) on the fertile soils of the flood plains of the Tigris, the Euphrates, the Nile, the Indus and the Yellow Rivers. These soils were so fertile that they could support crops in one place for longer and support a higher density of people. The rivers continuously replenished soil fertility by depositing alluvial silt during floods. They also supplied water for irrigation and a conduit for transport.

Continuous cultivation of the same soil, even fertile soil, requires special care (maintenance of soil organic matter and replenishment of soil nutrients). If not, the soil will become infertile and degraded. Similarly if animals are grazed continuously on the same land, and particularly if the number of animals is too great, the pastures will not regenerate and the soil will become degraded, sterile and subject to erosion by water and wind. Both of these misfortunes were evident in the early days of agriculture and regrettably are not unusual in parts of the world today. Neolithic farmers responded to this by using some of the basic agricultural techniques that have stood the test of time. They used manure and other fertilizers to maintain soil fertility, terracing to control erosion, fallow to rest the soil, and rotation with legumes to enrich soil nitrogen (Hughes, 1994). Even so, as the population increased and concentrated in settlements, more land had to be found to sustain crop yields and animal numbers. This was achieved by clearing forest. Indiscriminate clearing of forests, particularly on slopes, promoted soil erosion and downstream sedimentation that, later in the Bronze Age, choked irrigation canals and watercourses and land-locked coastal ports. This further reduced the crop and pasture yield on cleared land and even more forest was felled to compensate. Deforestation was also necessary to provide fuel and wood products for the developing civilizations.

The Rise and Fall of Civilizations

The trigger for the development of civilization came when humans first found they could produce more food than they needed for themselves. This meant there was time available for leisure and not everybody was needed to be involved in food production. Food surpluses led to trade and the increase in non-productive time meant other activities such as arts and crafts and other cultural activities could be developed. Human populations and particularly the density of humans living in close proximity increased. It was inevitable that hierarchical societies would develop and equally inevitable that the dominant groups would move to protect their positions. Thus, rulers conscripted police to enforce laws, priests promoted ideologies to enslave the poor and underprivileged, and armies were raised to steal resources from neighbours. Societies were segregated based on wealth and influence and the lower castes were usually slaves, or little better off than slaves. Also, because of the gross distortion of available wealth, the rich were profligate consumers of anything they could find including an outrageous consumption of forest products. This, however, is not all that different from modern times.

The civilization of the human race is arguably the greatest achievement in human evolution. However, when humans removed themselves from their close association with the land they lost the empathy with the land that the hunter-gatherers had developed in order to survive. This was the beginning of the ecologically flawed idea that humankind was separate from, rather than being intrinsically part of, nature. Only a minority of people was involved in producing food and these people were often the most underprivileged and least influential. The influential, used to surpluses and indeed dependent on them to maintain or extend their position, looked further afield when local supplies became inadequate. Population increases led to increased consumption of wood for fuel and shelter. Technological advances, particularly the smelting of metals, increased wood consumption (and therefore deforestation) further. The wealthy in the early civilizations were very dependent on wood to maintain their life styles and wood consumption, particularly for fuel for metal smelting and pottery, could be staggering.

A distinction needs to be made here between forest logging and forest clearing. Deforestation occurs when trees are felled and the land is cleared for a land use other than forestry. Forest cleared for agriculture and grazing is the most common example. Logging of forests is not deforestation if the area of forest is not reduced and the forest subsequently regenerates. However, during the golden civilizations, logging was done in a most destructive manner with no attention being given to regeneration strategies. Many forests became seriously degraded.

When forests were cleared or became degraded near to the centres of population, they were progressively exploited further away. Inevitably this reached the point at which the stronger communities were able to take (steal is probably a better word) the wood from weaker neighbours. This heralded the beginning of colonialism. Invasion greatly accelerated deforestation. Not only were forests needed to provide ships and other apparatus of war, but also invaders often destroyed the forests to deny timber to their enemies. Alternatively they burnt the forest to entrap their enemies. The citizens of vanquished cities transmigrated to the countryside and into the hills where they cleared more forest to grow crops.

A model for the rise and fall of civilizations is given in Fig. 1.1. This model best fits the cases for the civilizations of Mesopotamia and the Mediterranean but elements of it are universally applicable. The reasons why these civilizations ultimately declined are various but they all had the predisposing factor of poor soil management. They became weakened because they were unable to practise sustainable agriculture, which led to an inability to feed themselves. The forests were major casualties and these areas today are seriously degraded landscapes. The age of the golden civilizations of Sumeria, Babylonia, Assyria, Phoenicia, Egypt, Greece and Rome was a period of gross environmental mismanagement from which the regions have never fully recovered. Certainly climate change played a role, particularly in northern Africa, but there is no doubt that humans gave it very good assistance. It would be unreasonable to single out these ancient civilizations as environmental vandals without peer. Post-classical and modern societies often have done worse. Indeed there have been cycles of devastation and recovery of forests right across the history of the human race from the Bronze Age to the present. The ancient civilizations lasted for quite long times, many thousands of years in some instances. Also there is evidence that they understood some of the environmental problems that they were causing

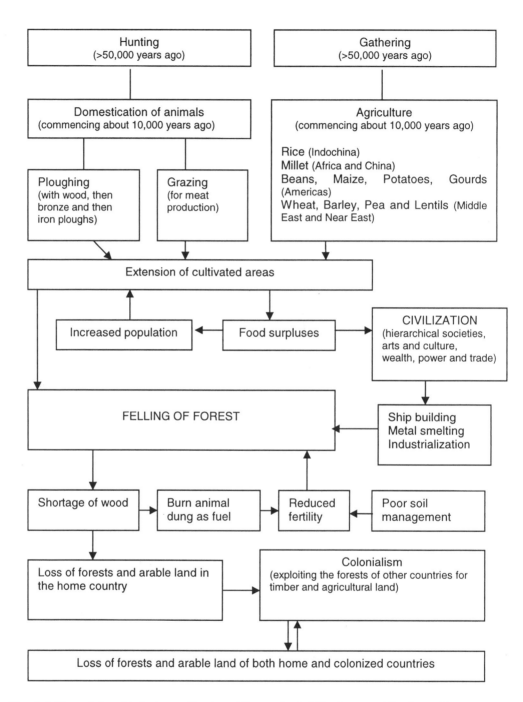

Fig. 1.1. The relationship between the rise of the ancient civilizations and the fate of forests (modified from G.B. Sweet, unpublished).

and they took steps to try and remedy them. Whether or not modern civilizations, with all their technological knowledge, will last as long as these ancient civilizations, is not yet known. Certainly modern civilizations have all of the same ingredients that were instrumental in the environmental degradation and the demise of earlier civilizations. Modern humans now have the knowledge to sustainably manage agricultural and forest systems and an understanding of the consequences of not doing so. This does not mean, however, that sustainable management will happen. There are plenty of present day examples to show that it is not happening. A more detailed discussion of individual civilizations follows later in this chapter.

The role of sedimentation in the rise and fall of civilizations is two-edged. Sedimentation is not necessarily a bad or destructive thing. The early Bronze Age civilizations developed on floodplains whose fertility depended on upland soils being eroded and carried downstream. This sedimentation would have occurred without the impact of humans but it is possible that Mesolithic and Neolithic humans assisted it by clearing forests upstream. Without this soil erosion these civilizations would likely not have developed. Civilization was spawned by soil erosion. However, with continuous felling of forests upstream and with poor soil conservation practices on cultivated land and overgrazing, the sediment loads became so large that they became an uncontrollable nuisance rather than a benefit. These same river valleys that supported the ancient civilizations still owe their inherent fertility today to soil erosion.

Over the last 10,000 years or so, humans have reduced the global forest cover from about one half of the global land surface area to about one quarter. Most of this deforestation has occurred in the temperate and sub-tropical regions and it is only in recent times that there have been significant incursions into the tropical and boreal forests. Also until recently, deforestation occurred close to where people lived. There has been a strong pattern of deforestation accelerating alongside increases in population and a pattern of forest recovery occurring after human populations move out of an area. Nowadays however, many more people live in large cities and transport has improved to the extent that forests of the most far flung regions of the world are readily accessible. It is not practical to comprehensively review all regions of the world. Therefore a few representative examples are given here as case studies. This commences with the civilizations of Mesopotamia and the Mediterranean, then looks at forests in Europe from the Dark Ages through to the 20th century and concludes by looking at the impact of colonialism on forests using the USA as an example. Williams (2003) provides a more comprehensive history.

Mesopotamia

Mesopotamia comprises all of the country dominated by the twin rivers, the Tigris and Euphrates. This covers all of present day Iraq and parts of Syria, Turkey and Iran. The fertile soils of the floodplain led to the region being called the Fertile Crescent. Settled agriculture and subsequent city-states developed here because the soil was fertile, suitable wild grasses were accessible for cultivation and the rivers provided water for irrigation. The productive soils were limited to the floodplain, which was surrounded by less hospitable and often dry countryside. Consequently crop productivity, particularly on the lower plain, needed to be supported by irrigation. Settled agriculture commenced at about 5550 BC from which arose a series of dynasties and empires. Sumeria developed first in the south of the floodplain and Babylonia followed in the middle of the floodplain. The Babylonian civilization came to an end after successive invasions from the Kassites, Elamites, Assyrians and eventually the Persians in 539 BC.

Initially the plain was peopled from agricultural villages in the uplands moving onto the plains. Written language was first evident at about 4000 BC and progressively these agricultural communities developed most of the trappings of civilization: hierarchical societies; the plough; the wheel; food surpluses; trade; building in stone; wood and wood-fired bricks and clay; wood-fired pottery; wood-fired smelting of copper ores to make tools and weapons; crafting with fine metals; sculpture; spinning and weaving in linen; and leather work. The temple was a main feature of the cities and this was usually mighty in size and required considerable labour and material in construction and maintenance (Hawkes, 1973).

Populations increased and the proportion of people directly involved in producing food became reduced. Consequently the productivity expected from a given area of cultivated land area increased. At first this was achieved. Irrigation canals were built and the fertile alluvium brought down in the floods prolonged crop productivity but not indefinitely. Siltation progressively increased and choked the irrigation canals and considerable labour had to be directed towards keeping the canals free of silt (Dasman, 1968). This dredging resulted in the canals being up to 10 metres above the surrounding fields (Hughes, 1994). Southwick (1972) reported that, after 3000 BC, silt from the Tigris and Euphrates Rivers filled in the Persian Gulf 180 miles from its origin. Crop yields were reduced and forests were felled to provide new land for cultivation as well as for providing wood. The felling of the forests exacerbated the problems on the plains. The upland areas exposed by deforestation or forest degradation were prone to soil erosion and even more silt moved into the river basins. The eroded soils exposed saline rocks and mineral salts which also moved down to the plain and soil salinization and irrigation with saline water further reduced crop yields. Salinization of the lower plain led to the more salt-tolerant barley replacing wheat as the staple cereal. Ultimately, the yields of barley fell to about 40% of their earlier yields. This was probably the main reason for the collapse of Sumeria and the centre of civilization in Mesopotamia moving north to the mid plains of Babylonia where there was a much lesser problem with salinity (Jacobsen and Adams, 1958). However, Babylonia eventually also succumbed to land degradation and low crop yields and ultimately the great Babylonian civilization was weakened to the extent that it became vulnerable to invaders.

The Mesopotamian civilizations were large consumers of wood. Perlin (1989) reviewed the use of wood in Mesopotamia. The scene is set in the Epic of Gilgamesh, one of the earliest stories ever written. The epic was Sumerian but was translated into Akkadian at the beginning of the Babylonian period (Saggs, 1995). An English translation is provided by Sandars (1960). The legend is set in about 2700 BC in Uruk in southern Mesopotamia where Gilgamesh, the ruler at the time, had to penetrate the surrounding cedar (*Cedrus*) forests to provide the timber to build his city. In order to do so he had to overcome the ferocious demigod Humbaba who was guardian of the forest. The presence of dense forests accessible to the river plains at this time is no legend. The forests of the Zagros range east of Babylonia were the first to be exploited and they provided oak (*Quercus*), cypress (*Cupressus*), pine (*Pinus*) and juniper (*Juniperus*). The rulers of Lagash, another city-state in Sumeria, progressively raided the cedar forests of the Amanus Mountains from about 2350 BC in order to build their grandiose cities. The third dynasty at Ur in 2100 BC was the pinnacle of Sumerian civilization and wood was extensively used for construction and to provide the fuel for smelting metals. Poplar (*Populus*) and willow (*Salix*) were taken from local forests; cedars were imported from the Amanus Mountains; juniper, fir (*Abies*) and sycamore (*Platanus*) were taken from northern Syria; and other timber was imported from northern Arabia and India. Later cedars were accessed from as far as the Lebanon range (Makkonen, 1969) and Crete (Perlin, 1989).

Timber became a precious commodity and wars were fought to secure access to forests. Wood was so scarce in Babylon that rented houses had no doors and tenants had to take their doors with them. Administrators took strict measures to slow down the consumption of firewood and to curtail unauthorized cutting of trees in the Babylonian forests. Firewood prices soared and this was reflected in the asking price for commodities that required heat in their manufacture (Perlin, 1989).

The Mediterranean

The forests in the Mediterranean are characterized by more than 5000 years of continuous and often intense co-existence with humans. Most of the Mediterranean was well forested in areas of adequate rainfall at the beginning of the Bronze Age (5000 years ago). Except for the edges of the Sahara in North Africa, there has been no significant change in climate over the Mediterranean for the past 5000 years. Therefore, deforestation events since then cannot be attributed to changes in climate.

Before human intervention, pines and oaks occupied most of the land below 1000 metres elevation. After human intervention much of this became covered by *maquis*, a cover of small woody shrubs, which after repeated destruction by fire, overgrazing or clearing is replaced by a community of low tough shrubs called *garigue*. With even more extreme disturbance, *garigue* becomes a grassland that is resistant to grazing. In the western and northern parts of the Mediterranean basin, deciduous forest with oaks, beech (*Fagus*), elms (*Ulmus*), chestnut (*Castanea*) and ash (*Fraxinus*) occurs in a zone between 1000 and 1500 metres. This zone is missing in the south and the east, possibly because it has been eliminated by persistent human removal over time or maybe it never existed in these parts. Above 1500 metres in undisturbed environments are coniferous forests containing pine, cedar, fir, and juniper (Hughes, 1994).

Significant deforestation and forest degradation occurred in the ancient Mediterranean and this has continued to the present day. This was particularly so in the south and the east and near to centres of population. Neolithic humans cleared some forests on the plains for agriculture and Bronze and Iron Age people extended this into the foothills. The forests on the upland slopes and mountain ranges were not cleared for agriculture but were extensively exploited for timber, often in a most destructive manner. Consequently much of the area of forest in the Mediterranean classified as forest today is degraded. Because of difficulties and cost of transport, accessible forests were the first exploited and the more inaccessible forests had a temporary reprieve. A substantial proportion of the Mediterranean forests only became accessible when railways were built in the 19th century (Meiggs, 1982). Another common element across the Mediterranean was overgrazing by cattle, sheep, pigs and goats. A particular menace was (and remains) the goat, which enjoys wandering into wooded areas destroying the young regeneration and reaching up into small trees to eat the foliage. Hughes (1994) quotes the poet Eupolis to demonstrate that goats eat almost everything available to them including trees.

'We feed on all manner of shrubs, browsing on the tender shoots
Of pine, ilex, and arbutus, and on spurge, clover, and fragrant
Sage, and many-leaved bind weed as well, wild olive, and lentisk,
And ash, fir, sea oak, ivy, and heather, willow, thorn, mullein,
And asphodel, cistus, oak, thyme, and savory.'

Continuous uncontrolled grazing is the major reason why many degraded and over-cut forests of the Mediterranean have not regenerated from classical times to the present.

The Cedars of Lebanon

The cedars of Lebanon were famous in the ancient world and are held in great respect, almost in glory, in the ancient literature. The Lebanese cedar (*Cedrus libani*) came from the Lebanon range which runs parallel to the eastern seaboard of the Mediterranean Sea between Beirut and Tripoli. Cedars were also found in the Taurus Range on the Cicilian plain and the Amanus Mountains in Syria but it was the cedars of Lebanon that received most acclaim. The cedars made Phoenicia a dominant maritime power and trading nation. The cedar was valued for its high quality durable wood that was close grained and easy to work. The trees were also large and suitable for roof beams in temple construction and in making large doors. People would go to great lengths to get the cedars of Lebanon and were prepared to transport them considerable distances. The Mesopotamian civilizations used them for building their temples as did King Solomon in the combined kingdoms of Judah and Israel. Cedars and other species of the Lebanon Ranges routinely were used in Egypt for shipbuilding, building construction and furniture. The Lebanon Range also contained a range of other tree species including evergreen oaks, cypress, junipers, firs and pines. Phoenicians, Egyptians, Hittites, Hebrews, Assyrians, and Babylonians all had a stake in using the cedars. When Alexander the Great sacked Persepolis, the Persian capital, in 331 BC, the wooden pillars of the royal palace were made of cedar from Phoenicia (Winter, 1974). Prior to human exploitation the cedar forests were extensive and magnificent and the dominant species over the Lebanon Range. As recently as the First World War, the remnants of the cedars of Lebanon were still being cut. Today the cedars are rare and found only in a small number of sparse, scattered and degraded populations (Meiggs, 1982).

Minoan Crete

Crete was well forested prior to the Minoan civilization, which commenced about 3000 BC and was very well developed by 1700 BC. Indeed the presence of these forests was a major factor in the development of the Minoan civilization. Cypress was prominent in the forest but fir, pine, oak, olive, ash, maple, black mulberry, tamarisk and possibly cedar were also found (Meiggs, 1982). The Minoans had a passion for building elaborate palaces and wood was widely used in construction. Wood was used for columns, beams, doors, windows, and furniture and also inside mud-brick walls. Minoans had a very good knowledge of the different properties and uses of wood from different species. Minoan Crete became a major nautical power and an important trading post and much wood was used in shipbuilding. Most wood, however, was used as fuel in making bronze, in cooking, making sacrifices, and in the preparation of lime for making plaster for the walls of the palaces. Because of the ready supply of wood, Minoan Crete was probably an exporter of both bronze and timber to the Near East.

Eventually deforestation of Crete caused timber to become scarce. Evidence for this was in the reduced use of timber in construction and its replacement with gypsum. There was a greater amount of recycling of bronze. Relatively more charcoal and less wood was used as fuel. Minoan Crete changed from being an exporter of wood to being an importer from the nearby Mediterranean mainland, particularly Messenia in south-western Greece. Timber became more and more difficult to obtain, and shipbuilding was reduced in quantity and quality. The Minoan fleet became smaller and less reliable, resulting in Minoan Crete losing its nautical superiority and advantage in trade. In 1450 BC Minoan Crete was overtaken by the Mycenaeans who subsequently used Crete as a sheep-grazing outpost. Sheep grazing ensured that degraded forests could not regenerate.

Mycenae

Mycenaean Greece occupied the southern section of the Peloponnese Peninsula. After 1450 BC, the Mycenaeans replaced the Minoans as the main traders with the Near East and this included the trading of timber. At this time Mycenaean Greece was well forested on the hills although the plains had been deforested to some extent by Neolithic and early Bronze Age agriculturalists. Oaks were prevalent on the plains, pines along the coast, firs in the hills and alder and poplar in the marshes (Loy, 1966). Mycenaeans became major exporters of bronze and ceramics, both of which required wood as fuel. Timber was used in building and vehicle construction and their strong shipbuilding industry gave them dominance in trade. The Mycenaeans, like their Minoan predecessors, had a penchant for palaces and large houses, which required large amounts of wood for their construction and upkeep.

The Iliad and the Odyssey are epics attributed to Homer and historically set in Mycenaean Greece. They both frequently refer to forests and to particular tree species. Homer clearly had a respect for woodsmen and craftsmen in wood and he knew the main uses of different woods. Homer frequently compared the falling of a warrior in battle with the felling of great trees. Meiggs (1982) gives many examples. A typical example from the Iliad is when Asios, killed by Diomedes, *'fell as an oak falls, or a poplar, or a tall pine, which craftsmen cut down in the mountains with their newly whetted axes to make ship-timbers.'* This quote, if it is to be taken literally, encapsulates several factors about forests in Mycenaean Greece. It shows that different species were recognized and named, that tree felling with axes was common practice, that trees were felled in the mountains (which may mean that the plains had already been cleared) and that timber was used in shipbuilding. Another quote from the Iliad says that Simoeisios, after being speared by Ajax, *'fell to the ground in the dust like a poplar that has grown up smooth on the edge of a great marsh, and its branches grow at the top. A chariot-maker cuts it down with his gleaming iron axe to bend a felloe for a specially fine chariot, and so it lies drying out by the river banks'* (Meiggs, 1982). This quote reinforces that poplars grew in or at the fringes of marshes but it also shows that there was a preference for knot-free timber for certain end-uses and that the principles of timber seasoning were understood. It could be a mistake to take Homer too literally as he wrote his epics at least 400 years after the events described and his information about Bronze Age Mycenae would have been passed down through the years. The use of the iron axe in the latter quote suggests some poetic licence. Also there was confusion about species identification, not only by Homer, but right across Mesopotamia, the Near East as well as in the Mediterranean (Thirgood, 1981). Homer mentions axes, usually bronze, quite often and with great respect. This respect transferred to the woodworker whose particular skill as a craftsman was recognized and who must have had an important and influential place in society. Homer does not mention saws. Although the Romans claim to have invented saws made of iron, there is good evidence of bronze and copper saws being used in Minoan Crete and ancient Mesopotamia and Egypt (Makkonen, 1969).

Mycenaean Greece followed the typical pattern of rise and decline shown in Fig. 1.1. The buoyant economy caused an unprecedented increase in population necessitating the clearing of forests further away to provide agricultural and pastoral land for increased food production and more timber for fuel and shelter. For example, authorities in Messenia gave tax concessions to settlers who would clear the more remote hilly areas of forests and bring them into agricultural production (Perlin, 1989). Consequently settlement moved into the hills. Woodcutters, bronzesmiths and potters moved north in the Peloponnese Peninsula away from the centres of population and into higher country where the forests remained. However, in due course, poor soil management took its toll. Crop yields were reduced due to declining soil fertility. Sparsely vegetated soils were prone to erosion by wind and water. This was particularly so on denuded hillsides where large amounts of soil eroded and were deposited as sediments downstream. The bare eroded hillsides greatly increased run-off after rain and promoted flooding.

Homer frequently refers to devastating forest fires and floods that uprooted trees and carried them along in the torrents. Forest fires would not necessarily cause significant deforestation unless agriculturalists and pastoralists moved in to occupy the burnt areas and it is likely that this happened to some extent. The clearing of upland forests may have exacerbated the floods and the consequent sedimentation of the rivers of the plains, but floods would still have occurred irrespective of forest clearing. It is easy to overstate the singular role of deforestation in causing floods and increasing sedimentation to streams. Any form of vegetation removal from upland sites will expose soil to erosion and increase the run-off and sediment load to streams. Deforestation, poor agricultural practice and overgrazing would all have contributed to the large scale soil erosion and sedimentation that occurred in the late Bronze Age of Mycenaean Greece. Inevitably the stage was reached in which there was not enough arable land to support the population and not enough readily available timber to support the wood-based industries. Mycenaean Greece received some respite by importing grain and timber from Troy but eventually, from about 1200 BC, Mycenaean Greece became impoverished, population numbers crashed and the inhabitants migrated to better sources of food, fuel and shelter or else retrogressed to a subsistence economy (Perlin, 1989).

Attica

Attica, the countryside around Athens, was well forested in the 6th century BC. Agriculturalists had already cleared some of the soils and probably most of the better soils. Theophrastus, writing in the 4th century BC but looking back through the centuries, recognized the common practice in agriculture of taking the best soils and leaving the worst soils to the forests when he wrote '*use your rich soils for grains and thin soils for trees*' (see Thirgood, 1981). The general pattern that developed was grain on the plains, vineyards and orchards on the foothills and middle slopes leaving forests of pine, oaks and firs on the mountains, which also served as summer grazing. At this early stage, however, the boundaries would have been quite diffuse and there still would have been significant areas of forest remaining on the plains. Olives (*Olea europaea*) were cultivated and elm and poplar were planted as fodder trees. At the end of the 6th century BC, timber was readily available and most of it would have been used for fuel. Also the Athenian navy was small at this time having fewer than 100 ships (Meiggs, 1982).

The situation changed from the beginning of the 5th century BC. Persia controlled the forests of most of northern Greece and all of Asia Minor and was particularly concerned with denying Athens access to shipbuilding timbers. However, with the ready access of local timber in and around Attica for shipbuilding and silver from the mines at Laurium to finance it, Athens increased the size of its navy and became the leading maritime power of the area. This was the most important factor in the victory of the Greek States over Persia at the battle of Salamis in 480 BC. However, the Persians inflicted enormous damage on the houses and buildings of Attica. Supremacy of naval power was again the key to the eventual victory of Athens over Persia in 469 BC (Meiggs, 1982) after which the Athenians were in a position to rebuild their sacked city with style and splendour. The golden age of Athens began. It was accompanied by the inevitable pattern of greatly increased wood consumption to support grandiose building construction (such as the Parthenon), to provide fuel (particularly for the silver mines of Laurium) and to provide shipbuilding timbers for a city-state with greatly increased population (Gomme, 1933) intent on maintaining supremacy of sea power (Perlin, 1989). By far most of the wood consumption was for fuel. Hughes (1994) estimates that 90% of the wood consumption over the whole of the classical Mediterranean was for fuel, particularly for metal refineries and pottery kilns.

Eventually and inevitably Attica became progressively deforested and this was reflected in an increase in the price of wood and the adoption of passive solar designs to maximize the penetration of winter sun into the living areas of the houses. Athenians imported timber from the nearby island of Euboea and further afield. Macedonia and Thrace became the major sources of shipbuilding timber and the Athenians established trading cities on the coast to access these forests. Athenians, after one unsuccessful attempt, eventually took control of Amphipolis, a trading city accessing the plentiful forests of Macedonia. This allowed Athens to supplement its dwindling timber supplies and also to prepare for the forthcoming protracted war with Sparta, the Peloponnesian war. Recognizing that Athens had maritime superiority, the Spartans fought back by targeting the forests. When the Spartans invaded Attica they cut down most of the trees and laid the landscape bare. This not only denied Athens access to local timber to maintain its fleet but also the bare hillsides became very badly eroded, the amount of run-off increased greatly and previously arable lowlands became unproductive mosquito-infested marshes. People living on the land that became marshes re-settled elsewhere, which placed additional pressure on the arable land base. The Spartans then captured Amphipolos in 422 BC, further denying timber supply to Athens (Thirgood, 1981; Perlin, 1989).

The war lasted from 431 BC to 404 BC. For most of this time neither party had a decisive victory and there was also a short truce in the middle of this period. However, the amount of timber required to sustain this protracted war was immense and Perlin (1989) gives a range of graphical examples of this prodigious consumption. Athens unsuccessfully invaded Sicily in 415 BC to get access to timber supplies of Italy and Sicily and in the process lost almost all of its fleet. However, Athens reached agreement with Macedonia to supply timber so that it could rebuild its fleet and subsequently destroy the whole Spartan fleet off the coast of Asia Minor. Sparta had an agreement with the Persian governor of Asia Minor to have access to the forests of Phrygian Mt Ida (Hughes, 1994) and Sparta rebuilt its fleet. In 404 BC, Sparta surprised the entire Athenian fleet while it was beached and its crew were on land looking for food (Perlin, 1989). They destroyed the whole fleet and Athens was defeated. There are numerous examples in the classical Mediterranean of security of access to forests being the key to maritime superiority, which in turn determined the balance of power (Hughes, 1994).

At the end of the Peloponnesian war, Attica was an environmental mess. There had been wide-scale deforestation and fertile soil on the hill country had eroded and left relatively impermeable sub-soils that shed more rainfall to marshes and the sea. Consequently the upland soils stored less water and previously reliable springs dried up. The dreadful state of Attica alarmed some 4th century BC Athenians. It is clear that some writers of the period appreciated some of the basic aspects of managing landscapes to retain their productivity. Plato (Critias, 111C) appreciated what Attica had been like and what it had become:

> 'The result is that Athens is now like one of the small islands, the skeleton of a sick body with barely any flesh on it. In those early days the land was unspoilt: there was soil high upon the mountains, and what we now call shrub had fields full of rich earth. There was abundant timber on the mountains and of this you can still see the evidence. Some of our mountains can now only support bees; it is not long ago that trees from these mountains supplied roof-timbers for the largest of our buildings, and the timbers are still sound. And there were many other tall trees cultivated which provided food in plenty for livestock. The year's rain did not, as now, run off the bare earth into the sea, but the water coming down from the hills was preserved underground and fed springs and rivers. One can still see sacred memorials where springs once existed.'

This is also the period in which Theophrastus (370 – 285 BC), a pupil of Aristotle and Plato, wrote his Aetiology of Plants and Enquiry into Plants, which was the most extensive and reliable treatise of the period on the classification, description and uses of plants and particularly trees. It is not until Pliny and his Natural History, completed in AD 77, that there was a work of comparable stature. It is also somewhat ironic that significant advances in the standards of agriculture were made during the environmental degradation of the 5th and 4th centuries BC. Superior breeds of plants and animals were selected, manures and mineral fertilizers were used, crops were rotated and a literature on agricultural practice developed. Perhaps this was a response to the shrinking of the arable land base necessary to support a burgeoning population.

The Romans destroyed Corinth in 146 BC and Greece came under Roman control. The cultured Greeks resented the less cultured but militarily superior Romans. Rome's record of environmental management is a poor one and Greece felt the impact of this. Rome sacked Piraeus in 86 BC and cleared most of the remaining trees on Attica (Meiggs, 1982). The Greeks had a relatively conservative ethic and recognized that the land needed to be rested. Rome on the other hand had a more aggressive approach, considering it could maintain soil fertility without the need to rest the soil. Rome replaced the small-scale peasant farm of the Greek city-states with large-scale, centrally-organized, slave-based farming units to feed Rome and its dominions (Thirgood, 1981). Indeed Toynbee (1935) attributes this as the main reason for the eventual collapse of the Greek city-states.

Macedonia

Macedonia was a relatively well-forested part of the Mediterranean. In the 4th century BC, Athens imported its wood from Macedonia because it had none of its own. Other powers in the region wanted control of the Macedonian forests and a succession of alliances and betrayals ensued to try to achieve this. Eventually Phillip of Macedonia, father of Alexander the Great, conquered Amphipolis in 356 BC and consequently had complete control of the supply of timber to the whole of Greece. Phillip used his abundance of timber as bribes and incentives to turn governments against each other and to further his own cause. Later the Macedonians used their forests for their own development and subsequently became the dominant power in much of the world known to them at that time. This continued until Rome conquered Macedonia in 167 BC. Macedonia became a Roman province in 148 BC and the Romans, recognizing the relationship between timber and the balance of power, prohibited the Macedonians from felling their own trees (Perlin, 1989).

Rome

Prior to Rome coming to prominence it was surrounded by forest. Indeed most of Italy and its neighbouring islands were extensively forested. Fir, spruce (*Picea*) and pine were found at the higher elevations and beech, oak, hornbeam (*Carpinus*) and linden (*Tilia*) at lower altitudes. When the Etruscans controlled Rome in the 6th century BC, supply of timber greatly exceeded demand and relatively little forest clearing had occurred to support agriculture. After the Gauls sacked Rome in 390 BC there was pressure to quickly rebuild the city and citizens were given permission to cut timber wherever they pleased providing they constructed their home within one year (Meiggs, 1982). There still remained, however, extensive forests around Rome. In 310 BC, the Ciminian Forest, about 100 km from Rome was reported by Livy as being so dense that it was impenetrable. The protracted wars between Rome and Carthage in the 3rd century BC required a navy and the ships were built in Rome made from timber, especially fir, probably from the Apennines. Carthage was finally defeated by the end of the 3rd century BC and Rome had acquired Spain, Corsica and Sardinia. In the 2nd century BC, Greece, Macedonia and Syria became Roman provinces. The Roman Empire at its zenith in the 1st century BC had some degree of control over all of the lands surrounding the Mediterranean and Western Europe as far as Britain.

During this colonial expansion, Rome changed its agricultural practice from subsistence farming on small properties to a trading enterprise based on broad scale agriculture and grazing, often using slave labour. The population of Rome greatly increased and lowland forests around Rome and elsewhere in Italy were cleared to provide land for agriculture and grazing. Timber was taken from the higher and more remote forests to meet the greatly increased consumption of wood for fuel and construction. When supplies became scarce, Rome moved into her colonies and repeated the exercise. Thus large areas of forest in North Africa and Spain were cleared to provide grain to Rome. When Rome ran short of timber in Italy, it imported timber from Spain, North Africa, Gaul and Britain. Even though there was considerable deforestation and environmental degradation at the hands of the Romans, Italy was spared the full brunt of this because Rome exported much its destructive behaviour to the colonies.

Many of the forests of Italy were under state control with systems of management and regulation. Forest guards not only guarded the forest from unauthorized activities, but eventually came to control cutting, harvesting, hunting, watershed protection and silviculture. Sacred trees and groves were set aside as reserves. Coppicing, thinning, pruning, plantation establishment, tree breeding, controlled pollination and introduction of exotic species all occurred to some extent (Thirgood, 1981). However, the insatiable appetite for food and fuel overcame any semblance of sustainable management.

The Roman colonies brought considerable wealth to Rome and the citizens of Rome responded with excesses. The dramatic increase in population meant a boom in private housing construction. Hellenistic influences in the 2nd century BC encouraged more elaborate public building construction and over the next several hundred years there was a passion for greater and more magnificent buildings. Timber species were fir, pine, oak, elm, chestnut, poplar and cypress. The architects of the day had good knowledge of their wood properties and uses in construction. Building activity was assisted by the relatively regular occurrence of disastrous fires in the city. Julius Caesar initiated an unprecedented building boom commencing about 50 BC, which continued after his assassination in 44 BC through to and following the death of Augustus in AD 14. Italy was still well forested at the commencement of the building boom and it was only in the later days of the empire that Italian forests came under significant pressure, particularly for fuel. The amounts of timber used in construction would have been small relative to the amounts of wood or charcoal used in cooking, heating and cremation.

Wealthy Romans used large amounts of wood to centrally heat their villas. Also Romans had a penchant for hot baths and the amount of wood required to support this habit was prodigious. Emperors competed for constructing the most elaborate public baths and towards the end of the empire there were over 900 public baths, some catering for more than two thousand bathers at a time. Whole forests close to Rome were cleared to support the habit and considerable effort was directed towards searching far and wide for wood to fuel the baths. Industries requiring wood as fuel developed in parallel with the rise of the empire. The amount of wood-fired bricks increased and wood fuel was used for smelting iron and bronze and for making pottery. Water pipes and aqueducts were made from lime-based concrete or clay, both of which required wood fuel. Glass manufacture developed in the 1st century BC and subsequently consumed large amounts of wood as fuel. Towards the end of the empire there were clear indications of shortage of wood for fuel. Charcoal was substituted for wood, fired bricks became thicker and used less mortar, glass was recycled, and there was a return to mud-brick construction. Fuel shortages progressed to the colonies. Shortage of local supplies of fuelwood around smelting operations in southern France and in Britain caused the smelters to relocate to better-forested areas (Perlin, 1989). Copper mining in Cyprus and iron smelting on Elba depleted the islands' forests although there was still significant amounts of forest on Cyprus at the end of the Roman Empire (Meiggs, 1982). Silver mining in Spain over four hundred years was particularly destructive of the forests, precipitating a shortage of silver and a dilution of the amount of silver in coinage. This was caused by shortage of wood and not shortage of silver ore (Perlin, 1989).

Rome, like Greece, suffered the general pattern of environmental degradation shown in Figure 1.1. Over-population and urbanization promoted deforestation, soil erosion, siltation, overgrazing and loss of soil fertility. A flood in Rome in 241 BC inundated the lower parts of the city and blocked up the sewers. This in itself was not necessarily an indication of deforestation. This, however, coincided with the time when the population of Rome began to increase, in turn accompanied by agricultural expansion. Records indicate that floods increased after that date (Hughes, 1994). Deforestation caused significant erosion in Italy. Silt from deforestation of the Tiber River catchment landlocked Rome's coastal port. The Po River catchment was originally covered in forests but the lowlands were cleared for agriculture, followed by the foothills and then some of the higher country. Consequently, by the time of Christ, marshes had developed near the coast and some coastal settlements had become surrounded by water. Modern deforestation of the headwaters of the Po has greatly contributed. Today, Ravenna (the coastal port in the times of the Roman Empire) is 10 km from the sea and the riverbed is up to 3 metres higher than the surrounding floodplain over 350 km of its course (Winters, 1974).

Agriculture expanded at the expense of forests. Rome offered free possession to anyone who would clear the forest and bring it under cultivation. The prevailing attitude was that progressively clearing the forests for agriculture was a sign of progress. Rome was particularly efficient at deforesting its colonies. For example, by the end of the 1st century AD, North Africa was sending sufficient grain to meet two-thirds of the requirement of Rome's population of approximately one million people (Meiggs, 1982). The cultivated area to support this enterprise was originally covered in good quality forest. Rome depended on its colonies to supply its food and if the harvests failed the consequences for Rome were dire. Towards the end of the empire, famines in Rome were not unusual.

Like their Greek counterparts, the Romans understood some of the basic principles of good agricultural practice and developed a literature in this area. Animal (including human) manures were used, composting was recommended, rotation with legumes was practised, limestone and chalk were used as fertilizers, marble was burnt to make fertilizer, soils were fallowed, weeds were controlled by ploughing, and hill country was terraced to prevent soil erosion (Hughes, 1994). Even so, the soils lost their fertility. Lucretius (translated by Humphries, 1968, see Hughes, 1994) recognized decreasing harvests.

'But the same earth who nourishes them now
Once brought forth, and gave them, to their joy
Vineyards and shining harvests, pastures, arbors,
And all this now our very utmost toil
Can hardly care for, we wear down our strength
Whether in oxen or in men, we dull
The edges of our ploughshares, and in return
Our fields turn mean and stingy, underfed,
And so today the farmer shakes his head,
More and more often sighing that his work,
The labour of his hands, has come to naught.
When he compares the present to the past,
The past was better, infinitely so.'

Columella (Rust 3.3) noted that no one alive in his day could recall when the harvest produced as much as four times the seed that had been sown. Clearly the agricultural principles espoused in the literature, particularly by Columella, were not followed. Also clearing for agriculture probably moved onto marginal lands with soils of lesser fertility. The frequent wars that the Romans instigated greatly disrupted agriculture. Armies lived off the land, targeted crops for destruction, and conscripted or killed farmers, thereby denying them the opportunity to maintain their terraces and manure the soil. Siltation studies show that catchment erosion was most rapid when there was war in that catchment. Declining agricultural production in Italy was supplemented by imports from the colonies and much cultivated land in Italy was given over to pasture. The inevitable consequence of agricultural decline was famine and depopulation (Hughes, 1994).

While it is true that good fertile soils support good forests, the reverse, that good forests must grow on good soils, is not necessarily true. Pliny partly recognized this when he said *'A soil in which lofty trees do brilliantly is not invariably favourable except for those trees: for what grows taller than a silver fir? Yet what other trees could have lived in the same place?'* (Hughes, 1994). Given time forests can support large trees even on relatively infertile soils. They do so by being very efficient at recycling nutrients. A significant proportion of the nutrients on the site are contained in the trees and when these trees are felled for agriculture, the nutrients are lost from the site. Roman farmers cleared good forests only to find that the soil was not good for permanent agriculture.

Towards the end of the Roman Empire both capital and labour became scarce in Italy and depopulation continued. Cultivated land and pastures were abandoned and reclaimed by the forest. The country was weak, disorganized and ripe for invasion by the Goths and the Vandals in the 5th century AD (Thirgood, 1981). This marked the end of the Roman Empire and the commencement of the 'Dark Ages'. There were many factors contributing to the decline of the Roman Empire. The reduction in agricultural productivity and the environmental degradation through deforestation, soil erosion and overgrazing were major contributing factors. A period of restoration of the forests began after the fall of the Empire.

The Mediterranean after Rome

Considerable regeneration of forests would have occurred during the Dark Ages. In the Middle Ages and beyond, however, cycles of depletion and regeneration of forests occurred up to and including the present. Overall, there has been a net decline in the forests since Roman times and today the forests and woodlands of much of the Mediterranean are very degraded, fragmented and much reduced in area. They are largely restricted to poor soils in the higher country. In classical times there was little concern about the plight of the forests and little appreciation of the consequences of over-exploitation. However, it would be unfair to blame the classical and pre-classical civilizations for the deforestation and forest degradation evident in the Mediterranean today. There were still relatively extensive tracts of forest in the better-watered and/or less accessible country at the end of the classical period. The Mediterranean remained a centre of civilization for many centuries after the fall of the Roman Empire. The centre of activity was first in the east around Constantinople but later moved to the west around the maritime republics of Pisa, Genoa and Venice and later Spain and Portugal. The rise of these maritime trading powers and the consequent increase in population and living standards re-imposed the need for clearing forest for agriculture and timber. It also heralded the beginning of a whole new era of colonialism, because other continents had now become accessible.

Many of the forests that were severely exploited recovered and indeed have been through several cycles of exploitation and recovery. These cycles happened at different times in different places. For example, deforestation of Crete was a factor in the demise of the Minoan civilization in 1450 BC and yet cypress imported from Crete was used for the construction of the Venetian fleet in the Middle Ages. In the 16th century, the Idhi mountain range in Crete was covered in cypress; a century later it was described as a barren spot (Meiggs, 1982). The city of Iraklion is located near the site of ancient Knossos, the major city of Minoan Crete. In the 17th century AD, Iraklion repeated the deforestation of the ancient Minoans such that no more local supplies of firewood were available (Clutton, 1978; Perlin, 1989). The forests on Cyprus provided timber for shipbuilding (Meiggs, 1982) and bronze smelting (Perlin, 1989) in the Bronze Age. This depleted the forests on Cyprus and the kings on Cyprus responded by conservatively managing their forests only to have subsequent rulers take advantage of their efforts and deplete the forests further (Hughes, 1994). Meiggs (1982) gives evidence describing the gradual clearing of the forests on the plains of Cyprus in the 3rd century BC, but noted that Cyprus was well forested in the 4th century AD. At the beginning of the 14th century, Cyprus was extensively deforested to grow sugar cane for two centuries (Westoby, 1989) after which the sugar trade found alternative sources of supply. Some recovery of the forest in Cyprus followed. In the 18th and 19th centuries Cyprus again increased its population and the forests were seriously depleted for fuel, and overgrazing by goats hindered their recovery. Cyprus today has well managed forests and sets a fine example to surrounding countries (Thirgood, 1981). Despite the deforestation and the degradation of the Po Valley by the Romans, parts of the valley were still well forested in the 5th century AD because they acted as shelter to the invading barbarians (Winter, 1974). After the decline of Rome the population of Italy collapsed and the forests regenerated. Good local timber was available to the Renaissance centres of Rome, Florence, Venice, Pisa and Genoa and the cycle of deforestation, soil erosion and overgrazing started over again. These city states prospered and their populations increased as did the proportion of the population not directly involved in producing food. Accordingly, they became big importers of food which they obtained by deforesting areas of the wider Mediterranean. Venice and Florence responded to the pressure on the forests by establishing silvicultural regimes that mark them as the pioneers of European forestry. North Africa was extensively deforested by the Romans but subsequently recovered following the decline of the empire. It then became depleted yet again on the establishment of a pastoral culture (Thirgood, 1981). Essentially deforestation over the Mediterranean occurred where populations increased and reforestation occurred where populations decreased and people moved out of the area.

These examples are testimony to the wonderful capacity that, given time, forests have to regenerate and recover after even the most destructive logging. This should not be used as an excuse to degrade forests with cavalier abandon. The forest degradation and deforestation in the Mediterranean have been associated with depletion and extinction of wildlife. The replacement of wildlife with the domestic goat and the feral cat has been disastrous. Arguably the most destructive agent in arresting or delaying the recovery of degraded forests has been overgrazing. It follows that the best way to promote regeneration and recovery of degraded forests over the whole of the Mediterranean is to protect the area from grazing.

There is no doubt that the ancients were responsible for extensive deforestation, overgrazing, soil erosion and scarring of the landscape. However, their capacity to degrade was limited by population, technology and transport. The worst period for environmental degradation in the history of the Mediterranean was the late 19th and 20th centuries. The region now supports very large and rapidly increasing human populations. Improved transport and technology have provided access to previously inaccessible forests. Wars aided deforestation in classical times and this was also the case in the 20th century where the First and Second World Wars were particularly hard on the forests of Italy and Greece. In the context of the forests of the Mediterranean having gone through a series of cycles of degradation and recovery, the forests are now (hopefully) at the end of a particularly intense destructive phase and looking forward to recovery. Over the last 30 years or so the pressure on the Mediterranean forests for fuel, agriculture and grazing has declined (Ball, 2001). There has been considerable progress made in forest management by some countries in the region.

Temperate Europe (with particular emphasis on England, France and Germany)

The Romans not only depleted forests in their vicinity but they also depleted the forests of Spain, the Rhone Valley in France, and in southern England and Wales (Westoby, 1989). After the fall of the Roman Empire, England had only 15% of its area as forest. Most of continental Europe, on the other hand, was still well forested until the Middle Ages. Dimbleby (1976), however, provided evidence of significant deforestation having occurred in temperate Europe in the Mesolithic and Neolithic. This suggests that some recovery of the forests had occurred prior to the onslaught of the Romans. It is probable that almost all of the forested areas in temperate Europe have come under the influence of human interventions at some time or other and that wilderness as such does not exist.

The Peasants' Revolt

Although the details of the history of forests in Britain, France and Germany differ, there are some common threads. Prior to the Middle Ages, the occupants of the land were somewhat similar to the hunters, gatherers and early farmers of the Neolithic. The forests were used as communal property for fuel, building materials, hunting and supplementing their food supply. The occupants lived in close proximity to the forests and had a strong conviction that they had a right to use the forest. Their customary rights to the forest developed over a long period. The attempted repression or extinction of these rights caused great dissent in all three countries over a period of a thousand or more years up until the time the rural culture was replaced by an urban culture in the 19th century. In other parts of the developing world where a rural culture still exists, the repression of customary rights to the forest remains a significant issue today.

Two ingredients came together to mount an onslaught on customary rights. The first of these was a power structure with the capacity to enslave or subjugate, and the second was recognition that the forests have value. The power structure that developed was one where the aristocracy (kings, barons, dukes, lords, princes, clergy or whoever) demanded that the peasant work for them first and for themselves last. The epitome of this was the development in medieval times of the feudal structure where the peasant or serf ostensibly was given a small parcel of land by his feudal lord who extracted a proportion of the harvest as rent. The peasant might have some rights to use cleared land for grazing and to use the forest for fuel and other needs. This reinforced the peasants' belief that they had inalienable customary rights to the use of the forest. Feudalism essentially bound the peasant to the land for life, with no chance of a reprieve. The second ingredient was the realization that the forests have value and the overriding value placed on the forests by the aristocracy was as a resource for hunting for pleasure and for food. Consequently the aristocracy set about appropriating the forest for their own pleasure, systematically eroding the customary rights of the peasant to the forests and instituting and enforcing draconian penalties for poaching and trespass. Also, because the aristocracy controlled the forests, they could provide timber and fuel to those who could pay their price and it also put them in a strong position during periods of shortage in wood supply. By the Middle Ages, much of the forest was in the hands of the aristocracy and the church. While this caused great bitterness among the peasants, it had the one advantage of protecting the forests from undue exploitation (Barton, 2002). The first serious attempts at forest management in Europe were to manage game rather than regulate timber supply. Forest history in Europe is dominated by the struggle between the aristocracy to extinguish customary rights and the peasants to keep them.

In Britain, William the Conqueror strengthened his hold over the royal forests by producing, after the Battle of Hastings, the 'Charter of the Forest'. This was a set of forest laws aimed at legitimizing his and future kings' rights to greater control over existing forest and to acquire new forest. The term forest in England in medieval times was used quite loosely. Effectively it was any unoccupied land that was not cultivated, on which the native vegetation of the district grew and which may or may not have included trees (James, 1981). William also commenced the Domesday Book, which in essence was an inventory of what he could claim, or intended to claim, as his own property. This included the forests. William's forest laws upset the English barons but it took a further 200 years before they could respond. A new 'Charter of the Forest' issued in 1225 was directed at reducing the power of the king over the forests but it had little effect on the plight of the commoners. The Charter was re-enacted more than 20 times over the next 150 years, which serves to emphasize the importance that the kings placed on the forests and the mistrust that the barons had in the kings. The power of the king over the forest fell away during the 16th century largely owing to indifference and corruption. Ownership and control of the forests still lay in the hands of the powerful few and attempts to enclose common land were met with great resistance from the tenant farmers until such time that customary rights were permanently extinguished.

The English tenant farmer relied on the forests for fuel, grazing land and food and in lean times this could make the difference between life and death. They resisted any attempts to curtail these rights. The Peasants' Revolt against the landed clergy in England in 1381 was one of the first examples of organized resistance. The legend of Robin Hood who 'steals from the rich to give to the poor' epitomizes the anger that the peasants felt about the unreasonable demands of their landlords and the erosion of their customary rights in the forests. Poaching was one of the main forms of resistance but occasionally it was outright forest destruction. In 1723 the British Parliament passed the 'black act' which created 50 new capital offences including hunting, stealing or wounding deer, unauthorized cutting of trees and even blackening the face as a disguise (Westoby, 1989). Retribution against offenders was swift and terrible and castration of offenders was not uncommon.

Similarly in France, royalty, the nobility and the church progressively took possession of the forests during the Middle Ages. As in England, the principle purpose was for hunting and the conservation of game. Again this had the unplanned but welcome outcome of conserving the forest. The feudal system that developed in the 12th century conferred communal privileges and rights but these were always at risk of being eroded by the nobility. Ordinances regulated forest use in France from the 12th to the 17th centuries. By the middle of the 17th century, however, forest devastation, poor management and crimes against the forest had progressed so far that Colbert recognized that 'France will perish for lack of woods'. He instituted the forest ordinance of 1669 to bring some order into the use of the forests and to prosecute those who did not comply (Fernow, 1907). The conflict between forest owners, gamekeepers, foresters and forest guards on the one hand and poachers and peasants exercising what they considered to be their customary rights on the other, progressed from the 12th century through to the beginning of the 20th century. Following the French Revolution in 1789, nearly 3 million hectares of forest were confiscated, becoming part of the national domain which the peasants interpreted as their common property. 'Theft' of timber and fuel and other 'crimes against the forest' increased. The forest code of 1827 was imposed to conserve the forest and to regulate yield but the peasants saw this as a further erosion of their rights to collect fuel and timber and to graze their animals. This led to violence over the next couple of decades. Westoby (1989) recounts a piece of savagery where the peasants would cleave a log with an axe, force the hand of the forester into the cleavage and then withdraw the axe. After the mid 1800s the number of crimes against the forest decreased. By the end of the 19th century the social order had changed to the extent that the customary rights of the rural poor were no longer an important issue for them, effectively extinguished.

In Germany after about 400 AD the people were mainly herders, gardeners and hunters. Outside of their immediate yard and garden all property was communal and their rights to the forest were at first unchallenged. The forests largely remained communal property until the 13th century but customary rights were eroded over time. As early as the 7th century there were tracts of forests controlled by the kings and the nobility. The overriding value of the forest to the nobility was for hunting but they also extended their income, influence and territory by clearing land for agriculture and grazing and by selling timber. It was at this time that the German word 'forst' originated, from which the English word 'forest' is derived. The word for normal woodland was 'silva' but it was called 'forst' if it was necessary to secure a property against the interests of other claimants (Kiess, 1998). From the 7th century the kings forbad trespass on their forests and in the 10th century they extended this to territory not belonging to them. They established banforests which under special forest laws enforced by forestmasters consolidated the right of the kings to hunt exclusively wherever they chose. The wooded areas were not continuous over the country but occurred as a mosaic with cleared land. The size of the forest, as in England, was often expressed as the number of pigs it could support. This demonstrates that the forests had some agricultural significance. The hunting rights of the sovereign remained supreme, however, and ensured that the forests, irrespective of ownership, were not devastated (Kiess, 1998). The kings granted land to barons, bishops and officials in return for hunting rights and the barons and bishops in turn declared the ban for their own use which ultimately meant that the commoners had restricted customary rights to hunt and fish in the forests. This practice of higher ranking persons granting land to lower ranks in return for service marked the beginning of the feudal system in Germany. By the time it trickled down to the lowest rank (the vassal) the rights to use the forests were somewhat reduced although not yet extinguished (Fernow, 1907).

The preoccupation of the nobility with the right to hunt helped to conserve the forest. As in England and France, however, the progressive whittling away of property and customary rights among the common folk led to resistance and sometimes outright rebellion. From the 12th century until the 18th century, squatters appropriated forested land without regard for property rights (Fernow, 1907). Again, as in England and France, penalties for trespass and poaching were severe. Westoby (1989) recounts an act by the Archbishop of Salzburg (in Austria) who in 1538 publicly executed a poacher by having him sewn into deer skins and thrown to the bloodhounds. The church (Catholic and later Protestant) was a major forest owner in Germany. There was a peasant rebellion in southern Germany in 1525 which, among other things, sought to restore the customary rights to hunt, fish and collect wood from the forest. This rebellion was savagely suppressed by the union of Protestant princes, the Swabian League, with the apparent approval of Martin Luther who strongly supported the property rights of the princes (Westoby, 1989).

After the Middle Ages, from the commencement of the 16th century, the feudal system deteriorated and more modern organizations of property and power took their place but the displaced feudal vassal still retained some residual rights to the use of the forest (for grazing stock and collecting litter and wood). From the 16th to 19th centuries there was considerable pressure on the forests. It was during this period that the foundations for silviculture and management were established and many restrictions on forest use were introduced. However, effective forest management was seriously impeded by the exercise of the residual customary rights of the rural landowner/tenant until these rights were effectively extinguished in the 19th century. The restrictions on forest use in Germany had a profound flow-on effect on political and socioeconomic thought. Karl Marx's path to socialism commenced with his concern over the laws relating to the theft of wood from the German forests.

The Fate of the Forests

England was extensively deforested by the Romans and consequently the availability of agricultural land has not been a major issue since. In continental Europe though, clearing forest for agricultural land was a major cause of deforestation after the Middle Ages. The same general principles applied to the deforestation of England, France and Germany as applied in classical times. Deforestation followed increases in population and living standards. The pressure was taken off the forests during the periods of low population following the sacking of Rome in the 5th century and the Black Death in Europe in the mid 14th century. For 200 years following each of these events the forests had time to recover.

Population and wealth increased in the 12th, 13th and first half of the 14th century. This was also an active period of deforestation, for farmland in France and Germany and for wood products in England. Germany was extensively deforested during this period and ordinances for strict felling and forest use were issued in response. The first prohibition of clearing in Germany was in 1165, followed by ordinances to limit the cut, preserve the best trees and to use only dry wood. In England in 1184, the Assize of Woodstock proclaimed forest laws, distinct from common law, which controlled forest use mainly to protect deer. Over the next 200 years forest legislation in England was mainly about the rights of the king or his barons to the game rather than for the care of the forest.

The hold of the kings and nobility over the forests had weakened considerably by the 16th century after which the emphasis on the use of the forests switched from hunting to the provision of wood. This marked the start of another period of pressure on the forests, which was most intense in the 17th and 18th centuries. This was a period of great devastation of the forests caused by increased fuelwood consumption (for cooking, heating, smelting metals, manufacturing glass, producing salt), building construction, pit props for mining, shipbuilding (England and France), fires (Germany), civil wars, famines, clearing land for agriculture (mainly Germany and France), poor forest management, theft of firewood and forest products, browsing of regeneration by game, lack of forest inventory, and corruption and ignorance of officials. In France, the forest area decreased from about 35% at the beginning of the 16th century to about 25% by the middle of the 17th century (Ball 2001). Following the 1762-1776 famine in France, tax concessions were given to farmers to encourage them to clear even more forest for agriculture. The shortage of firewood in England in the 17th century prompted the burning of wheat stubble, weeds and animal dung which reduced the fertility of the soil. The reduction in forest area and the poor condition of the residual forest in all three countries promoted a raft of regulation to limit forest use and the development of rudimentary forest management.

In England during this period (16th to 18th centuries) there were regulations to protect coppice, promote regeneration, retain underwood, limit forest clearing, conserve firewood, limit the use of wood as a fuel for smelting iron, deal with unlawful cutting, limit agistment of grazing stock to ensure food was available for deer, encourage the planting of trees, and to change ship design (James, 1981). Colbert's French Forest Ordinance of 1669 recommended replacing local regulation and customary use with a national policy and, following the French Revolution of 1789, even more forest land came under state control. Regulations on forest use included control of grazing, hunting, replanting, road building, conservation of firewood and restriction of felling and clearing. The Forest Code of 1827 imposed further limitations on use of French forests. In Germany the first regulations were to protect the forests for hunting but these later progressed to conserving timber. Regulations restricting forest use included imposing diameter limits for felling, using building codes to ensure that wood was used efficiently, encouraging alternatives to wood in building construction, having hedges and ditches instead of wooden fences, using poorer woods for fuel, using windfalls in preference to standing trees, prescribing economies in charcoal burning, substituting turf and coal for firewood, regulating the use of litter and protecting regeneration from grazing. Eventually no felling of trees at all, on private or public land was permitted without permission of a designated forester (Fernow, 1907). Devastation of forests in other countries in temperate and boreal Europe occurred over roughly the same time scale. Regulations to control forest use were introduced into Denmark and The Netherlands in the 18th century and into Norway and Sweden in the late 19th century (Ball, 2001).

The Rise of Silviculture and Forest Management

Pollen records demonstrate that the very early agriculturalists in Western Europe deliberately encouraged and may even have planted hazel and other species around their homes. They understood coppicing and pollarding, and also thinning and spacing to encourage the growth of favoured trees. The ancient Romans were perhaps the first to establish a body of silvicultural practice (coppicing, thinning, pruning, tree breeding, cross pollination, plantation establishment) but at the onset of the Dark Ages this information became lost and had to be rediscovered one thousand or more years later. Rudimentary ideas about seed trees and diameter limits for felling were established in Germany before the end of the 15th century but it was from the 16th century on that England, France and particularly Germany made progress in silviculture and management in direct response to the over exploitation of the forest occurring over this period. Silviculture and forest management progressed from trial and error in the 16th century through to scientifically based technologies in the 18th century and beyond.

In England, the preoccupation was growing oak, particularly to support the shipbuilding industry. The main silvicultural systems were coppice and coppice with standards. Coppicing occurs when cut stumps (stools) or their root systems vegetatively sprout (coppice). Coppice with standards is a combination of coppice with older and larger trees that originate from seed rather than stools. A statute in 1482 allowed the closure of forest areas for seven years to prevent grazing of regeneration and coppice. An Act in 1543 introduced thinning as a silvicultural tool and established that a minimum number of 12 trees per acre should be left. Towards the end of the 15th century oak and beech seed were being directly sown into the forest and the first oak plantation established from seedlings rather than sown seed was established in the mid 16th century. By the beginning of the 17th century, nurseries for raising oak and beech seedlings were more common. In 1664, Evelyn published his Sylva in which he recommended that cutting of coppice should follow a regular plan to ensure sustained yield. Evelyn's Sylva is the pivotal discourse on early silviculture and forest management in England. Later in the 17th century coppice was being grown on fixed rotations and in the 17th and 18th centuries there was a considerable increase in the area of forest set aside for timber production. These came from enclosing and managing existing forest areas and hedgerows. The forests were coppice alone or under standards of mainly oak, ash, birch, hazel and alder. However, in addition to coppice there were increasing areas of plantations being established.

The crown forests remained in poor condition into the 19th century because of poor management and corruption, damage from grazing stock and commoners still exercising their residual customary rights. Crown forests improved, however, under the tenure of the Commissioners of Woods and Forests in 1810 and the Forestry Commission in 1923. Ultimately it was realized that the oak plantations were not capable of producing timbers of a suitable size to the navy quickly enough and landowners turned to the faster growing conifers. Even though the conifers Norway spruce (*Picea abies*), silver fir (*Abies alba*), European larch (*Larix decidua*) and Corsican pine (*Pinus nigra*) had been introduced from continental Europe in the 16th to 18th centuries, an escalation in the planting of conifers did not occur until Douglas fir (*Pseudotsuga menziesii*), Sitka spruce (*Picea sitchensis*) and giant fir (*Abies grandis*) were introduced from North America. Scientific forestry and higher forestry education did not commence in England until the end of the 19th century. This was largely due to the influence of French and particularly German foresters (James, 1981).

The development of silviculture and forest management commenced earlier in France and Germany than in England. In Germany, the first example of a forest being divided into felling areas was in 1359 and the first undisputed evidence of a conifer plantation being established was in 1368. An oak plantation was recorded in 1491. Coppice and coppice with standards was developed by 1450 and cutting limits were imposed in 1488. The retention of seed trees after logging to allow regeneration was introduced in 1524 and the protection of seed trees against wind damage was achieved in 1565 by leaving groups (and later strips) of trees around the seed trees. By 1530, there was the recognition that thinning improved growth rates of the residual stand. By 1580, coppice was being grown in planned rotation lengths and in the 17th century a literature on seed and nursery management developed. In 1720, the shelterwood silvicultural system for protecting the young developing crop was introduced and from the mid 1700s, plantations of firstly oaks but later conifers were being established to enrich poor areas of native forest or to plant waste land. Agroforestry

(Waldfeldbau) was promoted in 1744 as a means of diversifying farm incomes. The first account of the scientific basis of thinning was in 1761 and in 1764 discounted cash flow analysis was first used for comparing management regimes. Volume tables were first produced in 1770 and yield tables in 1785. By 1788 the concept of a 'normal' forest and of sustained yield was well understood (Fernow, 1907). A 'normal' forest is a mathematical construct, an ideal that does not occur in reality. A normal forest is one with a range of age classes such that if the oldest age class is periodically removed, the next oldest age class will take its place over the period between this and the next harvest. By so doing a sustained yield and a constant age class distribution will be maintained. This is shown for a plantation or clearcut system in Fig. 1.2, but the same principles apply to an uneven-aged forest where the age class distributions are segregated vertically on the same area of ground rather than horizontally on different areas of ground.

There is a significant German literature on silviculture and forest management in the 18th century of which Carlowitz (1713), von Moser (1757) and Stahl (1772-1781) are perhaps the most significant contributors. Forestry textbooks were written and lectures on forestry presented at universities. Several forestry schools were established towards the end of the 18th century. The first was established in Ilseneburg in 1768 and the most influential in Zillbach in 1795, later transferred to Tharandt in Saxony in 1827. Students from this school established forestry schools at Nancy in France (1825) and in Spain (1848). Forestry schools were established in Austria and Russia early in the 19th century and in England and the USA in the latter part of the 19th century (Westoby, 1989). The German influence was strong in the early forestry schools and in the development of forest science in these countries (Lorentz in France; Brandeis and Schlich in England; and Fernow in the USA and Canada).

In the 19th century the forests of Germany became increasingly consolidated into the state and a lively discussion ensued about the relative merits of different silvicultural systems (selection, shelterwood and clearcutting). Selection systems are where single trees (or small groups) scattered all over the forest are removed more or less evenly and continuously. This supposedly mimics the 'natural' forest where large individual trees die and fall over to make place for natural regeneration. The shelterwood system is where the larger trees are felled progressively rather than all at once so that some protection is offered to the young regenerating crop. Clearcutting occurs when coupes of a specified size are clear felled at intervals to develop and maintain a range of age classes over the forest.

In France (as in England) forest management was more by restrictions on forest use than by active silvicultural intervention. Colbert's ordinance of 1669 enshrined in law the requirement to keep a certain number of seed trees for all forest types under all climates and all soils and to some extent this inhibited the development of silviculture for the next two centuries. This continued until French forest practice came under German influence. However, France was the first to recognize the role of forests in protecting soil and since the beginning of the 19th century, France has established large areas of plantations to stabilize previously deforested and overgrazed areas, particularly coastal sand dunes and mountain slopes (Barton, 2002). These forests now form an important part of the French forest estate. A School of Forestry at Nancy commenced in 1825 and still exists today. It promoted German style forest science. This was not always easy because it was sometimes challenged as being unpatriotic. The argument between the relative merits of coppice and selection (French) versus shelterwood (German) was a major issue (Fernow, 1907).

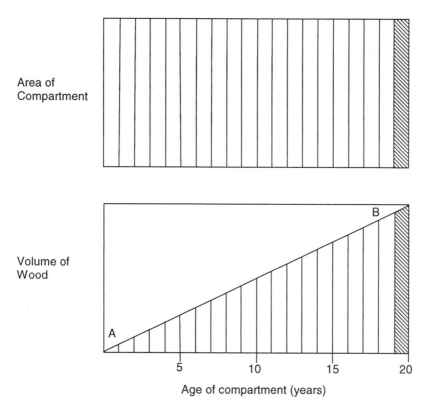

Fig. 1.2. A normal forest. This example is for a plantation where one compartment of rotation age 20 years is harvested each year and replanted in the following year. This will produce a sustained yield at rotation age and a constant age class distribution across the forest. (The line AB is actually a curve which is concave or convex depending on species, growing conditions and rotation age.)

Europe Today

German and French professors and foresters established forestry as both a science and a profession and presented it to the world. Even so, the development of scientific forestry in Europe was preoccupied with timber and game. Other roles of forests were not well understood or acknowledged. It was not until the 19th century that the role of forests in protecting soil and water was first appreciated in France. Even though the concept of sustained yield was well developed before the end of the 18th century, the concept of ecologically sustainable forest management, which includes all of the values of the forest, is a relatively modern concept that developed over the 20th century. Indeed, Hume and Kant, the 19th century philosophers wrestling with the relationship between humans and nature, still believed that clearing of forests improved climate. Environmentalism and the commencement of the conservation movement started in the late 19th century and foresters played an important role in this. The conservation movement had an important influence on the development of forestry in the 20th century and this will be discussed at greater length in later chapters.

The pressure on the forests of temperate Europe started to ease from the 19th century. The reasons are complex but there are several contributing factors. The economic situation saw a movement of people from the country to the city and there was no longer the pressure of the rural poor on their immediate forests. Iron replaced wood in ships and coal and later oil and gas replaced fuelwood for both domestic and industrial uses. Substitutes for wood were found, transport improved and sources of wood from outside Europe became more accessible and more economic. Population growth eased and living standards rose. Scientific forest management increased yields and more intensive plantation forestry produced wood on a smaller land base. Agricultural production became more efficient and also required a smaller land base. The pressure to deforest to provide land for agriculture no longer applied and indeed was reversed. Because of over-production in the agricultural sector in the later part of the 20th century, incentives were put in place to take agricultural land out of production. Perhaps one pivotal factor heralded the end of the wood age and the beginning of the coal age. This was the discovery in the midlands of England of a feasible way of smelting iron using coal instead of wood and charcoal. This, together with the harnessing of steam power, marked the beginning of the industrial revolution and placed England, which hitherto had been lagging, at the forefront of industrial progress.

Today Europe has a net annual increase in forest area and several countries report that the annual growth rate of their forests exceeds their harvest. For example, Sweden's forests were seriously depleted in the 17th century because of fuelwood requirements for its iron smelting industry. Today Sweden is a major exporter of wood products to the European community. Its annual increment, however, still exceeds its harvest. Growth ring analysis has shown that some of the forests of Europe are growing faster now than before. Improved forest management must be contributing to this but, ironically, two interacting environmental problems have assisted. Atmospheric concentrations of carbon dioxide have been steadily increasing since the beginning of the industrial revolution and this has the capacity to increase the rate of photosynthesis and therefore forest productivity, providing other factors such as the supply of water and nutrients can support the increase in productivity. Air pollution from coal burning in the industrialized parts of Europe has added nitrogen to forest soils. Perhaps it is this combination of increased carbon dioxide in the atmosphere supported by increased soil fertility that is responsible in part for the increased productivity that has been observed.

In many respects Europe in the 17th and 18th centuries represents the situation in many developing countries today and maybe is an example of what these countries might look forward to and indeed might aspire to. However, Western Europe imports wood from other countries including tropical countries (United Nations, 2001) and therefore has a responsibility to promote sustainable forest management in the countries from which it imports.

There has been a trend across Europe since the 18th century away from an intimate mixture of forest, pasture and agriculture to a landscape in which these areas have become separate, larger and more distinct (Kirby and Watkins, 1998). Also there has been the tendency for forest management to concentrate on one or a few productive species, predominantly conifers. Recently there has been a move by some to modify management to produce what is closer to the composition of the original forest on the site (a 'near natural' forest) even if this is at the expense of reduced productivity. The difficulty of course is agreeing on what was the original composition. Indeed the whole concept of what is and what is not natural is a confusing one and will be looked at in more detail later.

Colonialism

The colonial exploits of the ancient Greeks and Romans and the impact of this on the forests has already been discussed as has that of the Italian medieval city states. These were limited in their geographical spread. The European maritime powers after the Middle Ages 'discovered' and colonized the 'new world' which must have seemed to them to be an infinite resource ripe for plunder. The result was a new onslaught on the forests, and particularly the tropical forests which until this time had largely escaped wide-scale deforestation.

The Tropical Forests of the Americas

There is a long history of agriculture and forest clearance in South and Central America (Barraclough and Ghimire, 1995). A new phase in deforestation commenced with the invasions by the European maritime colonial powers of Portugal, Spain, France, Holland and Britain, all of whom had some part to play in the deforestation of tropical Central America, the Caribbean and South America from the 16th to the 19th centuries. The early explorers, being used to a temperate forest that had been subject to repeated bouts of deforestation, were awestruck by the size and magnificence of the forests that they found as well as by the wide variety of exotic plants and animals. The areas of forests were so large that they were considered to be infinite. The richness of the forest suggested to the colonizers that the soils were very fertile. Madeira, just 150 miles off the African west coast and 'discovered' by the Portuguese in 1420, was so densely forested that the Portuguese named it 'isola de madeira' or 'island of wood'. When Columbus 'discovered' Hispaniola in the Caribbean he reported that it had a thousand varieties of trees growing so high that they nearly reached the sky (Perlin, 1989). The wide-scale deforestation of the lowland forests of tropical America began at about year 1500. This was mainly for supplying sugar to Europe. The Caribbean Islands and the east coast of Brazil were particularly badly affected by the sugar trade. The sugar trade was an especially shameful episode in human relations (slavery and collapse of indigenous populations) but the effect on the forests was also a disaster. Large areas were cleared for sugarcane plantations and displaced persons were forced to migrate to outlying forested areas where they felled more forests in order to grow the basic essentials for survival. The development of sugarcane monocultures exposed as a myth the supposed fertility of rainforest soils. Most of the nutrients were contained within the deforested crop and the residual soil fertility was poor. Continuous cropping of sugar on the same area led to reduced yields and the need to clear more forests. Also the milling of sugarcane required very large amounts of fuelwood. Later in the 19th century the hill forests of Central America were cleared for coffee and bananas and in the 20th century for cotton and beef. However the greatest rate of deforestation of the tropical forests of the Americas was in the second half of the 20th century and it is still proceeding unabated. Central America had 40 billion hectares of forest in 1950 and this has since been more than halved (see Chapter 5).

The Tropical Forests of Asia

India and Southeast Asia had long periods of civilization, trade and forest exploitation prior to the arrival of Europeans. However, European colonists promoted deforestation by enforcing cash crops for export rather than subsistence farming. This had the additional effect of pushing the rural population to clear more forest land for subsistence farming. The general quality and utility of the timber was better in Asia than in tropical America (except for the Caribbean) and consequently there was a greater emphasis on the forests of Southeast Asia for speciality timbers. Teak was plundered by the British for shipbuilding and sandalwood was prized by the Chinese above all as a trading commodity. Consequently this species was hunted to almost extinction by traders with China. In contrast to South and Central America, the indigenous populations of Southeast Asia did not disintegrate and populations did not crash. Many countries of this region regained their independence from the colonizing powers shortly after the Second World War. However, the strong influence of outside countries (particularly Japan, China, and the European Union) continues in the region and the pressure on the forests remains.

The United States of America

The European colonial powers were also responsible for accelerating forest removal in Africa and in temperate climes of Canada, Australia, New Zealand and the USA. The USA is used here as an example.

The forests of the USA were strongly influenced by their indigenous peoples for the 8000 years or so following the last recession of the ice. They cleared land for their maize-based agriculture and periodically burnt the countryside to improve the habitat for game and wild plant food (Williams, 1989). Fire promoted an extension in the area of grassland and the development of fire-dependent forest types. In 1600, just prior to European colonization, forests and woodlands occupied about one half of the land area of the USA. In 2000, the figure was about one third (FAO, 2001). Timber exports to the colonial powers (especially shipbuilding timbers for England) contributed a little to forest loss, but the major cause of deforestation was for agriculture to feed the growing population of settlers. The new settlers also needed wood, particularly fuelwood and wood for fencing, but deforestation for wood products was relatively insignificant compared with deforestation to support agriculture. Even so, the extensive deforestation could barely meet the demand of the increasing population for wood. There was a burst of population growth during the 19th century and forest clearance kept pace with this. The population increased from about 5 million in 1800 to more than 110 million in 1920 during which time the area of land used for agriculture increased from about 8 million hectares to about 167 million hectares. By the end of the 19th century the forests were in a mess. Forests that had not been cleared for agriculture were left in a very poor and degraded condition. Forest removals had greatly exceeded forest growth and slash fires following clearing had escaped and created havoc. Wildlife populations had been seriously reduced and many animal species were threatened. However, there was a turnaround in the 20th century, particularly after the First World War.

While the trend in increasing population continues to the present day, the area of cropped land quite suddenly stabilized after 1920 and is much the same today. There are several reasons for this. During the 20th century the USA changed from a rural society of mainly subsistence farmers to an urban industrialized economy. The efficiency of agricultural production increased enormously through crop breeding, pest control, the use of fertilizers and a reduction in the use of draught animals. Consequently crop yields per hectare increased many times. In parallel with this there was an increase in the efficiency of the use of wood. Fuelwood comprised more than one half of the wood used during the 19th century and more than 90% of the total energy consumption at mid century. Towards the end of the 20th century fuelwood was about 3% of the total energy consumption of the USA with two-thirds of this being from industrial waste (MacCleery, 1994). The price of wood relative to competing materials increased and consequently wood substitutes became more common as less wood was used in construction. Wood processing plants became more efficient and generated less waste.

The standard of forest management improved by using scientific forest management backed by well-trained foresters and forest scientists. There has been significant investment in forest research and intensively managed and highly productive plantations have been established. The mean annual increment (growth rate) of US forests has increased considerably since the First World War. The USA has recorded an increase in forest area over the decade 1990-2000, despite the fact that it has the greatest *per capita* consumption of forest products in the world. However, like Europe, the USA imports timber products and therefore has a responsibility to promote sustainable forest management in the countries from which it imports. The US forests remain very vulnerable to fires, pests and diseases due in part to bans being placed on harvesting in publicly owned forests. In the latter part of the 20th century there has been a shift in log supply from public forests in the northwest to private forests and plantations in the southeast.

During the 19th century there was a growing concern about the adverse effects of forest clearing and poor forest management on biodiversity, water and soils. Early environmental writers (Thoreau, 1854; Marsh, 1864) prompted public concern which in the 20th century grew into an environmental movement whose aim was to protect the non-timber values of the forests and to reserve national forests. The environmental movement progressed during the 20th century from disorganized groups of concerned citizens to an organized and politically influential movement that has had a major influence on US forest policy and practice. Virtually all of the public forests (about 40% of total forests) are now protected areas.

Retrospective

Much of this chapter has been about historical deforestation and poor forest management. However, the chapter ends on a positive note. The developed world is growing forests faster than they are cutting them. The amount of forest in protected areas is increasing. There is greater emphasis on forest management for non-timber values. Scientific management is becoming increasingly sophisticated and has delivered real gains. The conservation movement has grown in strength and has greatly influenced forest management. Sustainable forest management has become a catchword for the future. However, most forests in the world are not sustainably managed and examples of destructive forest management are still easy to find. Even so there is reason to be guardedly optimistic about the future.

One factor that stands out when looking at the history of human interactions with forests has been the resiliency of the forests. By and large the forests recovered whenever humans left them alone and kept their grazing animals away. It appears that deforestation to provide land for agriculture has stabilized in the temperate (developed) zone. Perhaps there is hope that the current bout of deforestation in the tropical (undeveloped) zone will also stabilize as tropical countries become more developed and pass through the pain of the transition from a rural subsistence culture to an urban industrialized culture. This will be further discussed in Chapter 5.

References and Further Reading

Ball, J.B. (2001) Global forest resources: history and dynamics. In Evans, J. (ed.) *The Forests Handbook, Volume 1*. Blackwell Science, Oxford, pp. 3-22.

Barraclough, S.L. and Ghimire, K.B. (1995) *Forests and Livelihoods – The Social Dynamics of Deforestation in Developing Countries*. Macmillan, London.

Barton, G.A. (2002) *Empire Forestry and the Origins of Environmentalism*. Cambridge University Press, Cambridge.

Bechman, R. (1990) *Trees and Man – The Forest in the Middle Ages*. Paragon House, New York.

Boulter, M. (2002) *Extinction – Evolution and the End of Man*. Columbia University Press, NewYork.

Bush, M.B. (1997) *Ecology of a Changing Planet*. Prentice Hall, Upper Saddle River.

Calder, M. (1983) *Timescale – an Atlas of the Fourth Dimension*. Viking Penguin, New York.

Carlowitz, H.C. von (1713) *Sylvicultura œconomica*.

Clutton, E. (1978) Political conflict and military strategy as exemplified by Basilicasta's Relatione. *Institute of British Geographers Transactions* 3 (new series) 278 and 281.

Cohen, M.N. (1989) *Health and the Rise of Civilization*. Yale University Press, New Haven.

Cole, S. (1970) *The Neolithic Revolution*. British Museum (Natural History), London.

Collins, N.M., Sayer, J. and Whitmore, T.C. (1991) *Conservation Atlas of Tropical Rainforests, Asia and Pacific*. Macmillan, London.

Craw, R.C., Grehan, J.R. and Heads, J.H. (1999) *Panbiogeography – Tracking the History of Life*. Oxford Biogeography Series 11, Oxford University Press, Oxford, New York and London.

Dasman, R. (1968) *Environmental Conservation*. Wiley, New York.

Diamond, J. (1998) *Guns, Germs and Steel - a Short History of Everybody for the last 13,000 years*. Vintage, London.

Diesendorf, M. and Hamilton, C. (eds) (1997) *Human Ecology, Human Economy: Ideas for an Ecologically Sustainable Future*. Allen and Unwin, St Leonards, NSW, Australia.

Dimbleby, G.W. (1976) Climate, soil and man. *Philosophical Transactions of the Royal Society of London* B 275, 197-208.

Enright, N.J. and Hill, R.S. (1995) *Ecology of the Southern Conifers*. Melbourne University Press, Melbourne.

Evelyn, J. (1670) *Sylva or a Discourse of Forest Trees and the Propagation of Timber in His Majestie's Dominions* (3rd edition). Printed for John Martyn, Printer to the Royal Society, London.

FAO (2001) *Global Forest Resources Assessment 2000 - Main Report*. FOA Forestry Paper 140, Food and Agriculture Organization of the United Nations, Rome.

Fernow, B.E. (1907) *History of Forestry*. Toronto University Press, Toronto.

Fitch, W.M. and Ayala, F.J. (1995) *Tempo and Mode in Evolution - Genetics and Paleontology 50 years after Simpson*. National Academy Press, Washington, DC.

Flannery, T.F. (1994) *The Future Eaters – an Ecological History of the Australian Lands and People*. Reed New Holland, Sydney.

Futuyma, D.J. (1998) *Evolutionary Biology* (3rd edition). Sinauer Associates, Sunderland, Massachusetts.

Gomme, A.W. (1933) *The Population of Athens in the Fifth and Fourth Centuries BC*. Argonaut, Chicago.

Goudie, A. (2000) *The Human Impact on the Natural Environment* (5th edition). MIT Press, Cambridge, Massachusetts.

Hawkes, J. (1973) *The First Great Civilizations: Life in Mesopotamia, the Indus and Egypt*. Hutchinson, London.

Hindle, B. (ed.) (1975) *Americas Wooden Age – Aspects of its Early Technology*. Sleepy Hollow Restorations, Tarrytown, New York.

Hughes, J.D. (1994) *Pan's Travail – Environmental Problems of the Ancient Greeks and Romans.* John Hopkins University Press, Baltimore.

Humphries, R. (1968) *The Way Things Are.* Translation of Lucretius 1157-68, page 85. Indiana University Press, Bloomington.

Iversen, J. (1958) The bearing of glacial and interglacial epochs on the formation and extinction of plant taxa. *Uppsala Universiteit Arssk* 6, 210-215.

Jacobsen, T. and Adams, R.M. (1958) Salt and silt in ancient Mesopotamian agriculture. *Science* 128, 1251-1258.

James, N.D.G. (1981) *A History of English Forestry.* Basil Blackwell, Oxford.

Kiess, R. (1998) The word 'Forst/forest' as an indicator of fiscal property and possible consequences for the history of Western European forests. In Watkins, C. (ed.) *European Woods and Forests – Studies in Cultural History.* CAB International, Wallingford, pp. 11-18.

Kirby, K.J. and Watkins, C. (1998) *The Ecological History of European Forests.* CAB International, Wallingford.

Lee, R.B. and De Vore, I. (eds) (1968) *Man the Hunter.* Aldine Publishing Company, New York.

Lewin, R. (1993) *The Origin of Modern Humans.* Scientific American Library, New York.

Loy, W.G. (1966) *The Land of Nestor: a Physical Geography of the Southwest Peloponnese.* NAS-NRC Foreign Field Research Program Report No 34, Springfield, Virginia.

MacCleery, D.W. (1994) *American Forests – A History of Resiliency and Recovery.* Forest History Society Issues Series 3rd printing, Forest History Society, Durham, NC.

MacDonald, G.M. (2003) *Biogeography - Introduction to Space, Time and Life.* John Wiley and Sons, New York.

MacPhee, R.D.E. (1999) *Extinctions in Near Time - Causes, Contexts, and Consequences.* Kluwer Academic/Plenum Publishers, New York.

Maisels, C.K. (1990) *The Emergence of Civilisation - from Hunting and Gathering to Agriculture, Cities and the State in the Near East.* Routledge, London and New York.

Makkonen, O. (1969) Ancient forestry, an historical study - part 2. *Acta Forestalia Fennica* 95, 1-46.

Mannion, A.M. (1991) *Global Environmental Change: a Natural and Cultural Environmental History.* Longman, NewYork.

Mark, A.F. and. McSweeney, G.D. (1990) Patterns of impoverishment in natural communities: case studies in forest ecosystems - New Zealand. In: Woodwell, G.M. (ed.) *The Earth in Transition: Patterns and Processes of Biotic Impoverishment.* Cambridge University Press, Cambridge, pp. 151-176.

Marsh, G.P. (1864) *Man and Nature.* (ed D. Lowenthal). The Belknap Press of Harvard University Press, Cambridge, Massachusetts.

Meiggs, R. (1982) *Trees and Timber in the Ancient Mediterranean World.* Oxford University Press, Oxford.

Meyer, W.B. (1996) *Human Impact on the Earth.* Cambridge University Press, Cambridge, New York and Melbourne.

Moser, W.G. von (1757) *Principles of Forest Economy.*

Murdock, G.P. (1968) The current status of the world's hunting and gathering peoples. In: Lee, R.B. and DeVore, I. (eds) *Man the Hunter.* Aldine Publishing Company, New York, pp. 13-20.

Osborne, R., Benton, M. and Gould, S.J. (1996). *Atlas of Evolution.* Viking Penguin, London.

Perlin, J. (1989) *A Forest Journey - the Role of Wood in the Development of Civilization.* W.W. Norton and Company, New York and London.

Pielou, E.C. (1991) *After the Ice Age.* The University of Chicago Press, Chicago and London.

Pyne, S.J. (1982) *Fire in America - a Cultural History of Wildland and Rural Fire.* Princeton University Press, Princeton.

Pyne, S.J. (1995) *World Fire – the Culture of Fire on Earth.* Henry Holt, New York.

Rackham, O. (1980) *Ancient Woodland – Its History, Vegetation, and Uses in England.* Edward Arnold, London.

Robbins, W.G. (1985) *American Forestry- A History of National, State, & Private Cooperation.* University of Nebraska Press, Lincoln and London.

Roberts, N. (1989) *The Holocene: an Environmental History.* Basil Blackwell, Oxford.

Rosenzweig, M.L. (1995) *Species Diversity in Space and Time.* Cambridge University Press, Cambridge, New York and Melbourne.

Rushton, J.P. (1997) *Race, Evolution and Behaviour.* Transaction Publishers, New Brunswick (USA) and London.

Saggs, H.W.F. (1989) *Civilization before Greece and Rome.* Yale University Press, New Haven.

Saggs, H.W.F. (1995) *Babylonians.* University of Oklahoma Press, Norman.

Sandars, N.K. (translation) (1975) *The Epic of Gilgamesh; an English version.* Penguin, Harmondsworth, Mx.

Southwick, C.H. (1972) *Ecology and the Quality of our Environment.* D. Van Nostrand Company, New York.

Southwick, C.H. (1996) *Global Ecology in Human Perspective.* Oxford University Press, Oxford and New York.

Stahl, J.F. (1772-1781) *Onomatalogia forestalis-piscatoria-venatoria* (4 volumes).

Standford, C. (2001) *Significant Others - the Ape-Human Continuum and the Quest for Human Nature.* Basic Books, New York.

Stanley, S.M. (1986) *Earth and Life through Time.* Freeman, San Francisco.

Tattersall, I. (1993) *The Human Odyssey: Four Million Years of Human Evolution.* Prentice Hall, New York.

Tattersall, I. (1998) *Becoming Human - Evolution and Human Uniqueness.* Harcourt Brace & Company, San Diego, New York and London.

Thirgood, J.V. (1981) *Man and the Mediterranean Forest - A History of Resource Depletion.* Academic Press, London.

Thoreau, H.D. (1854) *Walden, or, Life in the Woods.* Published in 1992 by David Campbell, London.

Toynbee, A.J. (1935) *A Study in History.* Oxford University Press, Oxford.

Troeh, F.R., Hobbs, J.A. and Donahue, R.L. (1980) *Soil and Water Conservation for Productivity and Environmental Protection.* Prentice Hall, Englewood Cliffs, New Jersey.

Tyldesley, J.A. and Bahn, P.G. (1983) The use of plants in the European palaeolithic: a review of the evidence. *Quaternary Science Reviews* 2, 53-81.

United Nations (2001) Forest Products Annual Market Review. United Nations Economic Commission for Europe and the Food and Agriculture Organization of the United Nations Timber Bulletin Volume LIV No. 3, Published by United Nations, New York and Geneva.

Vera, F.W.M. (2000) *Grazing Ecology and Forest History.* CAB International, Wallingford.

Vrba, E.S., Denton, G.H., Partridge, T.C. and Burckle, L.H. (1995) *Palaeoclimate and Evolution with Emphasis on Human Origins.* Yale University Press, New Haven and London.

Watkins, C. (1998). *European Woods and Forests – Studies in Cultural History.* CAB International, Wallingford.

Westoby, J. (1989) *Introduction to World Forestry: People and their Trees.* Blackwell, Oxford and New York.

Williams, M. (1989) *Americans & their Forests – A Historical Geography.* Cambridge University Press, Cambridge.

Williams, M. (2003) *Deforesting the Earth – From Prehistory to Global Crisis.* The University of Chicago Press, Chicago.

Winters, R.K. (1974) *The Forest and Man.* Vantage Press, New York, Washington and Hollywood.

Wymer, J. (1982) *The Palaeolithic Age.* Croom Helm, London.

Chapter 2

Forests of the World

Factors Determining the Type and Distribution of Forests

The growth of plants (and animals for that matter) depends on photosynthesis, arguably the most important chemical reaction on earth. This is the chemical reaction that manufactures the basic substance from which virtually all living organisms are constructed. The reaction occurs in plant leaves, which are displayed to intercept light because the reaction requires solar energy. Carbon dioxide from the atmosphere enters the leaves via pores in the leaves (called stomata) and reacts with water from the soil to form carbohydrate (the basic substance of life) and oxygen. The reaction requires certain elements (nutrients) from the soil and occurs with the assistance of a special light absorbing pigment called chlorophyll. This is a gross oversimplification of a complex reaction but it will suffice for the purpose. The opposite reaction to photosynthesis is respiration where carbohydrate reacts with oxygen to produce carbon dioxide and water. This reaction releases energy and is necessary for maintaining metabolism in both plants and animals. It is also the chemical reaction of combustion and decomposition of carbohydrate organic matter.

It follows that for a plant to be growing (increasing in biomass), photosynthesis (A) must exceed respiration (R). For an actively growing tree a value for the photosynthetic respiratory ratio (A/R) of 10 would not be unusual. Photosynthesis minus respiration is called net photosynthesis (An) and in order for a plant to survive and grow, An would have to be positive. Forests are communities containing trees, understorey species, animals, soil microorganisms, and decomposing organic matter. In forest communities, A/R is less than that of single trees and may be greater than 2 for an actively growing forest like a plantation but close to 1 for a climax community such as a tropical rainforest. If A/R is less than 1, then the ecosystem is in decline. A forest ecosystem comprises all of the biotic components of a community as defined above plus the abiotic (physical and chemical) components such as the water and soil. The interactions in space and time among and between species in a forest ecosystem is the subject of Forest Ecology. This will not be pursued here but is well represented in the literature, such as by Kimmins (1997).

The stomata can open and close, thereby controlling the amount of carbon dioxide that enters the leaf for photosynthesis and, at the same time, the amount of water that is evaporated (transpired) from the leaf. Photosynthesis does not occur in the dark (at night) and therefore stomata shut at night to conserve soil water. Stomata open in the light (during the day) to let in carbon dioxide for photosynthesis. However, stomata may close or partly close during the day to conserve water if the plant senses low water content in the soil, or high temperature and low humidity in the atmosphere. The consequence of closing stomata during the day to conserve soil water is a reduction in the rate of photosynthesis and therefore growth. By understanding the factors that control photosynthesis and respiration, it is possible to define the environmental conditions that favour forest growth. Forests require light (but not too much) and warm temperatures (but not excessive). They need an adequate supply of water, nutrients and oxygen from the soil. It follows that the nature and location of forests depends primarily on the climate and soil, but it is also modified by disturbance.

Climate

The key climatic factors determining the nature and location of forests are the air temperature and the availability of water, and these two factors interact in determining the nature and location of forests. Temperature decreases with increasing latitude and altitude and is influenced by distance from the sea and by ocean currents. The availability of water depends on the balance between precipitation and evapotranspiration.

Figure 2.1 shows the relationship between climate (defined by air temperature and water availability) and the type and distribution of plant communities. Forest communities are shaded in this figure. Figure 2.1 shows that plant communities become more productive and complex as the climate becomes warmer and wetter and this is particularly so of forests. Forests grow under wet rather than dry conditions and there are more types of forest in warmer regions than in colder ones. The wettest and warmest climate of all supports the most biodiverse and productive forest of all, the tropical rainforest. Generally speaking, the warmer and wetter the climate, the higher is the productivity (Table 2.1) and the greater is the biodiversity (Fig. 2.2). Figure 2.2 shows that in the tropics as rainfall increases, so too does the number of tree species, tree height, the number of tiers and the number of epiphytes.

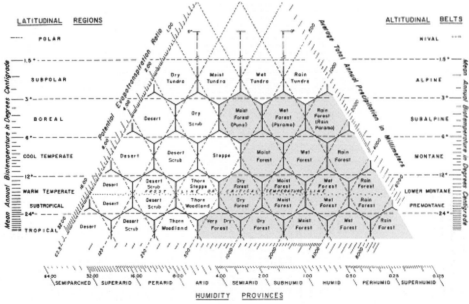

Fig. 2.1. Classification of the vegetation of the world based on temperature, rainfall and evapotranspiration (modified from Holdridge *et al.*, 1971).

Table 2.1. Productivity of different types of vegetation (modified from Leith, 1976).

Vegetation Type	Net Primary Productivity $(kg/m^2/y)$
Forests	
Tropical rainforest	1.75 – 2.8
Tropical deciduous forest	1.0
Temperate humid forest	1.0
Temperate dry forest	0.8
Boreal forest	0.65
Woodland	
Dwarf and open scrub	0.6
Tundra	0.16
Desert scrub	0.07
Grassland	
Tropical grassland	0.8
Temperate grassland	0.8
Desert	
Dry desert	0.003
Ice desert	0.0
Cultivated land	0.65
Fresh water	
Swamp and marsh	2.0
Lake and stream	0.5

Figure 2.1 shows that for any given temperature regime, there are a range of possible vegetation types depending on the availability of water. The availability of water depends on the balance between precipitation (P) and evapotranspiration (E). There are more forest types possible under a warmer than a cooler climate because of the influence of evapotranspiration. Potential evapotranspiration (Ep) is the evaporation from a continuous cover of vegetation having no water deficiency. In general terms this will occur if P is greater than Ep. Ep increases with increasing air temperature and decreasing relative humidity. If the vegetation has a water deficiency (P less than Ep), then the actual rate of evapotranspiration (Ea) will be less than Ep and Ep/Ea will be greater than 1. The greater the value of Ep/Ea, the drier the vegetation type (see right hand side of triangle in Fig. 2.1). The impact of evapotranspiration is the reason why, for example, 500-1000 mm annual rainfall might support very dry forest in the tropics, but moist forest in the cool temperate zone and wet forest in the boreal zone. If Ep/Ea is greater than 1 (water deficiency), only dry forest types can persist and then only in the warm temperate to tropical latitudinal regions (Fig. 2.1).

The effect of latitude and altitude are basically the same, both affecting temperature. A good example of the interacting effects of temperature and water availability can be seen in the forest types in Oregon, USA. There is a large range in temperature (<10°C to 30°C mean summer temperatures) and water availability (150 mm to 3000 mm annual rainfall) at various positions across the transect from the coast across the coastal range into the valley between the Coastal Range and the Cascades and then over the Cascades to the dry interior. The forest types segregate into those mainly constrained by low temperature, those mainly constrained by high water stress and those constrained by neither (Fig. 2.3).

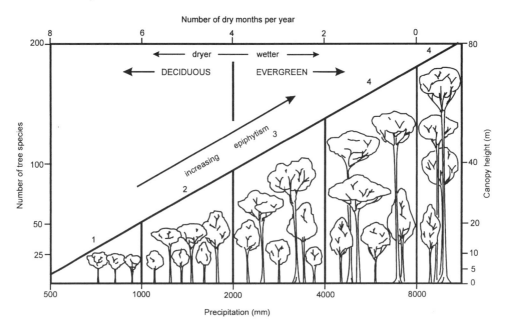

Fig. 2.2. The relationship between water availability and tree development in tropical forests (from National Research Council, 1982).

Soils and Topography

Soils anchor plants and provide them with water and nutrients. The ideal texture for a soil is a loam, which has both adequate water and air. Coarser soils (sands) have more air and less water and finer soils (clays) have more water and less air. The deeper the soil the better because this increases the store of available water and nutrients. Plants require a range of essential elements (N, K, Ca, Mg, P, S, Cl, Fe, B, Mn, Zn, Cu, Ni, Mo) from the soil in order to survive and grow. Soil organic matter is important because it assists in providing nutrients (particularly N) and beneficial soil microorganisms. The long-term health of forest ecosystems depends on the recycling of nutrients released from decomposing litter and the ability of forest ecosystems to recycle nutrients means that they can develop on quite poor soils. Littoral rainforest can even develop on coastal sand dunes. Soils with extremes of pH can be toxic to plants and promote nutrient deficiencies. Forests prefer and promote slightly acid soils. Compacted soils restrict root growth.

Usually soils at the top of a ridge are shallower and those near the valley floor are deeper. Aspects facing the sun (south in the northern hemisphere and north in the southern hemisphere) receive more radiation, and are warmer and drier. Soil on steep slopes is generally shallower, drier and more erodible than on less steep areas. Consequently, forest type and productivity can change from valley floor to ridge top, from one side of a ridge to another and across a change in soil type, which may be gradual or sudden. Within a local region, changes in soil type can cause quite marked changes in forest productivity and in species composition. A good example of a sudden change is in the Lamington National Park in south-eastern Queensland (Australia) where well-developed sub-tropical rainforest on soils derived from basalt suddenly changes to a low heath community on phosphorus-deficient soils derived from granite. Because of the strong relationship between soil type and productivity, soil survey is an important part of any plantation establishment programme.

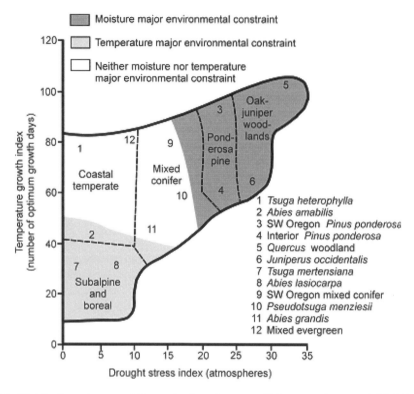

Fig. 2.3. Distribution of some of the major forest zones within an environmental field based on moisture (maximum plant moisture stress during the dry season) and temperature (from Cleary and Waring (1969) and Franklin and Dryness (1973)).

Disturbance

Large areas of forest have been modified by disturbance. Disturbances can be independent of humans (non-anthropogenic) such as windstorms, earthquakes, volcanism, fire from lightning strikes, landslips, floods, erosion, siltation, droughts and disease. Disturbances caused by humans (anthropogenic) include deforestation, reforestation, burning, grazing, and species dispersal. Humans can also contribute to non-anthropogenic sources. Forests are dynamic ecosystems in a state of flux with continuous cycles of germination, growth, maturity, senescence, decay and regeneration. Superimposed on this, however, are changes to the structure of the community following a disturbance. The term 'succession' is used to describe the sequence of change in community structure following disturbance. There is an intense debate in the literature about the finer details of forest succession (West *et al.*, 1981). Simply stated, the progress of change following disturbance moves through a sequence of community structures called 'seres' until finally a 'climax' community is reached where some sort of stability is imagined.

Reoccurring or persistent disturbance can prevent a particular sere from progressing towards its climax or can reverse its direction. For example, overgrazing in the Mediterranean has pushed forest communities to shrub communities and finally to grassland that is resistant to grazing. Also, deforestation and fire in Kalimantan has left large areas of *Imperata* grassland that are not regenerating to forest. Much of the cultivated and grazing land across the world that occurs in temperature and water availability zones suitable for forest would revert to forest if it was abandoned and not maintained in a periodically reoccuring state of disturbance.

There are many examples of forest types maintained by fire. Individual tree species may have adaptations to resist damage from fires, to re-sprout after fires, or to promote seed dispersal and germination after fires. Larch, Douglas fir, *Acacia* and many pines, particularly those with serotinous cones, may be favoured by fire. Jack pine (*Pinus banksiana*) is a very good example of a tree that is very well adapted to frequent burning. The Australian flora has been moulded by fire and has adaptations over the whole range from extreme fire resistance to the opposite where the plant is adapted to be flammable so that it will burn. *Eucalyptus regnans* forest of south-eastern Australia is a good example of the latter. These magnificent trees regenerate in the ash bed after very intense wildfires that essentially 'clear fell' the site. The seeds germinate in the ash bed and the young seedlings are favoured by open space and plenty of light. The forests are very productive and produce large amounts of ground cover as potential fuel for a wildfire. Usually the forests are too moist to burn, but on the occasions when dreadful fire weather follows a dry period (say once or twice a century) these very flammable forests burn and again reduce the area to an ash bed. The cycle recommences. If these forests could escape fire for long enough (hundreds of years) they would progress to a climax of cool temperate rainforest but the forests have adapted to promote their burning and therefore to hold themselves in this seral stage. Consequently these forests often are even-aged monocultures whose age is determined by the fire event that created them. This has implications for management. These forests can be successfully regenerated if large enough gaps are created such as occur in an intense wildfire. Consequently clear felling of small coupes will work whereas silvicultural systems that remove single trees or small groups of trees may change the forest to something else.

Classification of Forest Types

There is confusion in the literature about the definition of forest types. Angiosperms are called 'hardwoods' and conifers called 'softwoods.' However, some angiosperms (e.g. balsa, *Ochroma lagopus*) have soft wood and some conifers (e.g. yew, *Taxus* spp) have hard wood and so the terms hardwood and softwood are confusing and unhelpful. Also the term 'broad-leaved' is commonly used to describe non-coniferous tree species and 'needle-leaved' to describe coniferous species. However, some angiosperms have needle-like foliage (e.g. *Casuarina*) and some conifers have broad leaves (e.g. *Agathis*). The combined term 'deciduous hardwood' has often been used, particularly in northern hemisphere temperate countries, to distinguish angiosperms that are deciduous in winter from conifers. However, some conifers are deciduous (e.g. *Larix*) and a very large number of angiosperms in tropical regions are deciduous in the dry season. The distinction between evergreen and deciduous species is clear and unambiguous, as is the distinction between angiosperms and conifers. The terms hardwood, softwood, broad-leaved and needle-leaved are unsafe and will not be used in this chapter. However, these terms are endemic in the literature. Trees are either angiosperms or gymnosperms and the great majority of gymnosperm trees are conifers. Also there can be considerable confusion about common names. Australia serves as a good, although in no way unique, example. European colonists of Australia named Australian native species with familiar names such as ash (*Eucalyptus regnans*), beech (*Gmelina leichhardtii*), walnut (*Endiandra palmerstonii*), hickory (*Argyrodendron trifoliolatum*), maple (*Flindersia braleyana*), birch (*Bridelia exaltata*), elm (*Aphananthe philippinensis*), oak (*Casuarina, Grevillea*), cypress (*Callitris*) and pine (*Araucaria* and *Agathis*). The timber industry added to the confusion by naming species or species groups with names that are supposed to give them a marketing edge. For example, Tasmanian oak is a marketing name for a group of eucalypt species containing *Eucalyptus regnans*, which is also called ash. Blue gum represents a range of different species depending on region and the ultimate confusion occurs when *Eucalyptus tereticornis* is called blue gum in Queensland but red gum in other Australian states. The only way to be sure is to give a tree its scientific name, and this will be done throughout this chapter.

Forest Types and their Distribution

There are several classifications of forest type and distribution in the literature. The one used by FAO (2001a) has been adopted here because future reporting by the Food and Agriculture Organization of the United Nations (FAO) on forest resources will use this classification. Ecosystems of the World (edited by Goodall), which comes in many volumes, is another valuable source of information. FAO (2001a) divided the world's forests into ecological zones within domains (Fig. 2.4). The domains are tropical, subtropical, temperate, and boreal and are based on temperature variability throughout the year. These correspond largely, though not exclusively, with latitude. Ecological zones within domains have largely been determined on the basis of amount and distribution of rainfall and humidity. Forests do not occur where it is too dry or too cold. There is evidence, however, that control of overgrazing would extend the areas of forest into drier areas in some instances. Also there are large areas of non-forested land that occur in ecological zones that are warm and wet enough to support forest. This can be seen by comparing Fig. 2.4f, which shows actual distribution of forest, with Fig. 2.4a-e, which shows the total area of each domain. These areas of non-forested land were deforested to support agriculture, grazing and urban development and they would revert to forest if they were abandoned. The remaining tracts of continuous forest cover are confined largely to the tropics and to the boreal zones of the northern hemisphere (Fig. 2.4f).

Tropical Forest

Tropical forests contain 70% of the world's plants and animals, 70% of the world's vascular plants, 30% of all birds and > 90% of all invertebrates. Tropical forests straddle the equator where the climate is warm to hot and frosts are absent throughout the whole year. The main climatic factor that distinguishes forest types in the tropics is the length and severity of the dry season. With increasing distance from the equator, rainfall becomes seasonal and forests need to withstand progressively longer dry periods. High temperatures compound availability of water during dry periods. As the length and severity of the dry season increases, the forests pass through a sequence from tropical rainforest to tropical moist deciduous forest to tropical dry forest to scrub to thorn and ultimately to grassland. Through this sequence, the number of tree species per hectare is reduced, epiphytes and lianas become reduced and then disappear, trees become smaller and more scattered, buttressing is reduced and bark becomes thicker. Conifers are rare and angiosperms dominate.

Tropical Rainforest

Tropical rainforest (tropical moist evergreen forest) occurs in the warm, humid and wet regions straddling the equator. Temperature, relative humidity, rainfall and day length are relatively constant throughout the year. The main occurrences are: in the Amazon basin of South America; the Congo basin in equatorial West Africa; the west coast of Madagascar; parts of Mexico, Central America and the Caribbean; north-eastern Australia; New Guinea; and insular south-eastern Asia. Tropical rainforests are the most complex and biodiverse ecosystems of all and the level of endemism is high. The Amazonian rainforest is the most biodiverse ecosystem in the world. The Malay Archipelago has 25,000 to 30,000 plant species of which one third are trees (FAO, 2001a) and the Democratic Republic of the Congo has 11,000 plant species, 1100 bird species and 400 mammals (Tchatat, 1999). Because of the large number of families, genera and species it is difficult to select a few representative samples. Genera *Dipterocarpus* and *Shorea* are important commercial species in the south-eastern Asian tropical rainforests. There can be >200 species of trees per hectare in tropical rainforests and the number of species is so large that few people can identify them all. Timber species are often sold in assemblages without individual species identification. Trees are typically shallow rooted and buttressed and there is relatively little litter on the forest floor because it is very quickly decomposed. The soils are not necessarily fertile and can quickly lose their fertility if the forest is cleared. Most of the nutrients in the tropical rainforest are stored in the biomass in contrast to forests from cooler areas where most of the nutrients reside in the soil. Tropical rainforests are very efficient at recycling nutrients and if the trees are removed from the site or burnt on site, a large proportion of the nutrients can be removed from the ecosystem leaving the residual nutrients in the soil prone to leaching from the heavy tropical rains.

Tropical rainforests are home to some indigenous peoples who rely on them for their subsistence. They are a source of a range of non-wood forest products. Breadfruit, mangoes, avocados, brazil nuts, durian, jack fruit, edible oils, tannins, resins, bamboo, rattan, fodder, and spices are all products of the tropical rainforest.

Tropical Moist Deciduous Forest

Tropical moist deciduous forest is found adjacent to tropical rainforest. These forests still have a high annual rainfall, but with a distinct dry season. Extensive tropical moist deciduous forests occur north and south of the tropical rainforests in Africa and South America. In Africa the genera include *Marquesia, Berlinia, Laurea, Brachystegia* and *Acacia*. Most of the area, however, is wooded grassland and extensive clearing has occurred for agriculture. In South America the flora is very rich, with families Leguminosae and Myrtaceae well represented. The zone is a mosaic of grassland, tree savannahs and woodland with patches of more dense forest (FAO, 2001a). In India and south-eastern Asia,

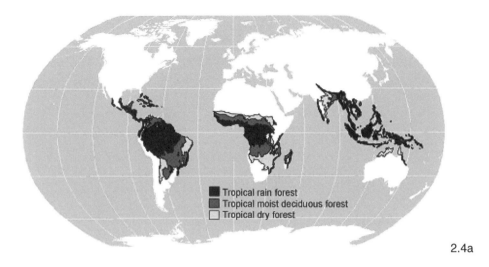

2.4a

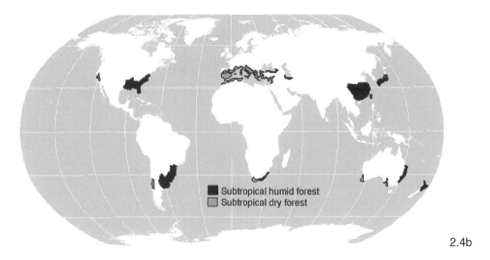

2.4b

Fig. 2.4 Ecological zones that are capable of supporting forest within a) the tropical domain, b) the sub-tropical domain, c) the temperate domain, d) the boreal domain, and e) mountain forests within all domains; f) shows the current distribution of forest (adapted from FAO, 2001a).

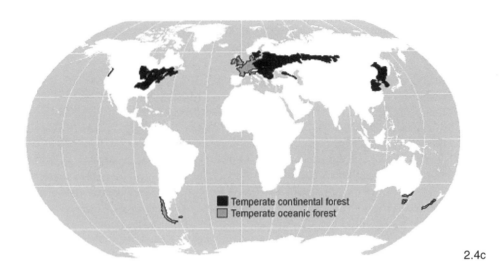

2.4c

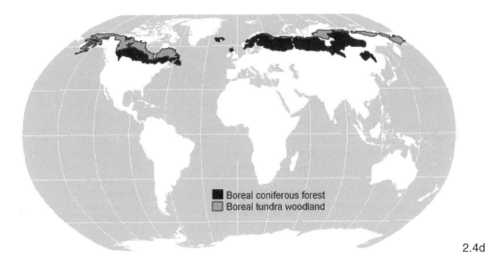

2.4d

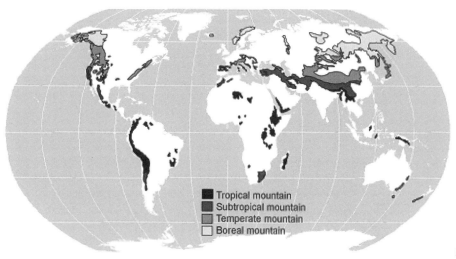

Tropical mountain
Subtropical mountain
Temperate mountain
Boreal mountain

2.4e

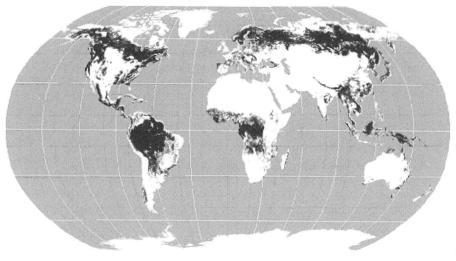

2.4f

there are many species from a wide range of families. These forests occur in monsoonal climates and are sometimes called monsoonal forests. *Tectona grandis* (teak) from Thailand and *Shorea robusta* (sal) from northern India are important commercial species. Angiosperms predominate but conifers are represented by *Pinus merkusii* in south-eastern Asia and *Podocarpus imbricata* and *Pinus latteri* in southern China.

In the tropics the deciduous habit occurs in the dry season where it is an adaptation to conserve soil water and thereby reduce water stress in the trees. This is in contrast to the temperate zone where the deciduous habit occurs in winter as an adaptation to protect the tree from freezing temperatures. Not all the species in the tropical deciduous forests are completely deciduous but most would shed some if not all of their leaves during the dry season. Many of the trees flower in the dry season and consequently deciduous forest, moist or dry, can be very colourful. Rubber (*Hevea brasiliensis*) is a deciduous tropical tree from Brazil, but which has been planted extensively in Malaysia.

Tropical Dry Forest

Tropical dry forests occur when the dry season becomes longer and more severe. There are extensive areas in Africa and also occurrences in northern Australia, India and eastern Brazil. The trees are smaller and more scattered, are often deciduous or semi deciduous, and tend to be spreading in habit. Leguminosae is an important family (*Amburana, Caesalpinia*, and *Mimosa* in South America, *Acacia* in Africa and Australia, *Curatella* and *Byrsonima* in Mexico and Central America). Dry deciduous dipterocarp forests and woodlands are common in south-eastern Asia as well as woodlands of *Tectona grandis* and *Pinus merkusii. Eucalyptus, Melaleuca* and *Callitris* are important species in northern Australia. Frequent fires and collection of fuelwood have reduced large areas of tropical dry forest to shrub and grass. Tropical dry forests are important for fodder, timber and fuel and because the trees are thinly distributed over grass they are also important for wildlife and for grazing. Tropical shrubland occurs when the climate is even drier and, although this zone does not support forest as such, it still may contain scattered trees.

Subtropical Forest

The subtropical domain is defined by FAO (2001a) as areas having mean monthly temperatures of over 10°C for more than eight months a year. Subtropical forests show some seasonal variation in temperature and day length and the two distinct types of subtropical forest, subtropical humid and subtropical dry, are distinguished by seasonal distribution in rainfall and humidity. Both of these forest types are close to the sea.

Subtropical Humid Forest

Subtropical humid forests are located on or near to eastern seaboards. They are uniformly warm and humid throughout the year without any pronounced dry season. The major occurrences in the northern hemisphere are in south-eastern USA and south-eastern China. In the southern hemisphere they occur in the east of South America and Australia. At their best these forests can be classified as subtropical rainforests where they have a similar structure to tropical rainforests. They are, however, less complex, less biodiverse, less dense, with fewer tree species, and with smaller trees with smaller and more waterproof leaves.

In the south-eastern USA the angiosperm species are evergreen oaks (*Quercus*), *Magnolia* (magnolia), *Carya* (hickory), *Liquidamber* (sweet gum), *Acer* (maple), *Ulmus* (elm), *Fraxinus* (ash), *Populus* (poplar), *Platanus* (sycamore) and *Cornus* (dogwood). The conifers include *Pinus taeda, Pinus echinata, Pinus elliottii* and *Taxodium distichum* (bald cypress). Extensive plantations of *Pinus taeda* and *Pinus elliottii* have been established in the area. In south-eastern China the forests are a mixture of evergreens and deciduous species. The forests are rich in oaks, laurel (*Cyclobalanopsis*), *Castanopsis* and conifers (*Pinus massoniana, Pinus taiwanenesis* and *Cunninghamia lanceolata*) but there are many other species. The extensive bamboo (*Phyllostachys*) forests of China are part of the subtropical humid forest zone. In southern Japan the species are mainly angiosperms (such as *Castanopsis, Cyclobalanopsis* and *Machilus*) but conifers (*Podocarpus, Tsuga, Abies* and *Pinus densiflora*) are also represented. *Torreya* which is a gymnosperm but not a conifer, is also present.

The subtropical humid forests of Brazil have been heavily exploited for timber and cleared for agriculture and grazing. Grasslands now dominate the region. The conifer *Araucaria angustifolia*, is a major constituent of any remaining forest. In eastern Australia the forests are predominantly eucalypts (such as *Eucalyptus grandis* and *Eucalyptus pilularis*) or closely related members of the Myrtaceae, and there are many species. In the absence of fire and on better soils these forests become subtropical rainforests of similar but simpler structure to the northern Australian tropical rainforests. The FAO classification of ecological zones places the North Island of New Zealand as sub-tropical humid. Local knowledge, however, would suggest that it is warm temperate. In the North Island of New Zealand the forests are a mixture of conifers (*Agathis, Podocarpus, Dacrycarpus, Dacrydium, Phyllocladus, Prumnopitys*) and evergreen angiosperms (*Weinmannia, Ackama, Metrosideros, Knightia, Elaeocarpus, Laurelia, Beilschmiedia, Litsea, Nestegis*).

Subtropical Dry Forest

Subtropical dry forests occur exclusively on western seaboards where the climate is characterized by a 'Mediterranean-type climate' of cool wet winters and hot dry summers. They are also further from the equator than what might initially be considered to be subtropical latitudes. There are occurrences in south-western Australia, Chile, western USA and the western Mediterranean (southern Spain, southern France, Italy, Greece, North Africa and the Mediterranean islands). The idea that the forests of Italy, southern France and southern Spain are subtropical would surprise many people in the region, especially during their winter. However, the effect of the warm ocean current adjacent to the Atlantic seaboard of Europe is remarkable. Toronto in Canada, which is not far from the boreal forest and has very intense winters, is on the same latitude (43.5° North Latitude) as Monaco on the Mediterranean Sea.

In the Mediterranean, the forest is usually olive (*Olea europaea*) or oak (*Quercus ilex*). In the USA, this forest type occupies parts of the coastal areas and plains of California and southern Oregon. Coastal fog is common and important in determining forest type and ecology. For example, in the north where the subtropical dry forests meet the temperate oceanic forests, *Sequoia sempervirens* (Californian redwood) occupies the seaward slopes together with *Pseudotsuga menziesii* (Douglas fir), *Thuja plicata* (western red cedar) and

Tsuga heterophylla (western hemlock). Further south are *Pinus* and *Cupressus* (cypress) on the coast and *Quercus, Pseudotsuga menziesii* and *Arbutus menziesii* (madrone) are further inland. Most of the lowland areas in this region have been converted to urban use or irrigated agriculture. At its drier extremities the forest is replaced by chaparral shrub land. Fire weather can be very severe and many of the forest types have been fashioned by fire. Even though this region is classified as a dry type on the basis of temperature and rainfall, it supports the tallest tree species in the world (*Sequoia sempervirens*) and the most successful conifer plantation species in the southern hemisphere (*Pinus radiata*). *Pinus radiata*, however, has a very restricted distribution in its native habitat. In south-western Australia, *Eucalyptus diversicolor* (karri) occurs on the better soils where the rainfall is adequate. Karri is a very tall tree and this demonstrates, along with *Sequoia sempervirens* in California, that the subtropical dry forest zone can support highly productive forests. (These tall trees occur at the very wet end of the sub-tropical dry climatic zone and perhaps are out of place in the FAO classification.) On the poorer soils, *Eucalyptus marginata* (jarrah), *Eucalyptus calophylla* (marri) and *Eucalyptus gomphocephala* (tuart) are the major species. In Chile, as in the USA, the subtropical dry forest lies north to south on the west coast and is adjacent to, but at lower latitudes than, temperate oceanic forest. The influence of the sea is important. The climax dry forest of *Cryptocarya, Lithraea, Peumus, Quillaja*, and *Beilschmiedia* has largely been degraded and replaced by thorn shrubs or converted to agriculture. Where rainfall is higher, *Nothofagus* and *Araucaria araucana* (monkey puzzle pine) are found.

Temperate Forest

Temperate forests are further from the equator than tropical and subtropical forests and are distinguished by marked differences in temperature and day length between the seasons. The domain is defined as having mean temperatures over 10°C for four to eight months. Temperate forests are divided into temperate oceanic and temperate continental types. Temperate oceanic forests are influenced by proximity to the sea and do not have the extremes of temperature experienced by their continental counterpart. Temperate continental forest occurs only in the northern hemisphere where large landmasses dominate the temperate latitudes. Temperate oceanic forests at their best are cool temperate rainforests that can rival tropical rainforests for grandeur and productivity. Tropical rainforests are productive because of high temperatures and abundant water. However the amount of radiation intercepted by tropical rainforests is limited by high humidity in the atmosphere and by cloud cover. Also the day length hardly changes in the tropics and warm night temperatures promote respiration, which reduces net photosynthesis. In the well-watered oceanic regions of the temperate zone, days are long during the growing season and cool night temperatures keep respiration in check. Examples of grand and productive temperate oceanic forests are the coniferous rainforests of the Pacific Northwest of the USA and the eucalypt forests of the wetter parts of Victoria and Tasmania in Australia. The temperate forests have a lesser number of species and are less layered than tropical forests. Sometimes the forests are dominated by only a few species, even just one.

Species composition differs between the northern and southern hemispheres. In the northern hemisphere, temperate oceanic forest occupies Western Europe, the north of Japan, and lowland areas of the Pacific Northwest of the USA and southern British Columbia. Temperate continental forest occupies the most of eastern USA and a large belt of forest extending from Western Europe across Asia below the boreal forest. They are characterized by a mixture of angiosperms (usually deciduous) and conifers. The angiosperms include *Quercus, Fagus* (beech), *Fraxinus, Acer, Carya, Ulmus, Juglans* (walnut), *Populus, Salix, Tilia* (lime) and *Betula* (birch) and the conifers include *Picea* (spruce) and *Pinus*. The species differ between the continents although many of the genera are common. *Thuya plicata* (western red cedar), *Tsuga heterophylla* (western hemlock) and *Pseudotsuga menziessii* occur in the oceanic forests in the west of the USA and *Quercus, Carpinus betulus* (hornbeam) and *Fagus sylvatica* are prominent in the oceanic forests of Western Europe. Temperate forests cover a large range in water availability, soils and temperature and the climate can be very variable. In Europe and the USA much of the temperate forest has been cleared.

In the southern hemisphere the temperate oceanic forests of New Zealand and Chile have some similarities, which reflect their common Gondwanan origins. The angiosperm genus *Nothofagus* (southern beech) is the main constituent in both countries and both have similar conifer genera (e.g. *Podocarpus* and *Dacrydium*). In Australia, fire has fashioned tall eucalypt forest (such as *Eucalyptus regnans, Eucalyptus globulus, Eucalyptus obliqua* and *Eucalyptus viminalis*). In the absence of fire on wet sites, rainforest species such as *Atherosperma moschatum* (sassafras), *Nothofagus cunninghamii* (myrtle), and *Acacia melanoxylon* (blackwood) can establish.

Boreal Coniferous Forest and Boreal Tundra Woodland

The boreal domain is defined as requiring no more than three months over 10°C. Boreal forest forms a large belt of almost continuous forest across Eurasia and Canada. It occupies the domain between the temperate and the polar. There is no equivalent in the southern hemisphere because there are no landmasses at the latitudes that support boreal forest in the northern hemisphere. These forests are sometimes called cool coniferous forests or taiga. They are dominated by *Picea, Abies, Pinus* and *Larix* (larch) but the angiosperms *Salix, Betula, Populus tremuloides* (aspen) and *Alnus* (alder) recolonize areas disturbed by fire, wind and human activity. Trees become progressively smaller from south to north as the climate becomes colder, the growing season becomes shorter and boreal coniferous forest moves into boreal tundra woodland. At the northern limit the trees are stunted and sparse, occurring in scattered groups on higher ground until the trees give way completely to tundra. A similar sequence occurs as treeline is approached with increasing altitude but this occurs over a much shorter distance. In North America the tree species approaching the tundra are *Picea glauca, Picea mariana, Abies balsamea, Pinus banksiana* and *Larix laricina*. In Eurasia, *Picea abies* and *Picea obovata* are the western species approaching the tundra and *Abies sibirica* and *Pinus sibirica* are the eastern species (Beazley, 1981).

Mountain Forest

There are extensive areas of mountain forests. Mountain forests usually are the last to be cleared and often the forests are restricted to the mountains in regions that have had a history of human activity. Mountain forests cover all of the domains from tropical to boreal but they are treated together here because their climate is different, and sometimes very different, to that of the lowland around them. Temperature becomes less and rainfall may increase markedly with increasing altitude. Consequently, mountain forests tend to be cooler types and sometimes wetter types than the domain in which they sit.

Africa

The Atlas Mountains in Tunisia, Algeria and Morocco reach up to 4165 metres and have a humid climate. They contain mainly *Quercus* and *Pinus* (*Pinus pinaster* or *Pinus halepensis*) on the lower slopes and *Cedrus atlantica* (atlantic cedar) and *Juniperus thurifera* at higher altitudes. The tropical mountains of Cameroon, Ethiopia, Kenya and the Kivu ridge support a great variety of distinctive vegetation types including bamboo (*Arundinaria alpina*) at the higher altitudes. In Madagascar, large areas of secondary grassland have replaced the original montane vegetation. In South Africa, the highveld reaches 3000 m and *Podocarpus* and *Apodytes* are found.

Asia

Treeline in the Himalayas is at about 4000 metres. The tropical lowland forest to the south of the Himalayas changes to an evergreen forest which may contain *Pinus roxburghii* (chir pine), above which is evergreen *Quercus* forest and, higher still, a coniferous forest of *Abies* and *Tsuga*. Montane rainforests occur in Myanmar, Malaysia, Indonesia and New Guinea and may contain some tropical conifers (*Podocarpus, Libocedrus, Araucaria, Phyllocladus*) at altitude. The mountains of the Arabian Peninsula are the only areas of the region where forest grows and, as altitude increases, *Acacia* and *Commiphora* are replaced by *Olea* and *Podocarpus* and then by *Juniperus procera*. There are extensive areas of subtropical mountain forests from the Mediterranean through to southern China. The Mediterranean forests have been significantly degraded and contain considerable areas of shrub and grasslands, but *Quercus, Abies cilicica, Cedrus libani* (Lebanese cedar) and *Pinus nigra* also occur. The mountains bordering the Caspian and Black Seas have well-developed forests of *Fagus, Carpinus, Acer* and *Quercus*. The northern Himalayas contain *Quercus, Pinus, Cedrus deodara* (deodar), *Picea, Abies, Juniperus, Betula* and *Rhododendron*. The subtropical mountain forests of China are dominated by *Abies, Picea, Pinus, Betula, Cupressus, Tsuga, Cunninghamia lanceolata* and with *Chamaecyparis* in Taiwan. The subtropical mountain forests of Asia are very diverse and the species list given here is by no means complete. There is a vast area of temperate mountain forest in Asia, to the north of the subtropical zone. *Pinus, Larix, Picea, Abies, Betula, Fraxinus, Polpulus, Acer, Tilia* and *Tsuga* are represented.

North and Central America

The tropical mountains of Central America contain temperate species such as *Quercus, Acer* and *Salix* but *Pinus* is also found. On wetter sites, Lauraceae is a prominent family. The subtropical parts of the Cascade Mountains, the Sierra Nevada and the Rocky Mountains contain *Pinus, Picea, Abies, Quercus, Populus tremuloides, Pseudotsuga menziesii, Tsuga* and the giant *Sequoia giganteum*. The temperate forests of the Pacific Coast Mountains are very productive and grand and are dominated by *Tsuga heterophylla* and *Abies amabilis* (silver fir). *Pinus, Picea, Abies, Populus tremuloides*, and *Pseudotsuga menziesii* are found in the temperate forests of the Rocky Mountains and species of *Fagus, Acer, Ulmus*,

Quercus, Betula, Picea, Pinus and *Abies* in the Appalachians. Forests at higher elevations in the boreal zone contain *Pinus, Betula, Picea, Abies* and *Populus*.

Europe

Europe has mountains in the temperate and boreal domains. In the temperate domain, beech forests with *Picea, Abies, Fraxinus, Acer* and *Ulmus* occupy the lower elevations and *Picea* and *Abies* the higher elevations. In the boreal domain the mountain forests contain mainly *Betula, Pinus, Picea, Abies* and *Larix*.

Australia and New Zealand

In Australia and New Zealand, mountain forests are mostly confined to the temperate domain. In south-eastern Australia (which includes Tasmania) the higher elevation forests contain tall eucalypt forests where it is wet enough. The main species are *Eucalyptus delegatensis, Eucalyptus viminalis, Eucalyptus regnans* and *Eucalyptus nitens*. *Eucalyptus pauciflora* is a treeline species. Cool temperate rainforest containing *Atherosperma moschatum, Nothofagus cunninghamii* and *Acacia melanoxylon* is also represented. In New Zealand the montane forests are mainly *Nothofagus* but conifers (such as *Podocarpus* and *Libocedrus*) are also significant.

South America

In South America the montane forests follow the length of the Andes for virtually the whole length of the continent. Tropical mountain forests are to be found in Colombia, Venezuela, Peru and Bolivia and they contain distinctive highland elements of the tropical forest on the adjacent lower country. There are many species. In the subtropical and temperate montane forests, the major genus is *Nothofagus*.

Mangrove Forest

Mangroves occupy the intertidal zones of inlets and estuaries on tropical and subtropical coastlines. They are a taxonomically diverse group, most belonging to four genera: *Bruguiera* and *Sonneratia*, and particularly *Avicennia* and *Rhizophora*. Mangroves are specifically adapted to survive and grow in the daunting marine intertidal environment. They have to be able to survive in a soil that contains no airspace and therefore almost no oxygen. They need to be able to absorb water and desalinize it before presenting it to the vital metabolic sites. They need to be able to persist in a very hot and desiccating environment when they are exposed at low tide.

Mangroves provide oxygen to their roots by having a variety of aerial roots. Air enters the aerial root through pores on the surface (called lenticels), which are connected to the root tissues below the water surface by a network of cells with a lot of airspace (called aerenchyma). *Rhizophora* has stilt roots, originating from the main stem well above the water line. *Avicennia* has pneumatophores, emerging vertically from a radial root close to but below the soil surface.

Mangroves desalinate water by a combination of excluding salt at the root surface, excreting salt through the leaves, depositing salt in metabolically less important tissue (such as in the non-functional components of the bark) and, at the cellular level, by depositing salt in the vacuole away from the sensitive metabolic sites in the cytoplasm. There is an energy cost in absorbing and transpiring fresh water extracted from salt water and consequently mangrove forests are not as productive as adjacent tropical land forests.

Mangrove species are a part of a marine forest ecosystem that can support epiphytes, insects, birds, reptiles, amphibians, mammals, decomposing microorganisms and a broad array of marine organisms including crabs and fish. Mangrove ecosystems have been an important source of construction timber, fuelwood, fodder, food, medicines as well as a habitat for shrimps and fish. Mangrove ecosystems are also important in arresting coastal erosion (Hogarth, 1999).

Plantation Forest

Plantations are usually (though not exclusively) of one species (monocultures) in groups of the same age (even-aged). Some argue that plantations should not be regarded as forests at all but rather as an extension of cropped agriculture. However, because plantations are trees growing over a rotation length of many years and supporting a more complex ecosystem with a wider range of community values than agricultural monocultures, most people regard them as forests. Certainly FAO includes plantations as forests when compiling their global statistics on forest areas. Plantations will be discussed in detail in Chapter 7.

Plantations have been established in both temperate and tropical areas. Natural monocultures are common in temperate forests but rare in the tropics. This suggests that planted monocultures (plantations) would be more suited to temperate regions than the tropics. This does not necessarily mean that plantations cannot be established in the tropics. Indeed, genus *Eucalyptus* and *Tectona grandis* (teak) have been very successful. It just means that it is a little trickier and more care needs to be made in species selection and silviculture. A successful plantation species has to grow fast, have adequate wood properties and be resistant to disease when grown as a monoculture. This restricts the number of suitable candidates enormously and is the reason why local species occurring in mixed forests are often unsuitable. The most successful plantation species are pioneer species. Pioneer species are the first trees to establish on bare ground caused by a disturbance such as fire, wind, road construction and deforestation. Pioneers grow under full sunlight and grow fast enough to out-compete weeds. Physiologists call pioneers 'sun trees' and trees that germinate and grow under shade, 'shade trees.' Sun trees have fast rates of photosynthesis at high light levels and consequently grow rapidly when there is plenty of light (Fig. 2.5). Shade trees have much slower rates of photosynthesis at high levels of light but are more efficient at photosynthesis at low levels of light. Shade trees are therefore better adapted to persist under the low light conditions of the forest understorey. Most prominent high quality furniture species are shade trees and attempts to establish them as plantations have been spectacularly unsuccessful. The mahogany (*Swietenia macrophylla*) plantations of Fiji and teak (*Tectona grandis*) plantations of Southeast Asia are exceptions. Even though a great number of species have been planted in plantations, it is likely that large-scale plantation enterprises will be limited to perhaps twenty pioneer species worldwide. Globally, the major plantation genera are *Pinus* (20%), *Eucalyptus* (10%), *Hevea* (5%), *Acacia* (4%) and *Tectona* (3%) (FAO, 2001a).

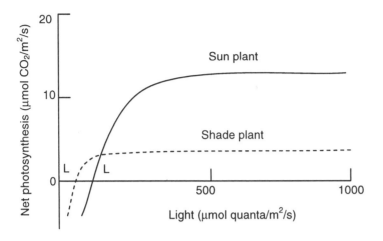

Fig. 2.5. A typical relationship between net photosynthesis and light (photon irradiance) for single unshaded leaves of sun plants and shade plants. L is the light compensation point (reprinted from Beadle and Sands 2004, copyright 2004, with permission from Elsevier).

There is a good argument for considering local native species first if they are suitable for the purpose. However, suitable native species may not exist because of the physiological constraints mentioned above and in this case an introduced species (exotic) could be the best choice. Exotics often grow faster in their introduced environment than their home environment. *Pinus radiata* (radiata pine) is a good example. It is the dominant plantation conifer in the southern hemisphere while of little consequence in its home environment of California and adjacent islands. Here it is an insignificant relict tree under threat. The reason for its success outside of its native range has been a matter of great debate. When genus *Eucalyptus* is planted outside Australia, it often grows faster and carries more leaf area. Eucalypts outside Australia no longer co-exist with the plethora of fungi and leaf eating insects that occur in Australia. Clearly planting exotics comes with some risk and this needs to be recognized and managed. Another risk associated with planting pioneer species is that they are very 'weedy' and can establish and dominate areas where they are not welcome.

Asia has by far the most plantations (62%), mainly in China, India, Japan and Thailand. The principal species in China are *Cunninghamia lanceolata*, *Pinus massoniana*, *P.koraiensis*, *P. tabulaeformis*, *P. elliottii*, *P. taeda*, *Eucalyptus* spp., *Populus* spp. and *Larix* spp. In India the major plantation genera are *Eucalyptus* and *Acacia* (*A. nilotica*, *A. catechu*, *A. tortilis* and *A. leucophloia*). In Japan, the major plantation species is *Cryptomeria japonica* (sugi) but *Chamaecyparis obtusa*, *Pinus densiflora*, *P. thunbergii* and *Larix kaempferi* are also represented. The major plantation trees in south-eastern Asia are *Tectona grandis*, *Hevea brasiliensis*, *Eucalyptus* spp. and *Acacia* spp. (*Acacia auriculiformis* and *A. mangium*). Europe has the second highest area of plantations (17%) mainly due to the contribution of the Russian Federation where the plantation genera are mainly *Picea* and *Quercus*. North and Central America have 9% of the world plantation area owing almost entirely to the contribution of the USA which has extensive plantations of mainly *Pinus taeda* in the south-east. South America has 6% of the world's plantation area, mainly in Brazil and Chile. The main plantation trees in Brazil are eucalypts and eucalypt hybrids (*E. grandis, E. urophylla, E. deglupta*) and *Pinus caribaea.* In Chile, *Pinus radiata* is the major plantation species. Africa and Oceania have the remaining 6% of plantations and *Eucalyptus, Acacia* and *Pinus* are the dominant genera (Turnbull, 2002).

Management of plantations can be intense with very high productivity. Eucalypt plantations in Brazil can exceed a mean annual increment of 50 cubic metres per hectare per year. For comparison, the mean annual increment of the world's forests on average is about 2 cubic metres per hectare per year. Not all plantations are managed intensively and mixed species and natural regeneration are other options for consideration.

How Much Forest is There

This is not a straightforward question to answer. FAO collects global statistics and until recently it got its information entirely by asking the various countries to report on their forest areas. This was unreliable because many countries could not (or chose not) to provide accurate figures. Recently, however, FAO has used remote sensing and geographic information systems to support country reporting and has been able to get a much better picture of the current state of the world's forests. This culminated in the FAO publication of the Global Forest Resources Assessment 2000 (FAO, 2001a), the most complete and accurate analysis available. Because of the changing patterns of deforestation and reforestation, the figures given here will be out of date before this book goes to print. FAO publishes its 'State of the World's Forests' every two years and the next Forest Resources Assessment will be in 2005. The distribution of the world's forest types is summarized in Tables 2.2 and 2.3. The definition of forest is land with a tree cover of more than 10%, an area of more than 0.5 hectares, a minimum height of 5 metres at maturity and not predominantly used for agricultural purposes.

Europe, which includes the Russian Federation, has the most forest area of any of the regions (27% of the world's forest area) but is third to South America and Africa in standing woody biomass. This is because the slow growing and relatively unproductive boreal forests stretching from Scandinavia to the Russian eastern seaboard dominate the forest area of Europe. The Russian Federation has 50% of its land area in forest, Sweden has 66% and Finland 72%, all well above the European average of 46% and the world average of 30%. European countries with a low forest cover include Iceland (0.3%), Ireland (10%), Moldova (10%) and the United Kingdom (12%). All countries in Europe, except Yugoslavia, had net increases in forest between 1990 and 2000. Europe is second to Asia in the area of forest plantations, largely due to the Russian Federation, which has 17.3 million ha.

South America has the highest percentage forest cover (51%) and the second highest forest area. This, together with the fact that South America has more than half of the world's tropical rainforest, means that the wood biomass of South America (43% of world total) is more than twice that of any other region. Brazil dominates not only South America, but also the world. Brazil has 113,676 million tonnes of wood biomass. The Russian Federation comes next with 47,423 tonnes. Brazil had 64% of its land area as forest in 2000 and was the largest deforesting country in South America losing 2.3 million hectares per year over the period 1990–2000. Some smaller countries in South America have very high percentage of forest cover: French Guiana has 90% and Suriname 91%. At the other extreme, Uruguay has only 7% forest cover and Argentina 13%. Uruguay, however, was the only country in South America to have a net gain in forest area over the period 1990–2000. Brazil and Chile also have extensive areas of forest plantations.

Table 2.2. The year 2000 distribution of forest by region and the change in forest area from 1990–2000 (from FAO, 2001b and 2003).

	Forest area 2000 $(10^6$ ha)	% of land area	% of world forest area	% of world woody biomass	Plantation area 2000 $(10^6$ ha)	Forest area change % per year 1990-2000
Africa	650	22	17	17	8	-0.8
Asia	548	17	14	11	116	-0.1
Europe	1039	46	27	14	32	+0.1
Nth & Cent. America	549	26	14	12	18	-0.1
Oceania	198	23	5	3	3	-0.2
South America	886	51	23	43	10	-0.4
World	3869	30	100	100	187	-0.2

Africa has 22% of its land area as forest, which is below the world average of 30%. Africa has the second largest wood biomass (17%) after South America because of the contribution of the productive tropical rainforest in west equatorial Africa. This distorts the fact that Africa is dry, with sparse or no tree cover over most of its area. No country in Africa had a net gain of forests over 1990–2000. Deforestation at rates exceeding 300,000 ha per year over the period 1990–2000 occurred in Sudan, Zambia, Zimbabwe, Nigeria and the Democratic Republic of the Congo. Almost all (98%) of Africa's forests are tropical. South Africa has 1.6 million ha of plantations.

Asia has a large area of tropical rainforest in the south-eastern mainland, Indonesia and New Guinea. It also has significant areas of drier tropical forest, subtropical forest, temperate forest and even a small occurrence of boreal forest. Asia has the lowest percentage of forest cover (18%) of any of the regions. Asia, however, has the greatest area of forest plantations, particularly in China, India and Japan, but also in Thailand, Turkey and Vietnam. Indonesia and Myanmar were the two greatest deforesting countries from 1990–2000, losing 1.3 and 0.5 million ha per year respectively. China, on the other hand, increased its forest area by 1.8 million ha per year over the same period. There was a net decrease in forest area of 0.1 % per year over the Asian region.

Oceania is dominated by Australia and consequently tropical dry types of forest predominate. However, Australia has some tropical rainforest in the north, subtropical humid forest in the east and temperate oceanic forest in the south-east and subtropical dry forest in the south-west. New Zealand has subtropical humid (warm-temperate) forest in the North Island and temperate oceanic forest in the South Island. The region is a net deforesting one. Australia lost most forest, losing 0.3 million ha per year over the period 1990–2000. New Zealand on the other hand, had a net gain of 50,000 ha per year over the same period. Both New Zealand and Australia have significant areas of forest plantations.

North and Central America have the full range of forest types from tropical rainforest to boreal tundra woodland. The region is dominated in area by the USA and Canada. Both countries have about the same area of forests and both about the same percentage of forest cover (Canada 27% and USA 25%). The forests are mainly subtropical and temperate in the USA, and temperate but mainly boreal in Canada. Consequently the wood biomass is greater in the USA (24.4 billion tonnes) than in Canada (20.2 billion tonnes).

Table 2.3. The percentage of forest types by regions in year 2000 (from FAO 2001b). Percentages in plain type are percentage area of the world and in italics percentage area of the region.

Ecological zone	Total forest	Africa	Asia	Europe	Nth & Central America	Oceania	South America
Tropical rainforest	28	24	17	-	1	-	58
Tropical moist deciduous	11	40	14	-	9	6	31
Tropical dry	5	39	23	-	6	-	33
Tropical mountain	4	11	29	-	30	-	30
Tropical total (% world)	47	28	18	-	5	1	47
Tropical total (% region)	*-*	*98*	*61*	*0*	*15*	*62*	*96*
Subtropical humid	4	-	52	-	34	8	6
Subtropical dry	1	16	11	30	6	22	14
Subtropical mountain	3	1	47	13	38	-	1
Subtropical total (% world)	9	2	42	7	37	7	5
Subtropical total (% region)	*-*	*1*	*23*	*5*	*16*	*30*	*2*
Temperate oceanic	1	-	-	33	9	33	25
Temperate continental	7	-	13	40	46	-	-
Temperate mountain	3	-	26	40	29	5	-
Temperate total (% world)	11	-	17	39	39	4	2
Temperate total (% of region)	*-*	*0*	*14*	*22*	*29*	*8*	*1*
Boreal coniferous	19	-	2	74	24	-	-
Tundra woodland	3	-	-	19	81	-	-
Boreal mountain	11	-	1	63	36	-	-
Boreal total (% world)	33	-	2	65	34	-	-
Boreal total (% region)	*-*	*0*	*2*	*73*	*40*	*0*	*0*
Total forests (% world)	100	17	14	27	14	5	23

Canada has no forest plantations to speak of but the USA has 16.2 million ha. Canada has recorded no net change in forest area over the period 1990–2000 but USA has increased its area by 3.9 million ha per year over the same period and is the only country in the region to record a gain. Mexico, Central America and adjacent islands carry tropical forests. Mexico has the worst record for deforestation in the region having lost 0.6 million ha per year over the period 1990–2000.

References and Further Reading

Beadle, C.W. and Sands, R. (2004) Tree physiology and silviculture. In: Burley, J., Evans, J. and Younquist, J.A. (eds) *Encyclopedia of Forest Sciences*, Elsevier Academic Press, Amsterdam, pp. 1568-1577.

Beazley, M. (1981) *The International Book of the Forest*. Mitchell Beazley Publication, London.

Burley, J., Evans, J. and Younquist, J.A. (eds) (2004) *Encyclopedia of Forest Sciences*. Elsevier Academic Press, Amsterdam.

Cleary, B.D. and Waring, R.H. (1969) Temperature collection of data and its analysis for the interpretation of plant growth and distribution. *Canadian Journal of Botany* 47, 167-173.

Evans, J. and Turnbull, J.W. (2004) *Plantation Forestry in the Tropics* (3rd edition). Oxford University Press, Oxford.

FAO (2001a) *Global Forest Resources Assessment 2000 - Main Report*. FAO Forestry Paper 140, Food and Agriculture Organization of the United Nations, Rome.

FAO (2001b) *State of the World's Forests*. Food and Agriculture Organization of the United Nations, Rome.

FAO (2003) *State of the World's Forests*. Food and Agriculture Organization of the United Nations, Rome.

Franklin, J.F. and Dyrness, C.T. (1973) *Natural Vegetation of Oregon and Washington*. Pacific NW. Forest and Range Expt. Station, USDA For. Service General Technical Report PNW 8.

Goodall, J.W. (ed.) (series, various years) *Ecosystems of the World*. Elsevier, Amsterdam.

Groombridge, B. and Jenkins, M.D. (2000) *Global Biodiversity – Earth's Living Resources in the 21st Century*. UNEP World Conservation Monitoring Centre, World Conservation Press, Cambridge.

Groombridge, B. and Jenkins, M.D. (2002) *World Atlas of Biodiversity – Earth's Living Resources in the 21st Century*. UNEP World Conservation Monitoring Centre, Universty of California Press, Berkeley.

Hogarth, P.J. (1999) *The Biology of Mangroves*. Oxford University Press, Oxford.

Holdridge, L.R. (1947) Determination of world plant formations from simple climatic data. *Science* 105, 367-368.

Holdridge, L.R., Grenke, W.C., Hatheway, W., Liang, H. and Tosi, J.A. (1971) *Forest Environments in the Tropical Life Zones*. Pergamon Press, Oxford.

Kimmins, J.P. (1997) *Forest Ecology - a Foundation for Sustainable Management*. Prentice Hall, Upper Saddle River, New Jersey.

Lieth, H. (1976) Biological productivity of tropical lands. *Unasylva* 28 (114), 24-31.

Nambiar, E.K.S. and Brown, A.G. (eds) (1997) *Management of Soil, Nutrients and Water in Tropical Plantation Forests*. Australian Centre for International Agricultural Research, Canberra.

National Research Council (US) Committee on Selected Biological Problems in the Humid Tropics (1982) *Ecological Aspects of Development in the Humid Tropics*. National Academy Press, Washington, DC.

Richards, P.W. (1996) *The Tropical Rainforest* (2nd edition). Cambridge University Press, Cambridge.

Tchatat, M. (1999) *Non-timber Forest Products - Role in Sustainable Management of the Rainforest in Central Africa. Local Program for Research and Development on Wetland Ecosystems in Africa*. Série FORAFRI Document 18.

Turnbull, J.W. (2002) Tree domestication and the history of plantations. In: Squires, V.R. (ed.) *The Role of Food, Agriculture, Forestry and Fisheries and the Use of Natural Resources*. Encyclopedia of Life Support Systems. Developed under auspices of UNESCO, Eolss Publishers, Oxford. (http://www.eolss.net accessed 12 March 2004).

West, D.C., Shugart, H.H. and Botkin, D.B. (ed.) (1981) *Forest Succession - Concepts and Application*. Springer Advanced Texts in Life Sciences, Springer-Verlag, New York.

Chapter 3

The Environmental Value of Forests

The Value of Forests

The value of forests means different things to different people. People may value forests for a range of reasons or may give prominence to one overriding value. People may differ over the values they place on forests and this can cause conflict. Likewise people may place different values on different types of forest. At the most fundamental level, forest dwelling communities value forests for their very survival. Some people place negative values on forests. Agriculturalists from ancient times to the present have seen forests as a barrier to development, as a harbourer of animals that prey on livestock and as a haven for plant pests and diseases. Urban developers often consider forests to be a costly nuisance. People who build their homes close to or in the forest may face a serious fire risk, albeit choosing to take this risk. Falling trees and limbs can cause damage to person and property. Tree roots can infiltrate drains and sewers and damage building foundations. Some find forests to be dark and claustrophobic. However, the overriding majority of people value forests and consider they provide a net positive benefit.

The benefits of forests may be tangible such as the forest products used by humans. These include wood used for construction and paper production; wood used for fuel (energy); and a wide range of non-wood forest products such as food, fodder, pharmaceuticals, dyes, tannins, cosmetics, essential oils, garden plants and resins. Tangible benefits are relatively easy to subject to economic analysis. Less tangible benefits are the environmental and social benefits of forests and the ecological services they provide. These include conservation of soil, water and biodiversity; recreation and amenity; aesthetic, cultural and religious values; protection from natural hazards; and climate amelioration. These less tangible benefits are not so easy to subject to economic analysis although natural resource economists are eagerly pursuing ways of doing so. This chapter will deal with environmental benefits of forests. The following chapter will deal with forest products.

Water Conservation

Water Quantity

The hydrologic cycle is shown at the global level in Fig. 3.1 and at the catchment (watershed) level in Fig. 3.2. Atmospheric water condensed in clouds falls as rain mostly over the oceans (79%) but also over the land (21%) where it enters the soil to provide water

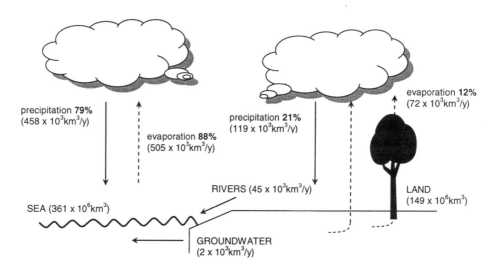

Fig. 3.1. Global hydrologic cycle (constructed from data of UNESCO, 1978).

for plants, runs off to streams over the surface or enters the groundwater. Some of the water from groundwater and streams finds its way back into the oceans. Water evaporates back to the atmosphere mainly from the ocean (87%) but also from the land (13%) either directly from the land surface or via plants in the process of transpiration (Fig. 3.1). Of the total volume of water (approximately 1.4 billion km³), only 2.5% is fresh and for plants, otherwise running off to streams over the surface or entering the groundwater. Most of this (approx. 70%) is tied up in polar and montane ice. Only about 0.05% of the fresh water is stored in the soil and available to plants. Solar energy acts as a giant still to ensure that fresh water is always being regenerated. Each year about 575,000 km³ is evaporated from land and oceans into the atmosphere (Fig. 3.1). If fresh water was evenly distributed over the earth, there would be no shortages to plants and animals. However, distribution is not equal and in certain parts of the world fresh water is the major limiting resource to plant and animal life.

Figure 3.2 is an idealized land surface in a forested catchment sloping down to a stream. The water balance is determined by accounting for all of the inputs (precipitation), outputs (evaporation and streamflow) and storage (soil water and groundwater):

$$Q = P - (E + \Delta S + \Delta G) \qquad (3.1)$$

where Q =   water yield (streamflow)
         P =   precipitation reaching the soil plus litter surface
         E =   evaporation of water from soil and litter, plus transpiration from leaves of water absorbed through the roots, plus evaporation of water intercepted by the canopy
         ΔS =   change in soil water storage
         ΔG =   change in groundwater storage

Rainfall intensity and duration is very uneven over scales of minutes to millennia. This results in droughts, rains and fluctuations in streamflow. It is the extreme events rather than the average that often creates most interest.

The water table is the surface that separates a zone of unsaturated soil above from saturated soil below. The unsaturated soil between the water table and the land surface contains air in its larger pores and water in its smaller pores. The force of adhesion of water to the pore walls increases as pore size diminishes. Consequently the smaller pores can hold water but the larger pores will drain to the water table. Plant roots require air to function and are confined to unsaturated soil. Plants, through the process of transpiration (evaporation of water from the leaf surface), can exert enough pull on the water held in the small pores to break the forces of adhesion and reduce the water content of the soil. Rainfall infiltrating the soil will replenish the water in the small pores until all of the small pores capable of holding water against drainage are full. The soil can hold no further water unless water is entering the soil at a faster rate than it can drain. Soil scientists say that soil in this condition is at 'field capacity.' The amount of water that a soil can hold at field capacity depends on its depth to the water table (or a water impeding layer): the deeper the soil the more water it will hold. The amount of water held at field capacity also depends on the texture of the soil: fine textured soils (clay soils) have relatively more small pores and therefore can hold more water than coarse textured soils (sandy soils), which have relatively less small pores. However the rate at which water can enter the soil (the infiltration rate) is greater in sandy soils than clay soils. Saturated soil, the soil below the water table, has its entire pore space filled with water and the water in the saturated zone is called groundwater.

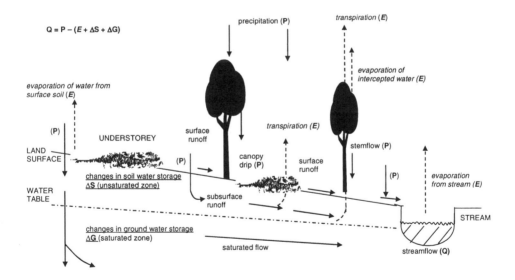

Fig. 3.2. Hydrologic cycle at the catchment level. Inputs are in plain type, outputs in italics, and changes in storage underlined.

Precipitation may be snow, sleet or hail but predominantly is rain. Rain falling in a forested catchment may fall directly to the ground or be hung up in the canopy. Anybody who takes shelter under a tree during rain can attest to the fact that initially the canopy provides good shelter. After a while, however, the canopy can intercept no further rain and water starts to drip from the leaves and flow down the stem. The amount of water intercepted by the leaves and which does not reach the ground can be substantial in forests. The precipitation reaching the ground surface (directly, by stem flow or by canopy drip) will enter the soil and will either add to the amount of water stored in the unsaturated soil or will flow (runoff) to groundwater and streams. The amount of runoff depends on slope, vegetation type, soil depth and texture, the amount of water already in the soil and, most importantly, the duration and intensity of the rain. Most flow to streams occurs as saturated flow from the groundwater or over the surface of an impermeable subsurface layer, but flow can also occur over the soil surface (overland flow) or even in the unsaturated subsurface under certain circumstances. Flow to streams at the expense of increasing soil water storage is favoured by intense rainfall, steep slopes and bare ground (no litter or soil humus). As the soil becomes wetter from prolonged rainfall and exceeds field capacity, the soil can no longer store water. In such circumstances all of the water will flow to streams or groundwater. If rainfall is entering the soil at a faster rate than it is draining from the soil profile, the groundwater may be recharged at such a rate that the water table may rise to the soil surface and overland flow may occur. Also, overland flow may occur as a result of intense rain on soils with low infiltration rates or over impermeable surfaces such as rocks.

Water is returned to the atmosphere by evapotranspiration which is the sum total of water evaporating directly from the soil surface, water taken up by plants from the soil through their roots and transpired through their leaves, and the direct evaporation of rain intercepted by the canopy. The water yield of a catchment depends on vegetation type. For many forested catchments, the additions and subtractions to soil water storage and groundwater storage averaged over a year balance out. Consequently the impact of ΔS and ΔG in equation 3.1 is small and water yield is mostly determined by P and E. Forested catchments usually have lower water yields than grassed, cultivated or otherwise disturbed catchments. This is because the amount of intercepted water is greater in forests and therefore less precipitation reaches the ground. Forests often transpire greater amounts of soil water as well. Forest trees have deep roots and can access water deeper in the soil. The fact that forested catchments have reduced water yields has been a source of controversy in catchment management, particularly when contemplating establishing tree plantations.

Different types of forest can produce different water yields. In general, evergreen forests (conifer and angiosperm) have canopies that intercept more water than deciduous angiosperm forests and therefore produce lower water yields. This happens because deciduous trees are without leaves to intercept and transpire water for several months in the year. Also water yield can depend on the stage of development of the forest: if the forest (including understorey) increases the size of its canopy, then the amount of both intercepted and transpired water will increase and the water yield will be reduced. Thinning and selective logging (removing a few trees) often has little effect on water yield because the residual vegetation can quickly increase its transpiring leaf area and resume near original values of transpiration and interception. Clearfelling a forest will increase water yield and if the site is deforested and transformed to another land use such as agriculture, the increase in water yield can be permanent. If the site is allowed to regenerate to forest, water yield ultimately will decrease back to forested levels. Forest plantations established on cleared land may initially increase water yield if weeds are controlled, but ultimately should reduce water yield. Any effect of clearing or replanting on water yield will be dependent on the area of disturbance involved. If the area in any year is small relative to the area of the catchment, the effect on water yield will be correspondingly small. If, on the other hand, a significant proportion of a catchment is cleared for agriculture or grazing, the increase in water yield can be substantial. Sometimes a wildfire will remove most, if not the entire, transpiring canopy in a catchment. Under these circumstances the increase in streamflow (and more particularly the decrease in water quality) can be substantial. Wildfires can also make soil surfaces water repellent, thereby reducing soil infiltration rates.

Streamflow can vary enormously throughout a year from no flow to full flood, with slope, soil type and land management practices having important effects. Streamflow can also vary diurnally (greater at night than at day) reflecting the greater water use by vegetation near streams during the day. If soil infiltration rates are slow owing to steep slopes, shallow soils of fine texture, soil compaction or the creation of impermeable surfaces (e.g. roads), runoff to streams may be accelerated to the extent that infiltration to the water table is diminished. This may reduce streamflow during the dry season. The impact of deforestation on flooding has often been exaggerated. Floods are caused by excessive rainfall and under these conditions all other hydrologic variables are swamped into relative insignificance. Because streamflow is less in forests, deforestation may cause a flood to start a little earlier and last a little longer. Also sedimentation to streams resulting from deforestation may reduce stream capacity and promote flooding. For these reasons deforestation on highlands is often blamed for causing floods on the lowland. This may or may not be true, depending on circumstances. Land management on the lowlands also plays a part: greater flood plain occupancy, greater channel constriction, more roads, more ditches and more non-absorbing surfaces will all contribute to lowland flooding. Reforestation of catchment uplands has several beneficial effects, but large-scale flood mitigation is not one of them. In general, forest cover in upland catchments will not prevent floods and deforestation of upland catchments is not the primary cause of floods.

If forests (deep-rooted with high evaporation) are replaced with pasture species (shallow rooted with lesser evaporation) then over time (perhaps many decades) the groundwater levels may rise. If the rising groundwater collects salts from the soil, the rooting zone of plants can become saline and salt may be exposed where the water table breaches the soil surface at lower parts of the catchment. For example, large areas of soils in southern Australia have become unproductive owing to soil salinity induced by clearing of forest. Replanting of forests can reverse this effect but the process takes a long time. Finding suitable species that are both copious transpirers and salt tolerant is a challenge.

Water Quality

A well-managed forest with intact understorey and well-developed litter and humus layers usually acts as a filter and produces better quality water with less turbidity and sediment than a catchment with an alternative land use. This will be discussed further in the next section on soil erosion. A case study from Canberra in Australia ably demonstrates the factors involved in water quality. The water supply to the city of Canberra is provided from two catchments. One is forested and the water requires no treatment. The other has a component of agriculture and grazing in the catchment and the water requires treatment for domestic consumption. However, in 2003, the forested catchment was extensively burnt and the water quality deteriorated markedly.

Soil Conservation

The conservation of soil can be considered in two complementary ways. The first of these is to maintain the fertility of the soil on a site and the other is to stop the loss (erosion) of soil from the site. Most of the land cleared in the past for agriculture and for urban development was on the better soils that would have originally supported forests. However forests can also exist on quite poor soils (e.g. boreal forests and forests on sand dunes) and on quite steep slopes. Consequently there has developed a culture that agriculture should have the first pick of soils and forestry should be left with the residue. This does not always make economic or environmental sense. Usually forests have better soils than adjacent non-forested areas on the same soil type. Young (1997) reviewed the evidence for this. Forests better preserve carbon and nutrients than non-forest soils and forest soils also tend to have better physical properties. It follows that reforestation of non-forested areas has the capacity to improve the soil and deforestation has the capacity to degrade soil. Deforestation, particularly in the tropics, can remove a significant amount of carbon and nutrients from the site. However, forest harvesting, as distinct from deforestation, does not appear to cause significant productivity loss in 'natural' forests (Attiwill and Weston, 2001). Forested areas usually suffer less soil erosion. Soil erosion is the process by which soil is removed from a site by agents such as ice and snow but mainly by water, wind, or human disturbance. Soil erosion clearly reduces the fertility, productivity and utility of the soil at the site from which it has eroded. The eroded soil is deposited elsewhere as sediments, which may be either a benefit or a nuisance.

Soil Erosion

The development of soil is a balance between the rate at which parent material is broken down to form new soil and the rate at which soil is removed by erosion. Topsoils are produced at a rate of 0.5 to 2.0 t/ha/y and soil loss under the best practice of cultivated agriculture is also 0.5 to 2.0 t/ha/y, which means that there is no net loss or gain of soil. Soil loss under poorly managed agriculture (and forestry) often will exceed the rate of soil development and there will be a net loss. Zimbabwe would require US$1.5 billion of fertilizers each year to compensate for natural nutrient loss from soil erosion. The USA loses US$18 billion per year in fertilizer nutrients to soil erosion (National Research Council, 1993). In relatively undisturbed forests on moderate or less slopes, the rate of soil development can exceed the rate of soil loss from erosion. Generally speaking, soils are developed under forests and lost under agriculture. Because topsoils are lost under intensive agriculture, productivity is reduced unless fertilizers are used. Otherwise more forests are cleared to compensate for lost productivity. This compounds the problem by increasing the contamination of ground and surface waters and increasing the overall rate of erosion by clearing more forest and exposing more land.

Any event that exposes bare mineral soil has the potential to promote soil erosion by both water and wind. Such events include fire, overgrazing, drought, pestilence (e.g. swarms of locusts), disease, deforestation, forest degradation, soil cultivation, weed control, road construction, forest harvesting operations, earth works and urban development. It follows that the key to controlling soil erosion is to avoid exposing mineral soil or to keep the time interval during which the soil is exposed as short as possible. Tying the soil together with a network of plant roots and promoting a protective cover of litter and humus on the soil surface as soon as possible after any disturbance can achieve this. The vegetative cover need not necessarily be forest, although the reduced runoff in forested catchments would provide an advantage. It follows that deforestation will promote erosion for the time the soil is exposed but erosion will be arrested when the site is revegetated by reforestation or by some other land use that provides a complete vegetative cover with a complex web of roots. If forest is replaced by cultivated agriculture, the potential for soil erosion by both water and wind is increased.

Erosion by Water

The factors determining soil erosion by water are described by the Universal Soil Loss Equation (Wischmeier and Smith, 1965)

$$A \quad = \quad RKLSCP \tag{3.2}$$

where A is the mass of soil lost per unit area, R is the rainfall factor, K is the soil erodibility factor, L is the length of slope factor, S is the slope gradient factor, C is the cropping management factor and P is a factor related to erosion control practices. Qualitatively this means that soil erosion by water is maximized when rain is prolonged and intense on erodible soils on long steep slopes that have no vegetative cover and where no attempts have been made to try to manage the problem.

There are various factors that make intensive agriculture particularly vulnerable to erosion by water: large areas of monocultures are often planted with little regard to the configuration of the terrain; there are extended periods in which the cultivated soil has little or no vegetative cover at all; heavy machinery exposes, compacts and reduces the water infiltration rate of soil; and roads provide paths for the runoff of water. Erosion can be significantly reduced by using minimum tillage, mulching, conserving crop residues, rotational cropping, mixed cropping and contour strip cropping. Row crops usually are most profitable in the short term but offer the least resistance to soil erosion. The reason that forests are effective at controlling soil erosion by water is a combination of several factors: forests have less surface runoff of water to streams; the surface of forest soils are generally covered by a protective layer of humus and litter; forests and their understorey form a complex root network that is effective in holding the soil together; and forests are often in remote areas and relatively immune from human disturbance. However, when forests are disturbed, particularly on steep country, the potential for soil erosion by water is substantial. The soils most vulnerable to erosion by water are poorly aggregated fine textured soils where the resistance to particle detachment by water is low and where water infiltration rates also are low. Sometimes the protective effect of vegetation is not sufficient to arrest erosion by water. On the very steep upper slopes of young and recently uplifted mountain ranges, such as the Himalayas and the New Zealand Alps, extensive soil erosion is inevitable and a fact of life. No management intervention can stop it. However, in most instances the management of vegetative cover is the key to arresting or slowing down soil erosion from water. The interacting effects of rainfall intensity, slope and ground cover are demonstrated in an agricultural system in Nigeria (Table 3.1). Soil loss increased with annual rainfall and with slope. Soil loss was greatest on the bare fallow plot. However, when maize was grown with a protective mulch, soil loss was zero even on the steepest slope.

Table 3.1. Soil loss (t/ha) during higher and lower rainfall years from different maize cultivation practices in Nigeria (derived from Lal, 1997).

Treatment	Slope (%)		
	1	5	15
Higher rainfall year (781 mm)			
Bare fallow	7.5	80.4	155.3
Maize without mulch	1.2	8.2	23.6
Maize with mulch	0.0	0.0	0.0
Lower rainfall year (416 mm)			
Bare fallow	3.7	75.8	73.9
Maize without mulch	0.4	2.8	17.1
Maize with mulch	0.0	0.0	0.0

A mixture of different roots is more effective at reducing soil loss than roots from a single species alone and the role of the understorey in arresting soil erosion is substantial. Forests are most effective in erosion control when they have a well developed understorey with a complex network of roots, leaf litter and humic horizons. The key factor in the relative efficiency of different land uses is the degree of intactness and completeness of the soil litter and humus layer, the complexity of the root web and the lack of disturbance in the system. This is well demonstrated by the range of land uses shown in Table 3.2.

Litter has been removed from some forests (for example Portugal) for centuries. This has gradually but relentlessly reduced the fertility of the soil as well as increasing the risk of soil erosion. Similarly collecting leaf litter for fuel under plantations in Nepal and collecting leaves under mahogany and teak plantations in the Philippines for producing mosquito repellents, chicken feed and fertilizers exposes the soil to the risk of soil erosion unless special care is taken. Care needs to be exercised to reduce soil erosion in forest operations. Residue retention between rotations of plantations not only conserves nutrients on the site but also reduces the potential for soil erosion. Cultivation and weed control are standard practices in plantation establishment but this comes at the risk of soil erosion which needs to be taken into account in steep terrain in areas of high rainfall. The practice of sowing a legume crop as soon as possible on the cleared land prior to plantation establishment, such as is practised in New Zealand, helps to reduce soil erosion as well as improve nitrogen nutrition. Soil conservation techniques such as contour ploughing, minimum tillage, cover crops, multiple cropping, mulching, terracing and building structures to control water flow can reduce soil erosion from water. For example, thousands of years ago, the Incas in South America constructed intricate terraces on the steep slopes of the Andes to intercept rainfall and prevent erosion. Today, intricate terraces support rice culture in Bali. This again demonstrates that erosion on steep slopes can be controlled without forests. Forest harvesting operations compact, deform, disperse and expose mineral soil. Soil erosion and sedimentation to streams can be substantial if steps are not taken to arrest this. The provision of untouched buffer strips around streams as sediment traps is a common requirement in logging codes of practice worldwide. Road construction is another forest operation that has the potential to cause significant erosion. Fire over a catchment can cause substantial erosion because it exposes mineral soil. Fire can eliminate the protective soil litter, destroy humus and soil organic matter, kill the roots and destroy soil structure. It can also increase runoff of water after heavy rain and this will further increase erosion of the exposed soil.

Table 3.2. The effect of forest type and ground surface cover on soil loss (t/ha/y) in the tropics (abridged from Hamilton and Pearce, 1987).

Mixed species uneven-aged forest	0.30
Forest plantations, undisturbed	0.58
Tree crops with cover crop or mulch	0.75
Shifting cultivation, cropping period	2.78
Tree crops, clean weeded	47.60
Forest plantations, burned or litter removed	53.40

Soil erosion by water can be from the soil surface (sheet erosion, gully erosion and stream erosion) or subsurface (tunnelling). Sometimes soil erosion on steep slopes can be very deep-seated and whole hillsides plus vegetation detach (mass wasting or landslides). The presence or absence of forests has little influence over the occurrence of deep-seated landslides. Erosion by water may be uniform and hardly noticeable. For example, sheet erosion may slowly but steadily remove layer after layer of soil until there is little left. More often, however, erosion by water is intermittent and uneven over the landscape and occurs in short bursts at times of intense rainfall on patches of ground with sparse or no vegetative cover. Surface erosion is potentially greater in the tropics than the temperate zone because rain is more frequent and intense, there is less litter on the soil surface, the soil has thinner humic horizons and there is less ground vegetation. Consequently logging operations in wet tropical forests are very sensitive to soil erosion, particularly on steep slopes.

Sites eroded by water can be rehabilitated by covering the soil surface with vegetation, litter and root system networks as soon as possible. Trees have a major role in rehabilitation because they are deep rooted and have strong roots. However, trees alone are not a panacea and should be used in conjunction with grasses, herbaceous species and shrubs in order to achieve a complex web of roots. Vetiver grass (*Vetiveria zizanioides*) grown in the tropics as a narrow dense hedge along the contours can be very effective at reducing erosion. Tree roots can assist in stabilizing areas prone to shallow slips and in preventing stream bank erosion. One of the many benefits of trees in agroforestry systems is erosion control. Rows of trees planted along the contours in agroforestry systems can trap soil moving down hill.

Erosion by Wind

Soil erosion in dry country can be by water following intense rainfall but erosion by wind is more common. Wind erosion is more uniformly spread over the landscape than water erosion and its consequences can be both spectacular and disastrous. For example, Leo Tolstoy witnessed the 'black storms' in the Ukraine in 1891, which caused crop failures resulting in entire villages starving to death. In the USA the natural vegetation of the prairies was removed and replaced by poorly managed agriculture on sub-marginal cultivated farmland. The resulting dust storms in the 1930s lasted for days, blocked out the sun, ruined machinery, inundated buildings and mechanically abraded plants, animals and structures. In dry country, vegetation cover is more fragile and recovers less well after damage. Droughts are often prolonged (and when the rains come they may be intense). The key to arresting soil erosion in dry areas is to maintain or improve the productivity of the soil so that the vegetation provides a more or less complete cover for all of the year and is protected as far as possible from the ravages of drought. This can be achieved by avoiding overgrazing, minimum tillage, not cultivating poor soils, managing the soil to increase its fertility and water holding capacity, mulching, conserving soil organic matter, water harvesting, avoiding soil salinity, avoiding fire and by constructing windbreaks and soil barriers. Annual grasses 'hay off' and die and the soil is vulnerable to erosion during this period. Trees and other perennials (in combination with grasses) have an advantage in controlling soil erosion in dry country because they are deep rooted and long lasting. Examples of the use of trees to combat soil erosion from wind include: extensive plantings in the African Sahel to slow down the desertification of overgrazed land; establishing the 6000 km long 'Green Great Wall' to protect Beijing from the advancing wind eroded soils of the loess plateau; and stabilization of advancing coastal sand dunes (in conjunction with coastal grasses) in many parts of the world (McKelvey, 1999).

Sedimentation

Eroded soil is either washed or blown to other locations where it is deposited as sediments. Sedimentation is a mixed blessing but the net effect is decidedly negative. Even before the soil is deposited as sediment it can have negative effects. Eroded soil carried in streams reduces the quality of the water for human and animal consumption by increasing the turbidity of the water and by carrying harmful substances such as insecticides, herbicides, fungicides, excess nutrients (particularly nitrogen and phosphorus) and industrial wastes. The reduction in water quality can harm or kill aquatic life and can limit its use in irrigation and industry. Wind blown soil reduces visibility and endangers health and water-borne sediments can raise the level of river beds and reservoir beds so that adjoining areas are subject to flooding. Groundwater levels can rise and wide areas can become waterlogged, developing into marshes. Reservoir capacities are reduced and the life of hydroelectric turbines and water pumps is also reduced. Water flows to industry and agriculture are impeded and irrigation canals become clogged and ineffective. Sometimes good soil is deposited where it is useful for agriculture, such as in the fertile flood plains of river systems (e.g. the Nile, Tigris/Euphrates, Ganges and Yangtze Rivers) whose continued productivity depends on erosion. However, very large amounts of eroded soil are deposited on stream and reservoir beds and in the oceans where it can be of no agricultural significance. The ancient civilizations on the fertile floodplain of the Tigris/Euphrates River developed on sediments from erosion but ultimately failed because they were overwhelmed by excess sediment (see Chapter 1).

The amount of soil eroded and deposited as sediments can be enormous. More than one billion hectares of the land surface of the earth are experiencing serious soil degradation as a result of water erosion (Oldeman *et al.*, 1991). The current global rate of soil loss is of the order of 75 billion tonnes per year (Pimental *et al.*, 1995) of which about 50% is human-induced. The world's rivers deliver about 20% (15 billion tonnes of sediment) to the ocean each year (Walling and Webb, 1996) and many more billions of tonnes settle on stream bottoms and/or silt up reservoirs behind dams. For example, China loses more than 2 billion tonnes of soil each year, most of which is deposited in the Yangtze River. The annual sediment yield from the upper Yangtze River basin alone is 0.5 billion tonnes (Dingzhong and Ying, 1996). It takes 3.5 billion cubic metres of water to flush 100 million tonnes of soil deposited in the Yangtze River to the ocean. This water could be used for more productive purposes (National Research Council, 1993). Erosion of highly erodible rocks and soils on steep slopes in areas of high rainfall can be spectacularly high. The annual sediment yield from a schist basin in a high rainfall area of the New Zealand Alps is nearly 30,000 tonnes per square kilometre (Hicks *et al.*, 1996), while each year 500,000 cubic metres of gravel is removed from the Waimakariri River in New Zealand for flood mitigation. The Netherlands flood because of sediment arriving from other countries upstream on the Rhine.

Upstream erosion causes downstream sedimentation and therefore better upland watershed management will assist in controlling downstream sedimentation. If natural erosion rates are not extreme, undisturbed forests with intact understoreys will reduce erosion and sedimentation to streams. Disturbing (e.g. logging) or removing forests may increase erosion and therefore sedimentation to streams. Greer *et al.* (1996) found that suspended sediment concentrations in the Segama River catchment in Sabah were elevated as a result of the mosaic of forest patches at various stages of recovery from logging. They remained lower, however, than in the lower catchment which contained commercial oil palm plantations, agriculture and degraded forest. Rapid regrowth of secondary vegetation following logging caused sediment loads to decrease to near undisturbed levels within three years. Forest establishment may decrease sedimentation if it replaces poor land use. For example, Dano (1990) showed that no-burning and reforestation increased water yield (9.5% and 11.5% respectively) and decreased sediment yield (59% and 72% respectively) compared to annually burnt *Imperata* grassland in the Philippines. Protecting stream banks and drainage lines during forest harvesting by leaving undisturbed vegetation (buffer strips) will reduce sedimentation.

Conservation of Biodiversity

What is Biodiversity?

'Biodiversity' is a contraction of 'biological diversity' and can be defined as the variability among all forms of life, past and present. It can be considered at three levels: (a) genetic biodiversity, which considers each individual as a genetic entity; (b) species biodiversity, which is essentially a count of the number of species; and (c) ecological biodiversity, which is variability within and between ecosystems. From a conservation management perspective, genetic biodiversity (and their relative abundances) is variation within a species, and tree-breeding programmes need to ensure that sufficient genetic diversity is conserved to protect the species against environmental variation and biological threats. Conservation management of species biodiversity is directed at identifying and caring for threatened and endangered species and reducing the rate of species extinctions. Conservation management of ecological biodiversity is directed at maintaining intact ecosystems to safeguard genes, species diversity and ecosystem processes.

Because species and particularly ecosystems cannot be precisely defined, boundaries are often indistinct. It is very difficult to draw the line between the number of entities and the degree of difference between those entities. Consequently the measurement of biodiversity is complex and never exact. Species richness (the number of species) is used frequently as a surrogate measure of biodiversity but it has its limitations (Gaston and Spicer, 1998). Most plants are angiosperms (>75%) and most animals are insects (>80%) and even though greater than 90% of all species of the past have become extinct, biodiversity has increased over time. Currently it is estimated that there are 13.5 million species with only 1.5 million of these described (Gaston and Spicer, 1998). In general, biodiversity decreases with increasing latitude and altitude. More than two-thirds of terrestrial biodiversity is in the tropics and 12 countries have been classified as 'megadiverse', containing between them about 80% of the world's species richness (Malagasy Republic, Australia, China, India, Indonesia, Malaysia, Thailand, Mexico, Brazil, Colombia, Equador and Peru). However humans have increased the extinction rate to perhaps 10,000 times greater than background and consequently human-induced loss of biodiversity is a crisis.

Biodiversity of Forests

Forests are the most biodiverse terrestrial ecosystems (Groombridge and Jenkins, 2002), particularly in the tropics and subtropics. Tropical forests have more than 50% of all terrestrial species and tropical rainforests are the most biodiverse terrestrial ecosystems of all. Biodiversity decreases pole-wards, cool-temperate and boreal forests having lower biodiversity than tropical forests. Indeed in places in the temperate zone where there has been a long history of agriculture and grazing, such as in Europe, a unique biodiversity has developed. Some biologists argue that reversion to large continuous tracts of temperate forest in Europe would reduce biodiversity and that maintenance of the current mosaic of forest, agricultural and grazing land is necessary to maintain current biodiversity. However, overall the clearing of forests and replacing them with agriculture, grazing or urban development have greatly reduced biodiversity. This biodiversity loss is alarming at best and catastrophic at worst but estimates of rates of extinction of species have been hampered by insufficient data. Again, distinction needs to be made between forest clearing (deforestation) and forest logging. Deforestation permanently removes the biodiverse forest and replaces it with a less biodiverse alternative land use. Forest logging, no matter how badly and irresponsibly done, at least allows the possibility of ecosystem repair over time. Deforestation almost invariably reduces biodiversity, whereas forest management for timber production is comparatively less damaging (Table 3.3). It follows that forest management for timber production will save biodiversity if it substitutes for deforestation (Burgman and Lindenmayer, 1998).

The impact of forest management for timber production on biodiversity depends on the forest type and the standard of management. There is little information and great ignorance. Forest 'mining' where little or no attention is paid to regeneration and where considerable damage is done to the residual stand and soil has a much greater potential to reduce biodiversity than forest management systems that are sensitive to the future welfare of the stand. Low intensity logging or low intensity shifting agriculture where the size of the disturbance is similar to natural disturbances such as storm damage and tree falls may have little effect on biodiversity. Enlightened sustainable forest management systems aim to reduce the biodiversity loss to background values (or even to increase biodiversity) and in some instances this claim can be backed up with credible evidence. Regrettably, much forest management globally is poor and probably is reducing biodiversity to an unknown extent. The challenge is to promote sustainable management systems that protect biodiversity. Forest management for conservation of biodiversity will be discussed in Chapter 6.

Table 3.3. Causes of extinction and past and present threats to endangered plant species in Australia (from Leigh and Briggs, 1992).

Threat	Number of presumed extinct species	Number of endangered species	
	Presumed cause	Past threat	Present and future threat
Low numbers	-	10	85
Roadworks	1	8	57
Weed competition	4	12	57
Grazing	34	51	55
Agriculture	44	112	50
Industrial and urban development	3	20	21
Fire frequency	-	10	17
Collecting	-	6	17
Mining	1	3	11
Forestry	-	10	17
Recreation	-	-	7
Dieback	-	-	7
Clearing	-	2	5
Railway maintenance	-	2	4
Salinity	-	-	4
Insect attack	-	-	3
Quarrying	-	-	3
Trampling by pigs and buffalo	-	1	3
Hydrostatic pressure changes in artesian basin	1	-	-
Drainage	-	1	2
Flooding	-	4	2
Other	-	-	2
Rubbish dumping	-	-	2
Vehicle damage	-	1	2
Erosion	1	-	1
Habitat degradation	-	-	1
Rock falls	-	-	1
Visitation	-	-	1
None identified	1	5	-
Unknown	-	12	6

Trees in rural areas, in groups not large enough or too scattered to be considered as forest, collectively are an enormous resource for conservation of biodiversity as well as a source of fuel, timber, food and medicines. Also, the trees and associated flora and fauna in streets and in private and public gardens can be very extensive and a valuable resource for the conservation of biodiversity.

Why Conserve Biodiversity

Fundamentally, humans, individually and collectively, operate out of self-interest in a competitive environment. It is unlikely therefore that the human race will look after biodiversity unless there is something in it for them. This may sound cynical but, from an evolutionary perspective, it is appropriate behaviour. If, for the case of argument, ethical, cultural, religious and philosophical considerations are put aside for the moment, it might be argued that the human race need only be concerned with those species that impact directly on their sustenance. Three cereals (wheat, maize and rice) represent about 50% of all human energy input. Twenty species of plants and five species of animals account for more than 90% of human sustenance and about 1000 species out of approximately 13.5 million (<0.01%) have some current economic value to humans (Solbrig *et al.*, 1994). It might be argued that, because the amount of biodiversity on the planet is so high, the loss of a significant portion of this will have little effect on the existence of humans on earth providing we look after these key species. However, this analysis, besides being heartless, is incorrect. What this analysis does show is that humans exist on a very narrow resource base and this in itself might be a cause for concern. There is considerable scope for widening the consumption base for humans.

Besides providing food, plants and microorganisms provide medicines. For example aspirin, the best-selling medicine of all time, is now synthetically produced but is based on extracts originally derived from the bark of the willow. There are more than 100 substances extracted from 90 species of plants used in medicines and about 5000 species of plants, many from forests, have been extensively investigated as sources for new drugs. The World Health Organization lists 21,000 names of plants that have been reported to have medicinal use although very few of these have been scientifically tested (Groombridge, 1992). Many ethnicities have a long history in the use of traditional medicines based on plants. Animals are also used for medicine. However, there is a flourishing market in animal (and sometimes plant) products from endangered species. Besides producing wood (for construction and fuel) and food, forests produce a range of other economically valuable products such as fodder, dyes, tannins, cosmetics, essential oils, garden plants and resins. These are termed non-wood forest products and will be discussed in more detail in Chapter 4. It is inevitable that more plants, animals and microorganisms will be found with medicinal and other uses and some of these may well be from currently undescribed species. Forest plants, and particularly rainforest species, are rich in chemicals, particularly alkaloids, which they produce to protect themselves from disease or insect attack. These chemicals produce useful drugs now and will do so in the future providing the species still exist. Forests may be very biodiverse but there is considerable ignorance about what is actually there.

The conservation of biodiversity is intimately associated with the integrity of ecosystems. Forests are complex ecosystems that provide a wide range of values to humans as discussed in this and the next chapter. There is evidence that some terrestrial ecosystems are dependent on a high diversity of plants, animals and microorganisms (Reaka-Kudla *et al.*, 1997). The integrity of these ecosystems depends on the involvement and interaction of biological entities at all levels. Loss of biodiversity may threaten the integrity and functioning of the whole. It is not known whether there is an element of redundancy in this and whether all organisms are necessary for ecosystem stability. Certainly some ecosystems with lower biodiversity appear to be stable (Kimmins, 1992). Indeed, the relationship between levels of biodiversity and ecosystem stability and resilience is obscure (Kimmins,

1997). The question of how much biodiversity we can lose before the human race is threatened is unknown. This highlights the lack of knowledge in this area. Certainly it would be wise to err on the side of caution. It follows that one of the best ways to conserve biodiversity is to conserve intact ecosystems. Indeed, the main forms of management for conservation of biodiversity in forests are to reduce habitat destruction and fragmentation, to provide representative protected areas of sufficient size and arrangement, and to promote sustainable forest management based on ecological principles (Chapter 6). The provision of protected areas alone is not sufficient to conserve biodiversity. A range of sustainable forest management options is required (Lindenmayer and Franklin, 2002). The object of forest management should be to mimic as far as possible the natural disturbances in the ecosystem (Hunter, 1999). The precautionary principle of 'if you don't know the effect of a human intervention, then don't do it' has considerable merit because extinction is irreversible. This can be taken to unreasonable extremes, however, when used as an excuse to not manage at all, an approach that can have worse consequences.

Many humans instinctively believe that the welfare of all other species is intimately connected to their own welfare. This is often a gut feeling but research into the complex web of interdependence of species within ecosystems strongly reinforces this. Humans also extend their self-interest to future generations in being concerned to hand the world on to their progeny in good condition. However, humans are capable of rising above self-interest and, on face value at least, are concerned about the conservation of other species without any apparent expectation of a reciprocal benefit. The concept that all biological entities have an intrinsic value independent of human values is difficult to explain but is widely accepted. Ethics, culture, philosophy and religion all play a part in our relationship with other species. Certainly Buddhism and Taoism strongly advocate conservation of biodiversity. Indeed, Hamilton (1993) considered that 'it is not the ecologists, engineers, economists or earth scientists who will save space ship earth, but the poets, priests, artists and philosophers.' The concept of 'deep ecology' espoused in the 1980's (Tobias, 1985) at its most extreme says that all individual living things are sacred and that humans have a moral obligation to respect this irrespective of whether this is to their advantage or disadvantage. While this has some appeal it must be said that it is most likely to be entertained by those humans who are already well fed, clothed and sheltered and plan to remain so. We eat our companion species, we use them for fuel and shelter, and plant and animal products dominate our living environment. Most humans would agree that it would be a betrayal of duty to be responsible for extinction of species. Even so, most humans welcomed the extinction of the smallpox virus and many would not be at all concerned with the demise of some invasive species that threaten us or even just annoy us. Also humans tend to place different values on different species: animals are generally favoured over plants and larger boutique animals (e.g. tigers and pandas) are favoured over less conspicuous animals such as undescribed insects. Another ethical issue is the use and abuse of genetically modified organisms. Genetically modified organisms contribute to biodiversity and potentially offer enormous economic advantages to the human race. However, once released they cannot be recalled and their impact on ecosystem composition and processes cannot be reliably predicted.

Exotics and Invaders

An exotic species is a species that has been introduced into an area in which it previously did not occur. Humans have been the agents of moving plants and animals around ever since the time of the hunter-gatherers. This raises the vexed question of how long does a species require to be established in a region before it is no longer an exotic but is accepted as a native. Introduced species change the nature of biodiversity but their impact (benign, negative or positive) is often not immediately apparent. Most attention is given to those instances where an introduced species out-competes other native species or preys directly on native species. This can result in crashes in population numbers and extinction of species. There are numerous cases from all over the world. For example, the indigenous vertebrate fauna of New Zealand is low in mammals and the birds evolved in the absence of predators. Consequently the birds are poor fliers or unable to fly at all. When rats were introduced into New Zealand in the 1700s and stoats and weasels in the 1880s, the birds were particularly vulnerable and several species became extinct, others becoming endangered. The Australian possum, a protected species in Australia, when introduced into New Zealand took a particular liking to eating the foliage of New Zealand indigenous forest species and to eating eggs and young birds. Populations soared to plague proportions and the indigenous forests are in decline as a result. The fungus *Endothia parasitica* introduced into the USA from Japan, Korea and China in the early 1900s has almost eliminated the American chestnut (*Castanea dentata*). There are of course examples where introduced species have brought great benefit. The moth *Cactoblastis cactorum* was introduced from Argentina to Australia in the 1920s to control the introduced and rampaging prickly pear (*Opuntia*) and this is one of the best known and most spectacular examples of biological control. However, *Cactoblastis* was introduced to the Caribbean in the 1950s and was discovered in the USA in 1989. Currently it is threatening native rare *Opuntia* species in Florida Keys and there is concern that it could spread to Mexico, the centre of endemism for *Opuntia*. There are many examples all over the world of introduced weeds out-competing and endangering native plant species. For example *Pueraria lobata* (Kudzu), introduced into the USA from China and Japan, now infests more than two million hectares of forest. However, the most destructive invasive species globally is *Homo sapiens* who is ruthlessly efficient at competing for habitat and preying on other species.

Uniformity

The opposite of diversity is uniformity. Humans employ intensive monocultural systems for providing their food and some of these are tightly bred with a narrow range of genetic variation between individuals. The biodiversity among grazing animals is similarly restricted. Plantation monocultures are becoming increasingly important for the production of wood and the proportion of genetically confined 'improved breeds' and clonal material is increasing. Clearly biodiversity is reduced when biodiverse forests are cleared and replaced by alternative more uniform land uses. Also the intensity of these systems in terms of energy consumption and the use of pesticides and fertilizers cause concern when they promote adverse off-site effects. There is a growing desire among some foresters to move towards more diverse and less intensive forms of forest management. However, the counter argument is that intense uniform systems are so efficient at production that they require a smaller land base and relieve the pressure on the broader landscape. These matters will be considered further in Chapter 7.

The Convention on Biological Diversity

The pressing need to conserve biodiversity was recognized internationally at the United Nations Conference on Environment and Development, which was held in Rio de Janeiro in 1992. More than 150 nations signed the Convention on Biological Diversity, which was ratified some 18 months later. The signatories agreed to undertake measures to conserve and sustain global biodiversity in a fair and equitable manner. It is noteworthy that the convention recognized that all species have intrinsic rights.

Recreation, Amenity and Aesthetics

Most people recognize forests as scenic and beautiful places and are concerned that this aesthetic quality is safeguarded. For some it is enough just to know that the forests are there and they are satisfied with pictures on calendars and documentaries on television to demonstrate to them that this is so. Others go one step further and visit the forests but rarely venture further than the road or the immediate vicinity of the picnic ground. Still others go right into the forests for days or weeks and enjoy in full what the forest has to offer. People usually consider scenery as static, almost like photographs. Forests, however, are not static but are dynamic systems that are constantly changing. It is unrealistic and ecologically unsound to expect a forest manager to 'freeze in time' an appealing forest vista. It is impossible to do so. Also it does not necessarily follow that the most scenic forests are those in the best ecological condition and *vice versa*. For example, forests with many over-mature and dying trees may look untidy. However, the process of death, decay and rebirth is an integral part of a forest's growth and survival. Forests may pass through stages of more scenic and less scenic during their development. For example, forests in a regenerative phase with many small trees and few larger trees are generally considered to be less appealing than 'old growth' forests where there is a preponderance of large trees. Forests that have been cleared either by fire, wind or human disturbance may look quite ugly but can subsequently develop into something quite attractive. Young plantations may look rigid and contrived but older thinned plantations can be very appealing. Old growth forests with very large trees can be very spectacular and awe inspiring and to some they have a spiritual value. Religions through the ages have set aside reserves of trees that they consider have particular significance.

A responsibility of forest management is to reduce any adverse visual impacts. This requires attention to the visual impacts of forest roads (route, shape, cuts and fills), forest shapes (edges, lines), forest operations (felling, planting, residue management, soil exposure and erosion, windrows, burning, weeds) and species (selection and distribution). The design of forest landscapes to please and not offend is an important and well-established discipline in its own right (Lucas, 1991). Usually it is not the forests in themselves but rather the placing of forests in the landscape that people find attractive. Large continuous tracts of forests are less appealing than forests broken up by water, tree-lines, rock outcrops or land cleared for agriculture or grazing. Humans are creatures of the forest edges.

Forests are used for a wide range of recreational uses: bush walking, sight-seeing, bird-watching, camping, nature observation, hunting wildlife, picnicking, photography, fishing, mountain-biking, use of off-road vehicles, cross-country skiing, canoeing, running, growing and concealing illicit drugs, car rallies, and this is not a comprehensive list. Not all recreational uses are compatible and some are undesirable. It can be a challenging task to manage forests in such a way that the legitimate aspirations of those who wish to use them for recreation are taken into account. Their conflicting interests have to be considered alongside the need to minimize damage to the forests.

Human empathy with forests extends to creating treed landscapes in urban areas. Indeed towns, suburbs, parks and streets with extensive use of trees add greatly to the value of the land. The trees and associated flora and fauna in streets and in private and public gardens can be a very valuable resource for amenity and recreation and indeed the most common experience of 'nature' for most people on earth.

Other Benefits of Forests

Mountain forests can protect humans from natural disturbances such as avalanches, rock falls and landslips. Forests are the home of some indigenous peoples who depend on them for food, fibre, fuel and shelter. Forests also have a critical role in arresting the climate change resulting from increased carbon dioxide concentrations in the atmosphere. Trees in urban landscapes may be continuous and extensive enough to be considered as urban forests. These benefits will be discussed in later chapters.

References and Further Reading

Attiwill, P.J. and Weston, C.J. (2001) Forest Soils. In Evans, J. (ed.) *The Forests Handbook, Volume 1*. Blackwell Science, Oxford, pp. 157-187.

Bruijnzeel, L.A. (2001) Forest Hydrology. In Evans, J. (ed.) *The Forests Handbook, Volume 1*. Blackwell Science, Oxford, pp. 301-343.

Burgman, M.A. and Lindenmayer, D.B. (1998) *Conservation Biology for the Australian Environment*. Surrey Beatty and Sons, Chipping Norton, NSW, Australia.

Chow, V.T., Maidment, D.R. and Mays, L.W. (1988) *Applied Hydrology*. McGraw Hill, New York.

Dano, A.M. (1990) Effect of burning and reforestation on grassland watersheds in the Philippines. In: Zeimer, R.R., O'Loughlin, C.L. and Hamilton, L.S. (eds) *Research Needs and Applications to Reduce Erosion and Sedimentation in Tropical Steeplands*. International Association of Hydrological Sciences Publication No. 192, IAHS Press, Wallingford, pp. 53-61.

Dingzhong, D. and Ying, T. (1996) Soil erosion and sediment yield in the Upper Yangtze River basin. In: Walling, D.E. and Webb, B.W. (eds). *Erosion and Sediment Yield: Global and Regional Perspectives*. International Association of Hydrological Sciences Publication No. 236, IAHS Press, Wallingford, pp. 191-203.

Gadow, K. von. (2000*) Sustainable Forest Management*. Kluwer Academic Publishers, Dordrecht.

Gaston, K.J. and. Spicer, J.I. (1998) *Biodiversity - An Introduction*. Blackwell Science, Oxford.

Greer, T., Sinun, W., Douglas, I. and Bidin, K. (1996) Long term natural forest management and land–use change in a developing tropical catchment, Sabah, Malaysia. In: Walling, D.E. and Webb, B.W. (eds) *Erosion and Sediment Yield: Global and Regional Perspectives*. International Association of Hydrological Sciences Publication No. 236, IAHS Press, Wallingford, pp. 453-461.

Groombridge, B. (1992) *Global Biodiversity: Status of the Earth's Living Resources*. Report compiled by the World Conservation Monitoring Centre, Chapman Hall, London.

Groombridge, B. and Jenkins, M.D. (2000) *Global Biodiversity – Earth's Living Resources in the 21st Century*. UNEP World Conservation Monitoring Centre, World Conservation Press, Cambridge.

Groombridge, B. and Jenkins, M.D. (2002) *World Atlas of Biodiversity – Earth's Living Resources in the 21st Century*. UNEP World Conservation Monitoring Centre, University of California Press, Berkeley.

Hallsworth, EG (1987) *Anatomy, Physiology and Psychology of Erosion*. John Wiley and Sons, New York.

Hamilton, L.S. (1993) *Ethics, Religion and Biodiversity*. The White Horse Press, Cambridge.

Hamilton, L.S. and Pearce, A.J. (1987) What are the soil and water benefits of planting trees in developing country watersheds? In: Southgate, D.D and Disinger, J.F. (eds) *Sustainable Resource Development in the Third World*. Westview, Boulder. Colorado, pp. 39-58.

Hawksworth, D.L., Kirk, P.M and Dextre Clarke, S. (eds) (1996) *Biodiversity Information - Needs and Options*. CAB International, Wallingford.

Helms, D. and Flader S.L. (eds) (1985) *The History of Soil and Water Conservation*. The Agricultural History Society, Washington, DC.

Hicks, D.M., Hill, J. and Shankar, U. (1996) Variation of suspended sediment yields around New Zealand: the relative importance of rainfall and geology. In: Walling, D.E. and Webb, B.W. (eds) *Erosion and Sediment Yield: Global and Regional Perspectives*. International Association of Hydrological Sciences Publication No. 236, IAHS Press, Wallingford, pp. 149-156.

Hladik, C.M., Hladik, A., Hladik, O.F., Linares, H., Pagezy, H., Semple, A. and Hadley, M. (eds) (1993) Tropical forests, people and food. In: *Man and the Biosphere Series Volume 13*. UNESCO and the Parthenon Publishing Group, Paris.

Hunter, M.L.J. (ed) (1999) *Maintaining Biodiversity in Forest Ecosystems*. Cambridge University Press, Cambridge.

Kimmins, J. P. (1992) *Balancing Act - Environmental Issues in Forestry*. UBC Press, Vancouver.

Kimmins, J.P. (1997) Biodiversity and its relationship to ecosystem health and integrity. *The Forestry Chronicle* 73, 229-232.

Lal, R. (1977) The soil and water conservation problem in Africa: ecological differences and management problems. In: D.J. Greenland and R. Lal (eds) *Soil Conservation and Management in the Humid Tropics*. John Wiley and Sons, New York, pp. 143-149.

Leigh, J.H. and Briggs, J.D. (1992) *Threatened Australian Plants - Overview and Case Studies*. Australian National Parks and Wildlife Service, Canberra.

Lindenmayer, D.B. and Franklin, J.F. (2002) *Conserving Forest Biodiversity - A Comprehensive Multiscaled Approach*. Island Press, Washington.

Lucas, O.W.R. (1991) *The Design of Forest Landscapes*. Oxford University Press, Oxford and New York.

McKelvey, P.J. (1999) *Sand Forests*. University of Canterbury Press, Christchurch, NZ.

National Research Council. (1993) *Vetiver Grass – A Thin Green Line Against Erosion*. National Academy Press, Washington, DC.

Oldeman, L.R., Hakkeling, R.T.A. and Sombroek, W.G. (1991) *World Map of the Status of Human-Induced Soil Degradation: An Explanatory Note*. ISRIC, Wageningen.

Pimental, D., Harvey, C., Resosudarmo, P., Sinclai., K., Kurz, D., McNair, M., Crist, C., Shpritz, L., Fitton, L., Saffouri, R. and Blair, R. (1995) Environmental and economic costs of soil erosion and conservation benefits. *Science* 267, 1117-1123.

Reaka-Kudla, M.L., Wilson, D.E. and Wilson, E.O. (eds) (1997) *Biodiversity II - Understanding and Protecting our Biological Resources*. Joseph Henry Press, Washington, DC.

Solbrig, O.T., van Emden, H.M. and van Oordt, P.G.W.J. (eds) (1994) *Biodiversity and Global Change*. CAB International, Wallingford.

Swanage, M. (ed.) (1996) *Environmental Science*. Australian Academy of Science, Canberra.

Tobias, M. (ed.) (1985) *Deep Ecology*. Avant Books, San Diego.

Troeh, F.R., Hobbs, J.A. and Donahue, R.L. (1980) *Soil and Water Conservation for Productivity and Environmental Protection*. Prentice Hall, Englewood Cliffs, N.J.

UNESCO (1978) *World Water Balance and Water Resources of the Earth*. UNESCO, Paris.

Walling, D.E. and Webb, B.W. (1996) *Erosion and Sediment Yield: Global and Regional Perspectives*. International Association of Hydrological Sciences Publication No. 236, IAHS Press, Wallingford.

Wischmeier, W.H. and Smith, D.D. (1965) *Agricultural Handbook No.282*. US Government Printing Office, Washington, DC.

Young, A. (1997) *Agroforestry for Soil Management*. CAB International, Wallingford.

Chapter 4

Forest Products

Wood is Good

Competitors with wood as a building material have used the image of deforestation and forest degradation in an attempt to discredit the use of wood and to promote their own product. For example, the US steel industry had a concerted advertising campaign using the slogan 'build your house with six used cars and not one acre of trees.' The American Plastics Council used the slogan 'save a tree – use PVC.' This advertising is misleading and a very good case can be presented to show, providing forests are managed sustainably, that wood is environmentally more friendly than its competitors. Building in timber uses less energy than its competitors. Wood both, as a building material and as a fuel, produces less carbon dioxide.

Wood is constructed by the process of photosynthesis. The chemical reaction of photosynthesis can be simply stated as:

$$\text{carbon dioxide} + \text{water} + \text{solar energy} \rightarrow \text{carbohydrate} + \text{oxygen} \qquad (4.1)$$

The chemical reaction for respiration, decomposition and combustion of carbohydrate is the reverse of photosynthesis and both energy and carbon dioxide are liberated.

$$\text{carbohydrate} + \text{oxygen} \rightarrow \text{carbon dioxide} + \text{water} + \text{energy} \qquad (4.2)$$

Since the industrial revolution the concentration of carbon dioxide in the atmosphere has been increasing to the extent that it is considered to be causing global climate change. The consequences of global climate change are thought to be so serious that it prompted an international response, the Kyoto Protocol, in an attempt to limit carbon dioxide emissions. Global climate change and the Kyoto protocol will be discussed in greater detail in Chapter 6. The increase in atmospheric carbon dioxide concentration mostly comes from the combustion of fossil fuels such as coal (carbon) or oil/gas (hydrocarbons) but also from the decomposition or burning of vegetation (carbohydrates) produced by photosynthesis (equation 4.2). Fossil fuels are non-renewable resources (at least over the time scales being considered here). However, combustion and decomposition of vegetation (biomass) can be carbon dioxide neutral or better, providing sufficient vegetation is re-established so that the products of equation 4.1 equal or exceed that of equation 4.2. Consequently the use of biomass as a fuel, and this includes fuelwood, is an environmentally preferable option to the use of fossil fuels.

87

Wood biosynthesis uses solar energy at no cost, and additional energy sources are only required for its transport and conversion. Other building materials require significant energy inputs in processing and most of this energy comes from non-renewable fossil fuels. Figure 4.1 gives a comparison of the energy per unit of dry weight required to extract, manufacture and deliver a range of building materials to a residential building site in the USA. Aluminium uses enormous amounts of electrical energy while wood requires the least energy. For this reason China is closing some high-energy metal processing operations, preferring to import further processed metal, thereby making more power available to urban consumers. Wood is even more attractive from an energy perspective than shown in Fig. 4.1 because about 75% of the energy requirement for solid wood products comes from burning the wood residues arising during wood conversion and only 25% needs to be provided from external sources. Also the weight of concrete, bricks and steel used in construction is considerably greater than that of wood for the same purpose. A recent report by Meil *et al.* (2004) compared the energy required to construct a wood-framed house with that required to construct a steel-framed house in Minneapolis and a concrete house in Atlanta. They found that most of the energy associated with construction was used in the extraction and manufacture of the materials. The steel house used 17% more energy and the concrete house 16% more energy than the wooden house.

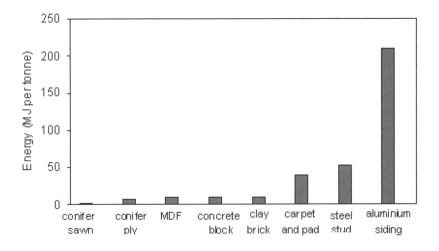

Fig. 4.1. The net energy requirements for extraction, manufacture and transport of various building materials to a residential building site in USA (compiled from data given by Koch, 1992).

Because wood in construction uses less energy, the substitution of wood for alternative materials in construction can save energy and reduce carbon dioxide emissions. Buchanan and Levine (1999) looked at global carbon stored and carbon emitted in forest products where processing energy came from fossil fuels (Fig. 4.2). This ignores past and current production and shows storage and emission functions starting from zero and assuming a continuous rate of production and consumption. Carbon storage reached a steady state after 40 years but accumulated carbon emissions increased indefinitely as the burning of fossil fuels continued. The emitted carbon equalled the stored carbon after about 150 years. Most of the storage was in solid wood products while most of the emissions came from the pulp and paper industry. Clearly the only way to eliminate carbon emissions in forest products industries is to replace fossil fuels with renewable sources of energy that do not have a net flux of carbon dioxide to the atmosphere. Buchanan and Levine (1999) showed that a 17% increase in wood usage in the New Zealand building industry could reduce carbon dioxide emissions by 20% from the manufacture of all building materials.

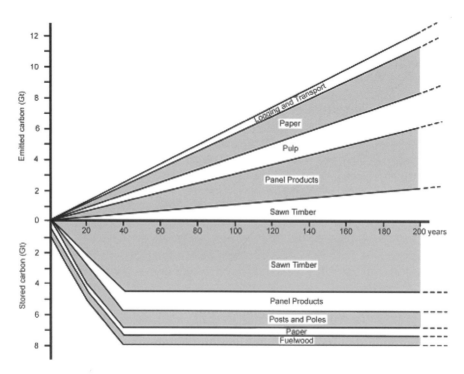

Fig. 4.2. Global carbon storage in forest products and resulting cumulative emissions of carbon from fossil fuel processing energy (adapted from Buchanan and Devine, 1999).

Wood for Energy

The regular use of fire and the use of tools by hunter-gatherers heralded the beginning of the increase in energy flow attributable to the human race. Prior to this, humans or pre-humans were like any other omnivorous animals that received and contributed their energy via their own metabolism (somatic energy). The somatic energy of humans can be quantified as a Human Energy Equivalent (HEE), which is about 10 megajoules per person per day (Newcombe, 1971). The energy use per average hunter-gatherer probably doubled as a result of the use of fire and in early agricultural humans increased to about 5 HEE as a result of fire, the use of wood as fuel, and the somatic energy of domestic animals. Since the commencement of the use of fossil fuels, energy consumption has risen dramatically. Energy consumption in England, Germany and the USA in the late 1800s rose to near 30 HEE because of the industrial revolution (Cook, 1971). In year 2000 the HEE calculated from consumption of fossil fuels alone (British Petroleum 2002) and population data (United Nations, 2001) was approximately 20 for the world and near 100 for the USA. The human population has increased from about 5 million at the beginning of the age of agriculture to about 6 billion in year 2000. Consequently the total energy consumption of modern humans is about 5000 times that during the early stages of agriculture. Almost all the energy use from the time of the hunter-gatherers until the present day has been solar in origin from the combustion of fossil or current biomass constructed from photosynthesis. Oil and coal reserves originate from biological organisms laid down in the Carboniferous, Permian, Jurassic and Cretaceous. Oil and gas are mainly from marine organisms but coal is mainly from forests, particularly swamp forests. Present day humans have developed a life style that depends on the utilization of these reserves. There is an enormous time gap between the development and the utilization of these reserves and it is all too easy to forget that coal is a gift of the forest.

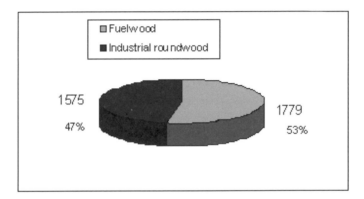

Fig. 4.3. World production of industrial roundwood and fuelwood in year 2000 (millions of cubic metres) (compiled from data given in FAO, 2003).

Energy may be classified as non-renewable (fossil fuels and nuclear power) and renewable (hydro, geothermal, wind, solar, landfill gas, waste incineration, and biomass). Excluding hydro power, about 11% of current world primary energy consumption comes from renewable resources, mainly biomass (IEA, 2002). Woody biomass can be converted to solid, liquid and gaseous fuels suitable for domestic, commercial and industrial uses. Wood can be burned directly, otherwise densified to form pellets or briquettes, carbonized to form charcoal, fermented to form alcohol or gasified to provide combustible gases.

Throughout the ages the major use of wood has been to provide energy (Chapter 1) and this remains the case today. More than 50% of the volume of wood removed from the world's forests annually is used as fuelwood or charcoal (Fig. 4.3). There is negligible international trade in fuelwood and fuelwood dominates total forest removals in Africa (88%) and Asia (79%) (Fig. 4.4). In developing countries wood and charcoal comprise between 30 and 80% of total energy consumption. Sub-Saharan Africa is particularly

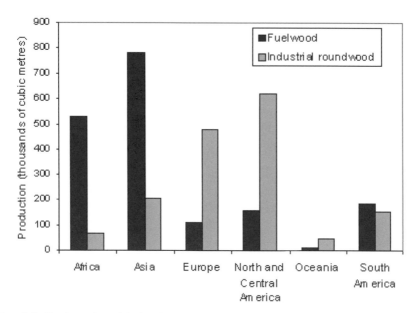

Fig. 4.4. Fuelwood and industrial roundwood production in various regions in year 2000 (compiled from data given in FAO, 2003).

dependent on fuelwood. For example, in Ethiopia, more than 90% of the inhabitants depend on fuelwood for their domestic energy needs, demand outstripping supply to the extent that 15% of the total energy requirement comes from burning agricultural residues and animal dung (Eschete, 1999).

The proportion of fuelwood in the developed world is greatly under-estimated in Fig. 4.4 because it does not account for the large amount of wood waste used as fuel in wood processing plants. The average proportion of the raw wood arriving at wood processing plants used as fuel is shown in Table 4.1. If waste from industrial operations amounts to 40% (and this is conservative) and all this was available for energy conversion, the total fuelwood for energy in Fig. 4.3 would rise to 70% of global wood production, and industrial wood products would be reduced to 30%. For example, kraft pulping, which is the dominant chemical method of making paper, has a yield of only 50%. The other 50% is largely broken, solublilized fragments of lignin. This carbonaceous residue, known as black liquor,

Table 4.1. The average percentage of the wood arriving at wood processing plants that is used as fuel (reprinted from Hakkila and Parikka, 2002, copyright 2002, with kind permission of Springer Science and Business Media).

Product	Fuel component of raw material	Fuel % by volume
Chemical pulp	Bark, screening residues from chips, black liquor	50-60%
Mechanical pulp	Bark, screening residues from chips	10-20%
Lumber	Bark, sawdust, slabs, cull logs, etc	15-60%
Plywood	Bark, log ends, waste plies, dust, screening residues, cull logs, etc.	40-75%

is concentrated and burnt to generate both process steam and electricity, enough for a modern pulp mill to be totally energy sufficient. In 1992 in the USA, 26.5% of the energy obtained from wood came from black liquor (Zerbe, 2004).

Wood as a Solid Fuel

Fuelwood

Wood was the main source of energy prior to the industrial revolution and today this is still the main use for wood (Fig. 4.3). Wood as a solid fuel has both advantages and disadvantages. Properly managed it is an environmentally friendly source of energy. It is a 'grow your own' energy source at individual, community and national levels. Fossil fuels are unevenly spread around the world and security of access to fossil fuels is not a foregone conclusion for countries without their own resources. Indeed the uneven distribution of fossil fuels is a politically charged issue capable of bringing down governments and provoking war. Virtually all cultures have an affinity to a wood fire, particularly an open fire, irrespective of its low efficiency. Also in many rural areas, fuelwood can be collected at no cost (but considerable labour). However, wood is an awkward commodity to store and transport; it produces smoke and particulates that may be harmful to health; and the amount of heat given off when it is burnt (net energy value) is low compared with fossil fuels (Table 4.2).

The energy value of wood does not vary greatly between species on a weight basis, but wood varies in basic density and consequently more dense woods provide more heat per unit of volume. The effective energy value of wood is very dependent on moisture content and wet wood is not an efficient fuel (Table 4.2). Lignin, resin, terpenes and waxes have higher energy values than cellulose and hemicelluloses and consequently conifers in general have higher energy values per unit of weight than non-coniferous species, branches have higher energy values than stem wood, and bark and leaves have higher energy values than wood (Hakkila and Parikka, 2002). Conifers, however, are often less dense than non-coniferous species.

Table 4.2. Net energy value of wood and other combustible fuels (MJ/kg) (averages from various sources).

Charcoal	29.5
Wood (oven dry)	19.0
Wood (air dry)	16.5
Wood (wet)	9.5
Anthracite coal	35.5
Bituminous coal	25.0
Air dry dung	17.0
Lignite coal	12.5
Peat (air dry)	10.5
Butane	49.0
Liquid Petroleum Gas	47.0
Kerosene	46.5

Charcoal

The production of charcoal from wood differs from combustion (equation 4.2) in that the wood is heated in an enclosure in the absence of air or under restricted air supply. In traditional operations, technology is simple and capital investment is low: the enclosure can be as basic as a pit in the ground or a simple kiln constructed out of bricks or earth. Industrial operations can be more capital intensive and the enclosure can be a more technologically advanced kiln or retort, but the processes are the same. The production of charcoal requires an external energy source until pyrolysis reactions become self-sustaining. Also there is some gaseous loss of carbon. Consequently there is a net loss of energy in producing charcoal. For example, charcoal from *Acacia bussei* at a distillation temperature of 500°C, yielded 23% of the dry wood weight and an energy yield of 38% (Hollingdale *et al.*, 1999). In some traditional operations charcoal yield can be as low as 10% of the weight of the wood. The loss of energy in manufacture must be balanced against the advantages of producing a fuel that has twice the energy value of wood per unit of weight, is less costly to transport, is smokeless, is easier to store, does not rot and is immune to insect attack. It is more efficient to transport the more compact and energy-intensive charcoal within urban communities than to burn wood. Charcoal is used in industrial applications. For example in Brazil, charcoal from eucalypt plantations is an important primary fuel in the smelting of iron.

As the temperature of the wood charge in the kiln increases, the wood absorbs heat and releases water vapour. The temperature stays at or slightly above 100°C until the wood is dry. Further heating to 270°C causes the wood to absorb further heat and the wood begins to break down, releasing some carbon monoxide, carbon dioxide, acetic acid and methanol. From 270°C to 290°C the exothermic decomposition of wood commences (wood releases heat) and various gases and some tars are released. From 290°C to 400°C the decomposition continues and the combustible gases - carbon monoxide, hydrogen and methane - together with carbon dioxide and the condensable vapours water, acetic acid, methanol and acetone are produced. From 400°C to 500°C the residual entrapped tar is driven off and the fixed carbon content of the charcoal rises to about 75% (FAO, 1985). During the cooling process the kiln must remain sealed from any air entry as this would reduce charcoal yield.

Liquid Fuel from Wood

Arguably the greatest challenge in finding alternatives to fossil fuels is to find satisfactory liquid fuels to replace petroleum-based fossil fuels. There are obvious advantages in having a fuel in liquid form. It is portable, easily transported and is the predominant fuel for most of the transportation sector. Ethanol can be produced by fermentation of sugars derived from plants and plant residues. Ethanol production from sugarcane, sugar beet and cereal grains is a well-developed and long-standing technology. Brazil is perhaps the best example where 30% of its motor fuel requirements come from ethanol derived from the products of the sugarcane industry. In Brazil, annual production of ethanol amounts to about 14 billion litres per year. In USA, production is about 12 billion litres per year, but it is growing at about 20% per year. In Brazil production is mainly from sugarcane, and in USA, mainly from maize.

Sugarcane and sugarbeet have free sugars and cereal grains have starch, both of which are relatively easily fermented. Wood is a lignocellulosic material and is more difficult to ferment. It requires several steps (mechanical, thermal, chemical or biochemical) to break down the cellulose and hemicellulose to sugars for fermentation while leaving lignin solids as a by-product (Wayman and Parekh, 1990). Usually the technologies are acid or enzymatic hydrolysis but the technology is still in the developmental stage. Ethanol from sugarcane and grain is still the only feasible option for large-scale ethanol production.

Table 4.3. The annual biomass production with potential for ethanol production (adapted from Wayman and Parekh, 1990; 160 litres of ethanol is equivalent to one barrel of oil).

	Million tonnes per year	Million litres ethanol equivalent
Cane and beet molasses	38	11,000
Cane and beet juice	-	5,000
Bagasse, surplus to fuel	24	7,500
Grain, dedicated	23	8,000
Grain, low grade	80	27,000
B starch	116	52,000
Straw, chaff, stover	3,300	1,000,000
Cassava, cull	2	1,000
Cassava, tops	45	14,400
Potato, cull	12	3,800
Jerusalem artichoke, tops	3	1,000
Forest logging residues and non-commercial harvest	360	125,000
Plantation forests	60	24,000
Municipal waste	250	37,000
World total	4,313	1,316,700

The amount of stored energy derived from global terrestrial photosynthesis is more than an order of magnitude greater than global energy consumption. The present standing terrestrial global biomass, which is mainly in forests, is much the same as proven fossil fuel reserves (Zsuffa, 1982). Biomass is a renewable resource that is carbon dioxide neutral (carbon dioxide consumption in growth equals carbon dioxide emissions from combustion). Fuel ethanol burns cleanly and does not produce the harmful emissions of petrol. Ethanol can be used as an additive to fossil fuels to improve fuel performance. Ethanol blends with petrol but requires an emulsifier to form a stable additive with diesel. The production of fuel alcohol would assist in reducing agricultural and forest waste and utilizing surplus crops. Currently fossil fuels are less expensive than fuel ethanol. However, as the price of oil increases there will come a time when ethanol becomes competitive and then its production should increase substantially. Large land areas would be required if ethanol was to become the preferred liquid fuel of the future. Such areas are available. For example, 1% of the total land area of the USA would be required to be dedicated to providing biomass to produce sufficient ethanol to supply 10% of USA motor fuel requirements. Ethanol production could be from a variety of biomass including municipal waste, waste from forest operations and waste from wood processing plants. Currently straw and chaff have the greatest potential, but these are lignocellulosic materials like wood and face the same technological challenges as wood. The potential world biomass for fuel ethanol production is shown in Table 4.3. If economically viable production systems for ethanol from lignocellulose become available in due course, the establishment of dedicated energy plantations for ethanol production would be a logical development.

Fuel ethanol should not be confused with 'wood alcohol' which is the colloquial name for methanol obtained from the destructive distillation of wood. Wood alcohol is produced when wood is gasified (see below) and the resultant hydrogen and carbon monoxide are recombined over a heated catalyst. However, currently almost all methanol is made from natural gas (mainly methane) which is much less expensive and easier to use as a feedstock. Therefore, ethanol, rather than methanol, is the prime candidate as the biomass fuel for the future.

Gaseous Fuel from Wood

The production of gas from wood is an old and mature technology but, except in particular circumstances, it is uneconomic compared to using alternative fuels. The process of gasification is similar in many respects to the process of charcoal production. Wood is gasified when it is heated to greater than 450°C and by the controlled combination of heat, moisture and air, the gases carbon monoxide and hydrogen can be produced in appropriate proportions to synthesize methanol. Alternatively gasification with air can yield 'producer gas' (typical composition 22% carbon monoxide, 18% hydrogen, 3% methane, 6% carbon dioxide and 51% nitrogen) to provide or supplement heating of industrial boilers, or it can be compressed and used as a low grade fuel for motor vehicles. Producer gas from burning wood, charcoal or coal was used as an alternative to petrol during World War 2 in periods of acute petrol shortage. It is a motor fuel of last resort because of its low calorific value (4-6 MJ/m^3). The calorific value of producer gas, however, can be greatly enhanced if oxygen is used rather than air (typically 40% carbon monoxide, 40% hydrogen, 3% methane and 17% carbon dioxide). Enhanced producer gas can be used to make ammonia, methanol and diesel fuel by catalytic processes. Today there are better filters for removing tar from the gases before they are used in internal combustion engines and this reduces maintenance costs (Zerbe, 2004). Sweden has developed a gasification system that uses a range of biomass feedstock to provide a combination of heat and power for a district heating system and Canada has a developed a gasification system that uses a wide variety of feedstocks (biomass, peat and municipal waste) to produce biogas to replace the oil currently used in commercial boilers (Lee, 1996). A survey of commercial gasifiers is given in Reed and Gaus (2001).

Biomass for Energy in the Developed World

Prior to industrialization, the developed world of today used fuelwood as its major source of energy. A major driver of development and industrialization was the replacement of fuelwood with more convenient and more efficient fossil fuels. However the amount of biomass for energy is now increasing in developed countries for domestic, commercial and industrial applications. Fossil fuels are non-renewable resources and, although recoverable reserves have been stable in recent times owing to better extraction technology, the price of fossil fuels will inevitably increase, making renewable energy alternatives more competitive.

Biomass-based fuels are environmentally friendly alternatives to fossil fuels and nuclear power, but the renewed interest in biomass as a source of energy in the developed world has more to do with price and future energy security. Whether biomass-based fuels have a significant role in the next 30 years is arguable. For instance, while oil may be close to peak production and facing decline, the other alternative is the tapping of large gas reserves, which are significantly more environmentally friendly than the burning of oil (although still non-renewable). Current gas production and marketing is regional, but there remains the opportunity to build supertankers for the inter-regional transport of liquefied natural gas that would set-back the emergence of biomass by a generation (The Economist, 2004a). Also, coal will probably remain a major fuel in the generation of power. Coal had fallen into disfavour because it was seen as a filthy fuel, but current advances in technology promise much cleaner coal-fired plants in the future (The Economist, 2004b).

The USA is by far the largest user of biomass for energy in the developed world but only 3-4% of its consumption comes from biomass (Hakkila and Parikka, 2002). The use of wood has always been important in home heating in developed countries rich in forests, such as in Sweden and Finland. Finland gets 19% of its primary energy from wood. The use of wood for home heating in the developed world generally has increased, partly because of environmental concerns. However, the use of wood has become competitive mainly because of the improved efficiency of furnaces and the increased price of oil. Increasing the proportion of renewable energy sources in the developed world is likely to be an integrated approach using a combination of different renewable energy options. For example, the European Union in its post-Kyoto energy policy strategy, plans by 2010 to have 500,000 photovoltaic roof and facade systems installed, 10,000 MW of wind farms and 10,000 MW of biomass installations for combined heat and power plants (Dengg *et al.*, 2000).

Biomass for Energy in the Developing World

Fuelwood may meet the whole range of energy requirements in parts of the developing world. About 2.3 billion people depend on fuelwood and other forms of biomass to meet their daily domestic energy requirements. Wood is also used in industrial and commercial operations, including the generation of electricity. Eckholm (1975) published a leaflet called 'The other energy crisis, firewood.' This and other studies predicted dire shortages of fuelwood in the developing world to the extent that massive deforestation and starvation were considered to be inevitable. This prompted a flurry of activity in the 1980s that culminated in 1985 in the development of an international donor-funded initiative, the Tropical Forest Action Plan (TFAP). The objectives of this plan were to decrease wood demand by using improved stoves or alternative fuels, to produce more fuelwood from existing forests and to establish plantations specifically for fuelwood. There was considerable investment in this plan but, with the benefit of hindsight, it appears that the crisis was exaggerated (Arnold *et al.*, 2003). There does not appear to be any fuelwood shortages at the national level. Fuelwood prices have failed to rise and farmers generally are not interested in planting trees to produce a low value product. The plantations were unable to meet the establishment costs if the wood was used as a fuel in the farmers' homes. In order to cover the costs of plantation establishment, the wood had to be used for a higher value alternative use. The irony of this is that fuelwood plantations were established on land that previously had supplied fuelwood and no longer could do so (Saxena, 1997). Consequently many of the fuelwood planting projects promoted in the 1980s have failed. Those that succeeded as fuelwood plantations often supplied fuelwood to industry rather than to the domestic market.

The predicted broad scale deforestation from fuelwood gathering did not eventuate either. There has been and continues to be substantial deforestation in the developing world but this is caused mainly by clearing for agriculture and generally not from gathering fuelwood (see Chapter 5). Gathering of fuelwood, however, may cause deforestation in some specific areas of high population density where the forests are communal and unregulated, and also in semi-arid areas and in highland areas near tree lines (see Chapter 5). Also, the rapid rise in charcoal production to service urban markets has the potential to cause deforestation because charcoal kilns need to be located close to the source of wood. A study in West Africa concluded that charcoal production was the main reason for forest loss in those areas where charcoal production was concentrated (Ninnin, 1995). Trees outside of the forest can provide a significant proportion of fuelwood supplies. Fuelwood comes from isolated trees outside the forests, from trees felled in forest clearing operations, from wood waste from wood processing plants and also from within the forests. In the forests woody litter and sometimes live branches are taken. Whole trees are taken under conditions of scarcity. Even so, most of the wood taken from tropical forests is used as fuel.

There is no overall shortage of wood for fuel. Forests in developing countries have an unused annual increment of about 6 billion cubic metres which is greater than three times the world consumption of fuelwood. The amount of wood that is burnt *in situ* when the forests in developing countries (mainly tropical forests) are cleared for agriculture and grazing, would alleviate all scarcity if it could be directed towards the area of need. The reason for scarcity, where it occurs, is distribution and transport as well as inequitable barriers to access. Projections of fuelwood and charcoal consumption from 1970 to 2030 are shown in Fig. 4.5. Asia is the main consumer of fuelwood, but its consumption is predicted to decline from the 1990s to 2030, as is total world consumption. Consumption of fuelwood in Africa and South America is predicted to continue increasing. Consumption of charcoal is predicted to increase in all regions, especially in Africa. The combined aggregate of fuelwood plus charcoal is still rising, albeit at declining rate, although much less rapidly than the equivalent growth in population (Arnold *et al.,* 2003).

Income and prices are the most important factors determining fuelwood demand and the reason for the move away from fuelwood shown in Fig. 4.3 is probably the rise in real incomes. There is the tendency to move up the energy ladder as incomes increase (Table 4.4).

Even though there do not appear to be national shortages, there still are rural regions where fuelwood is scarce and is becoming scarcer. Communities are responding to this by using less fuelwood and/or spending more time in gathering it. Changes in forest tenure have also displaced the rural poor from sources of fuelwood to which they previously had access. Indeed in the drier parts of Africa, poor nutrition can be caused by shortage of fuelwood rather than shortage of food. Typically in areas of scarcity, searching for and gathering fuelwood can occupy a large amount of time: four hours or more a day would not be unusual. The fuelwood gatherer is usually a woman (unless some form of mechanized transport is involved) and consequently solutions to scarcity must be considered alongside improving the welfare of women. In rural areas there is a maximum distance (about 15 km) that a woman can scavenge for firewood beyond which the family may need to relocate. The area around cities that depends on fuelwood can be stripped clean of fuelwood (in combination with clearing for agriculture) for a considerable distance beyond the urban fringe. However, as cities become larger and wealthier, there is a move away from fuelwood towards charcoal and other fuels and the problem becomes less intense. It is generally uneconomical to source wood further than about 100 km from a centre of population. For greater distances it is more efficient to transport charcoal, which provides greater energy per unit of weight.

Eating more uncooked food is not a solution to fuelwood scarcity. Cooking is necessary to make starchy foods digestible, to remove toxins and kill parasites, to boil drinking water and to preserve food. Reducing the number of meals per day is not a solution either. Experience has shown that health problems occur when the stomach is loaded up with large amounts of not very digestible food. Undercooking of food may save fuelwood but this can also lead to health problems if it does not remove the toxins or kill the parasites. Cooking over an open fire is common and this wastes about 94% of the heat. Considerable effort has been invested in developing and promoting fuel efficient stoves that can reduce fuelwood consumption by at least 50%. This has been only partly successful because many cannot afford the improved stoves. Also, an open fire provides light as well as heat, keeps away insects and provides a social centrepoint. For example, despite concerted government initiatives in India to promote the use of improved stoves, a recent survey over six states showed that less than 10% of houses contained improved stoves (Gundimeda and Köhlin, 2003). It is very easy and inexpensive to at least double the fuel efficiency of an open hearth simply by enclosing the fire with bricks or constructing a simple mud-baked oven. Poor uptake of improved stoves confirms that scarcity of fuelwood or alternatives is not a universal problem in the developing world.

In some parts of the world where fuelwood is unavailable or too expensive, agricultural and animal residues are burnt as fuel. This removes nutrients that would otherwise be returned to the soil and consequently soil fertility is reduced, crop yields fall and forest may need to be cleared to provide additional land for cultivation. Cow dung is burned in Sub-Saharan Africa, Ethiopia, Iraq, Bolivia and Peru but particularly in India where 300-400 million tonnes of cow dung are burned annually (Eckholm, 1975). The potential production lost due to declining soil fertility resulting from burning cow dung has been estimated as 20 million tonnes of grain each year. However, the burning of dung does not necessarily mean that there is a shortage of fuelwood. The incorporation of dung into the fields requires labour and associated costs. Farmers may well choose to forego the increases in soil fertility in favour of using dung as a low cost and easily obtainable fuel in preference to fuelwood.

Traditionally, fuelwood has been regarded as free and rural people collected it at no cost from wherever they could. This still remains the case in rural areas of developing nations and a significant proportion of consumption does not enter the cash economy. However, considerable amounts of fuelwood are traded as well and selling fuelwood can be a major source of income for the rural poor. The low price, however, gives little incentive to farmers to grow trees specifically for fuelwood. Price is the main factor regulating the choice of fuel, but households may still choose fuelwood even when it is more expensive than alternatives because of security of supply and because it requires no expensive stove (Boberg, 2000).

The trend for developing countries to move up the energy ladder is inevitable and appropriate. However, the developed world is now looking at increasing the amount of biomass in their energy mix and, in the long-term, developing countries may choose to do so for the same reasons. Ultimately the best option will be to increase biomass supply rather than decrease demand because most energy alternatives are finite and environmentally inferior.

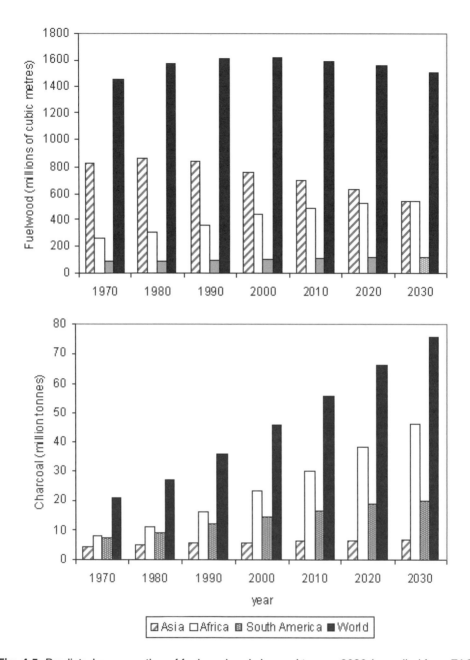

Fig. 4.5. Predicted consumption of fuelwood and charcoal to year 2030 (compiled from FAO projections given in Broadhead *et al.*, 2001).

Table 4.4. The relationship between income (USD per month) and energy use (kg oil equivalent per capita per month) in urban areas in 12 developing countries (unpublished data of Barnes *et.al.*, 2002, in Arnold *et al.*, 2003).

Income Class	Monthly income	Fuelwood	Charcoal	Coal	Kerosene	LPG	Electricity	Total
Low	8.59	3.63	3.28	2.38	1.33	0.15	0.60	11.59
Mid-low	15.51	2.57	2.66	3.21	1.73	0.42	0.82	11.59
Middle	25.02	2.10	2.20	2.83	1.50	1.25	1.15	11.15
Mid-high	41.94	2.62	2.54	0.67	1.14	2.09	1.77	10.82
High	116.95	1.66	1.79	0.00	0.60	3.70	4.15	11.62

Forests Managed for Biomass Production

Any argument that consumption of fuelwood should be curtailed to reduce deforestation is flawed. The gathering of fuelwood is not causing significant deforestation in current times and, properly managed, forests for biomass should increase the forest area rather than decrease it. The use of biomass as fuel should be encouraged rather than discouraged. Fuelwood can be obtained from native forests and plantations that are managed primarily for timber production. Fuelwood can also come from biomass for energy plantations established specifically for the purpose. Potential sources of fuelwood from the forest are much wider than for industrial wood. They can comprise forest residues, trees that are too small for industrial use, pre-commercial thinnings, mortality, trees that are of too low a quality for industrial use and even litter from the forest floor. Consequently, taking biomass for energy from the forest has the capacity to remove more material and a greater range of material from the forest than would be the case under conventional management for timber production. Care needs to be exercised because the maintenance of soil fertility depends on the cycling of nutrients and carbon and if these are removed from the site in biomass, soil fertility may be reduced. Foliage, small branches and litter contain a greater concentration of nutrients than stem wood and these should be left in the forest if at all possible. The return of the ash to the forest following combustion would alleviate nutrient loss to some extent. Also, any impact on conservation of biodiversity would need to be considered (Hall, 2002). On the plus side, the more thorough cleanout in a combined biomass and timber management regime would create a reduced risk of fire and damage from insects and disease (IEA, 2002). The challenge is to integrate the management and silviculture for biomass production with that for higher value timber production in a sustainable manner where environmental values are safeguarded.

In the developing world there are few situations where farmers grow trees specifically for fuel. The extensive areas of energy plantations, the wide-scale uptake of improved cooking stoves and the proliferation of gasification units envisaged under the Tropical Forest Action Plan have not eventuated. Probably the best strategy for the rural farmer is to grow multipurpose tree species where fuelwood supply is just one component of an overall land use system that meets broader objectives. Agroforestry systems are particularly attractive because they can provide food, fuel and fibre in a manner that provides a broader range of products in a way that reduces risk while caring for the land, reducing soil erosion and maintaining soil fertility. The most successful schemes have been where the rural farmers have been part of 'community-based participatory forest management' where the rural community shares in the planning, management and benefits. This is particularly important for managing land that is held in common. Community forestry will be considered in more detail in Chapter 8. Technologies need to be simple, decentralized and inexpensive. Subsidies, for example in seedling distribution, can assist as well as can the removal of legal disincentives to plant.

In the early 1980s the National Academy of Sciences developed a list of potential species for fuelwood and other products for the developing world (Table 4.5). Some of these species are very aggressive and care needs to be taken that they do not become weeds when introduced as an exotic species into a new habitat. Local species should be preferred if they are suitable.

Table 4.5. Potential fuelwood species (collated from National Academy of Sciences, 1980, 1983).

Humid tropics	Tropical highlands	Arid and Semiarid Regions	
Acacia auriculiformis	*Acacia decurrens*	*Acacia brachystachya*	*Haloxylon persicum*
Albizzia falcataria	*Acacia mearnsii*	*Acacia cambagei*	*Parkinsonia aculeata*
Bursera simaruba	*Ailanthus altissima*	*Acacia cyclops*	*Pinus halepensis*
Calliandra calothyrsus	*Alnus acuminata*	*Acacia nilotica*	*Pithecellobium dulce*
Casuarina equisetifolia	*Alnus nepalensis*	*Acacia saligna*	*Populus euphratica*
Coccoloba uvifera	*Alnus rubra*	*Acacia senegal*	*Prosopis alba*
Derris indica	*Eucalyptus globulus*	*Acacia seyal*	*Prosopis chilensis*
Eucalyptus brassiana	*Eucalyptus grandis*	*Acacia tortilis*	*Prosopis cineraria*
Eucalyptus deglupta	*Eucalyptus robusta*	*Adhatoda vasica*	*Prosopis juliflora*
Eucalyptus pellita	*Eucalyptus tereticornis*	*Ailanthus excelsa*	*Prosopis pallida*
Eucalyptus urophylla	*Gleditsia triacanthos*	*Albizzia lebbeck*	*Prosopis tamarugo*
Gliricidia sepium	*Grevillea robusta*	*Anogeissus latifolia*	*Sesbania sesban*
Gmelina arborea	*Inga vera*	*Azadirachta indica*	*Tamarix aphylla*
Guazuma ulmifolia	*Melaleuca quinquenervia*	*Balanites aegyptiaca*	*Tarchonanthus*
Hibiscus tiliaceus	*Melia azedarach*	*Cajanus cajan*	*camphoratus*
Leucaena	*Robinia pseudoacacia*	*Cassia siamea*	*Zizyphus mauritiana*
leucocephala	*Sapium sebiferum*	*Colophospermum*	*Zizyphus spina-christi*
Maesopsis eminii		*mopane*	
Mangroves		*Combretum micranthum*	
Mimosa scabrella		*Conocarpus lancifolius*	
Muntingia calabura		*Dalbergia sissoo*	
Pinus caribaea		*Emblina officinalis*	
Psidium guajava		*Eucalyptus*	
Sesbania bispinosa		*camaldulensis*	
Sesbania grandiflora		*Eucalyptus citriodora*	
Syzygium cumini		*Eucalyptus*	
Terminalia catappa		*gomphocephala*	
Trema		*Eucalyptus microtheca*	
		Eucalyptus occidentalis	
		Haloxylon aphyllum	

Biomass for energy plantations will increase in importance. Some (a minority) rural farmers in the developing world have been successful in making a profit from selling fuelwood from their woodlots and this may increase as markets become more sophisticated. Plantations will become increasingly important for supplying fuelwood to industry in the developing world and particularly to the charcoal industry. In the developed world biomass plantations are a particularly attractive option for the rehabilitation of degraded sites and for reclaiming land that has been released from agricultural production. The most common species for energy plantations are poplars (*Populus*), willows (*Salix*) and eucalypts (*Eucalyptus*). 11,250 hectares of plantation growing at 10-15 tonnes per hectare per annum could produce enough biomass to fuel a 30MW power station which could supply enough electricity for 30,000 houses (IEA, 2002). Fuelwood is a low value product and the economics of biomass plantations is marginal. In order to be economical for biomass, the plantations need to be highly productive and intensively managed as coppice systems. Plantations for biomass should become an increasingly attractive proposition as the price of fossil fuels increase, if trading in carbon credits eventuates and if it is given incentives by agricultural, forestry and energy policies. Probably, however, most woody biomass will continue to be obtained as an additional product to the management of forests for timber production. Management and silvicultural regimes will need to be modified to achieve this dual role (Andersson *et al.,* 2002).

Wood Products

The Structure and Properties of Wood

Trees grow by increasing in height and diameter. In order to do this they have to create new cells at the growing tip (the apical meristem) and around the circumference of the stem where the bark meets the wood (the vascular cambium). The vascular cambium produces cells (phloem) outwards that subsequently develop into bark and inwards (xylem) that subsequently develop into wood. The primary function of the stem of a tree is to provide a pathway: for the transport of water and dissolved nutrients from the soil through the roots to the leaves (this occurs mainly in the wood); and for the movement of sugars and carbohydrates produced by photosynthesis in the leaves to other parts of the tree (this occurs mainly in the bark). Consequently wood has to be somewhat like a system of pipes to facilitate its role in transport and the cells in wood are elongated axially to achieve this function. Wood also has to be strong enough to hold the tree and its canopy erect, while withstanding external forces such as wind. Consequently the elongated cells of wood in trees undergo secondary thickening and have thicker and more anatomically complex walls than cells of smaller herbaceous plants. In much of the literature a distinction is made between conifers and hardwoods. An argument was given in Chapter 2 that the use of the term hardwoods is ambiguous and misleading and that more correctly they should be called angiosperms. FAO circumvents these difficulties by classifying wood products as coniferous or non-coniferous. Except for a few species such as palms and bamboos, all forest trees that are angiosperms (non-coniferous) are dicotyledons (or more recently called by some sub-class Magnoliidae). Conifers and dicotyledons have distinctively different wood anatomies.

The predominant cell types in wood are fibres and vessels (in dicotyledons), tracheids (in conifers) and ray cells (in both). Tracheid, fibres and vessels are axially elongated cells (Fig. 4.6). Tracheids in conifers perform both transport and structural functions and they are connected to each other by bordered pits (essentially holes in the wall) to facilitate sap flow. Fibres in dicotyledons are shorter than tracheids and perform only the strength function while the elongated tube-like cells, called vessels, perform the transport function.

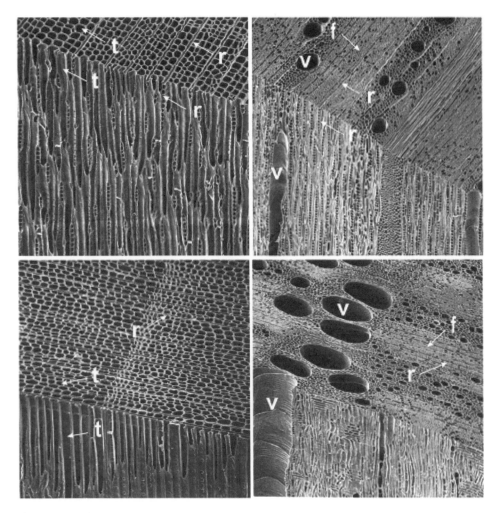

Fig. 4.6. Scanning electron micrographs of the conifer *Dacrydium dacrydioides* (left hand side) and the dicotyledon *Quercus robur* (right hand side) showing a transverse and tangential longitudinal face (top row) and a transverse and radial longitudinal face (bottom row). v =vessel; r = ray; t = tracheid; and f = fibre (scanning electron micrographs from Brian Butterfield; magnifications 90x left upper, 70x left lower, 50x right upper and lower).

The difference between dicotyledon and conifer wood is easily distinguished by the presence or absence of vessels. Tracheids, fibres and vessels lose their living contents (protoplasm) quite soon after differentiation and from then on are dead cells. Only the outer part of the wood (the sapwood) functions in transport and, in order to do this, the elongated transport cells (tracheids and vessels) need to be in contact with or close to a living cell over some portion of their length. These living cells (parenchyma) are oriented in radial rows called rays. Rays cells may be in a single radial column (uni-seriate) or stacked into multiple columns (multi-seriate). Most often they are lens-shaped when viewed end on (tangentially) (Fig. 4.6). There is a zone of wood adjacent to the vascular cambium, called the sapwood, which is active in transport. As the tracheids and vessels become older and further displaced from the physiological centre of activity (the vascular cambium), the adjacent ray cells die and the transport function is no longer supported. The walls and lumens (the space inside the walls) of the tracheids, fibres and vessels become progressively clogged up with extractives and the cell to cell pit connections break down. Differences in the amount and nature of extractives are responsible for differences in colour and durability between the woods of different species. This inner wood that is no longer involved in transport is usually darker and is called heartwood. In small, young and actively growing trees, sapwood may occupy all or most of the wood, but in older larger trees sapwood may be confined to a small ring adjacent to the vascular cambium.

There is very great variety in the way the cells are organized and this gives rise to the wide variety in the macroscopic appearance of the wood of different species and in the way that the wood behaves in service. In temperate climates radial tree growth effectively ceases in winter. Conifers lay down thinner walled cells in the early part of the growing season (earlywood) and thicker walled cells in the latter part of the growing season (latewood) and dicotyledons display different patterns of fibre and vessel distribution across the growing season. Consequently there is a distinct pattern of annual rings across a transverse section and counting the rings near to the base of the tree will give the age of the tree. In climates where there is no distinct cold season, trees grow throughout the whole year and do not have annual rings. However, they still may have growth rings owing to differences in water availability and temperature over weeks, months or years. The wood that is formed early on in the life of the tree (near to the centre or pith) is called juvenile wood. This wood has shorter tracheids and fibres with thinner walls and greater microfibril angles. Consequently it is not as strong. This has implications in wood utilization, particularly in plantations where trees grown over short rotations have high proportions of juvenile wood. In sawnwood, knots are found where branches meet the stem. Knots weaken the timber and mostly are considered to be unsightly. If the lower branches are cut off close to the stem, wood formed after this will be clear of knots. This is the rationale behind pruning trees to confine the knots to a central knotty core in order to produce clear wood outside the knotty core.

Tree stems are round but sawn timber is square or rectangular. The appearance of the face of a sawn board will differ depending on whether it is closer to being aligned radially (quarter sawn) or tangentially (back sawn) (Fig. 4.7). In quarter sawn boards the growth rings are near to perpendicular to the face of the board and the rays are parallel to the face of the board. Quarter sawn boards expose the sides of the rays and consequently are prominently figured. The prominent flecks in quarter sawn oak for example are caused by the prominence of the rays. In back sawn boards the growth rings are approximately parallel to the face of the board and the rays are perpendicular. The rays are less prominent because their ends rather than their sides are shown (Fig. 4.6). Boards between quarter sawn and back sawn where the rings intersect the face at angles of 30° to 60° (rift sawn) will have an appearance somewhere between that of quarter sawn and back sawn boards.

Wood is an anisotropic material, which means that its mechanical properties are not equal in all directions. This has advantages in that strength is maximized where the fibres or tracheids run parallel to the long edge of the board. However, the anisotropic nature of wood also creates problems, particularly during the drying of wood. Wood shrinks when it dries. Shrinkage along the length of the board is small in comparison to shrinkage across the face and edge of the board. Tangential shrinkage is greater than radial shrinkage and, because growth rings are circular and board faces are flat, this can cause the boards to dry in an uneven manner. The tendency is for the curvature of the growth rings to straighten in drying. A quarter sawn board will shrink less across the face than a back sawn board but may shrink unevenly across its thickness. A back sawn board may cup across its face. A square section may become a diamond, a circular section an ellipse and a rift sawn section a parallelogram (Fig. 4.7). Differences in shrinkage patterns along a board may cause the board to crook, bow, or twist (Fig. 4.7). These differences along the board are often caused by the presence of reaction wood, which is composed of abnormal cells produced in response to uneven loading such as in a leaning tree. Reaction wood has greater longitudinal shrinkage than normal cells. Small cracks (checks) may occur on the ends and faces of boards during drying. These problems can be minimized with careful and controlled drying (seasoning) of the wood.

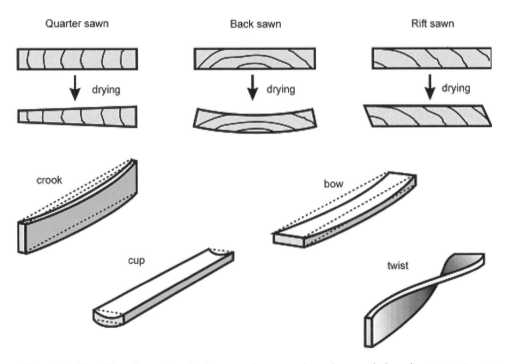

Fig. 4.7. Possible dimensional changes in sawn boards on drying (movements are exaggerated for effect).

Wood is a lignocellulosic material that, unless burnt, ultimately will decompose under the influence of wood-degrading fungi, insects, bacteria and marine borers. This is, of course, desirable because decomposition is an integral part of the cycling of nutrients and carbon in forest ecosystems. Forest floor litter has to decay. If it did not do so, woody material would continue to accumulate on the forest floor and the productivity of the system would grind to a halt. Wood decay may also be a boon if it is done in a controlled manner in order to pulp the wood prior to paper making. Wood decomposition, however, is a costly nuisance when it come to using solid wood and engineered wood products as a building material. Chemical treatments have been used for centuries to reduce the rate of wood deterioration. The most commonly used preservatives have been oil-based creosote and pentachlorphenol and water-based arsenicals. All of these are under public scrutiny because of proven or suspected toxicities and have been banned in certain countries. The safe disposal of treated wood remains an unresolved problem. There is considerable research into finding environmentally friendly and economically viable alternatives (Goodell *et al.,* 1993). Another solution is to adopt a life style that does not require timber to last so long and this seems to be happening to some extent. The life expectancy of homes, structures, decks, and outdoor wood products is not as great today as it has been because of an increasing tendency to redesign and rebuild.

Global Production of Wood Products

Production of coniferous and non-coniferous industrial roundwood is shown in Fig. 4.8. Coniferous industrial roundwood comes mainly from the temperate developed world and non-coniferous industrial roundwood comes mainly from the tropical developing world.

The production of industrial roundwood has increased over the last 40 years and continues to do so, although at a lesser rate. The production of coniferous industrial roundwood has always been about twice that of non-coniferous roundwood. Predicting future trends is a risky business and predictions of future production vary quite widely

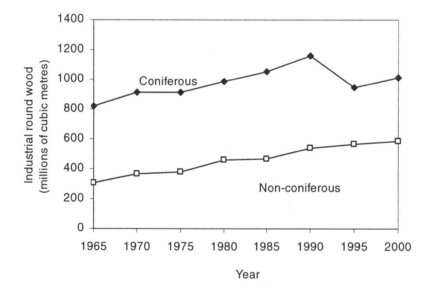

Fig. 4.8. The world production of coniferous and non-coniferous industrial roundwood (compiled from FAO statistics on-line).

from supply to demand driven. There is a range of factors that impact on future demand and supply. Increased consumption resulting from increased population and living standards will be tempered by environmental pressures, the use of alternative materials to wood, and increased efficiencies in processing. Plantations produce a small but increasing proportion of industrial roundwood. Currently plantations occupy 4.8% of the world's forest area (FAO, 2003) and 48% of the plantations have been established to provide industrial wood (FAO, 2001). Plantations can be very productive and their contribution to overall wood production can be greater than their proportional areas. New Zealand is a striking example. New Zealand's forest industry supplies 1.1% of world and 8.8% of Asia Pacific's forest products trade from just 0.05% of the world's forest resource and an annual harvest area equivalent to 0.0009% of global forest cover. Virtually all of New Zealand's timber production and trade comes from plantations (NZFOA, NZFIC, MAF, 2004).

Wood can be used in the round (post and poles), as sawnwood (lumber), as engineered wood products (particleboard, medium density fibreboard, laminated veneer lumber, plywood, oriented strand board), and can be pulped to produce paper and cardboard. Also wood can be used as a composite with other materials such as plastic and cement.

Sawnwood and Engineered Wood Products

Time trends for sawnwood production are given in Fig. 4.9 and for plywood, particleboard and fibreboard in Fig. 4.10.

The volume of sawnwood production is more than twice that of all wood-based panels combined, but sawnwood production is decreasing with time and wood based panels are increasing. Sawnwood is the traditional primary wood product from the tree but its production is wasteful, it can behave badly in drying, it is often not durable, it is variable in its properties, its strength is determined by its weakest part, and pieces are limited in width. These problems can be overcome to some extent by gluing veneers together (plywood, laminated veneer lumber) or by breaking the wood down into small pieces and then reconstituting it as particleboard, fibreboard, or oriented strand board. These processes (except for plywood production) also have the advantage that they are adaptable to continuous flow technologies and to large scale operations where the cost of labour is minimized. Even so, solid wood still holds a prominent position for framing and for appearance grade uses. The fine furniture industry is based on solid wood combined with veneer faced panels.

The production of plywood, and especially particleboard and fibreboard, is increasing with time. More than one half of the fibreboard production is medium density fibreboard (MDF). It is likely that wood-based panels will continue to expand at the expense of sawnwood.

Plywood is made by gluing together layers of veneer so that each layer has its grain perpendicular to the adjacent veneer. The number of layers is always odd so that the face and back veneers have the grain running in the same direction. The core is the centre section of the sheet and may be laminated solid wood (as in doors), particleboard, MDF or veneer. Plywood having a veneer core is the strongest and the more veneers for the same thickness the stronger the plywood. The veneers are peeled on a large rotary lathe in which case they form a tangential longitudinal section with the rays perpendicular to the face. Accordingly the boards do not have a pronounced figure. A more pronounced figure can be obtained by slicing the veneers, in which case boards closer to a radial longitudinal section can be obtained (Fig. 4.6). Sliced veneers are expensive to produce and are only used for high value decorative end uses. Plywood is strong and dimensionally stable. It is used in construction and also for high value end uses when the face veneer is of high quality. It can be made suitable for outdoor applications by using waterproof adhesives. Laminated veneer lumber (LVL) is similar to plywood in that it is made by gluing veneers together except that in LVL the grain of the veneer is in the same lengthwise direction for all layers. Because of this, LVL can be made in a continuous rather than a batch process and longer pieces can be made. LVL has the advantage over solid wood in that it has uniform properties, is largely defect-free and can be fabricated into large (long, wide and thick) sizes. It is strong when edge loaded as a beam or face loaded as a plank. The manufacturing process is suitable for making large pieces from small trees. Waterproof glues can be used for applications requiring moisture resistance. LVL is a structural material with cutting, fastening and connection requirements similar to solid wood. Usually it is not used for appearance applications.

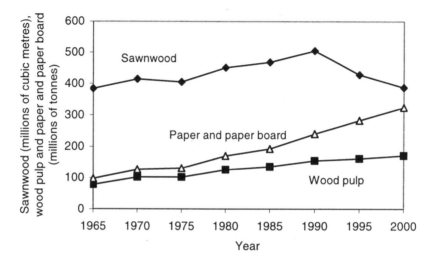

Fig. 4.9. World sawnwood, wood pulp and paper and paperboard production (compiled from FAO statistics on-line).

Particleboard (chipboard) is a panel made from wood chips bound together under heat and pressure with urea formaldehyde for utility cabinet manufacture or with the more water resistant phenol formaldehyde for industrial applications such as underlay flooring. It can be faced with a quality veneer for appearance or with a melamine laminate for a hardwearing surface. Particleboard will not bow or warp like solid wood or plywood but it can swell when exposed to water. It can be machined but, because of the size of the particles, there is always a risk of the surface tearing out. Also the inconsistent surface does not paint very well. Fibreboards are panels formed from wood fibres that are reconstituted under heat and pressure. They come in a range of densities from low density such as caneite (160 – 450 kg/m^3) to masonite (about 1000 kg/m^3) but the most versatile is MDF (around 800 kg/m^3) which is formed by combining conifer fibres with wax (paraffin wax) and resin (urea formaldehyde) and applying high temperature and pressure. MDF was developed in the 1970s and large scale production commenced in the 1980s. It has replaced particleboard for many applications. It has a smoother surface, a more even and consistent structure and better moisture tolerance than particleboard. Because MDF has no grain as such, it can be machined (cut, drilled, sanded, filed and carved) without damaging the surface and also can be painted. It does not bow or warp and has a very wide range of uses including furniture, cabinets, doors, panel mouldings, and floors. MDF can also be faced with decorative veneers or melamine. Both particleboard and MDF contain formaldehyde and the possibility of health risks remains a contentious issue. Oriented strand board (OSB) is made from large chips or flakes of wood that are formed from layers oriented at right angles to each other which gives strength to the panels. OSB can be cut with saws but cannot be milled satisfactorily. It is used for structural applications where a strong panel is required that doesn't need finishing.

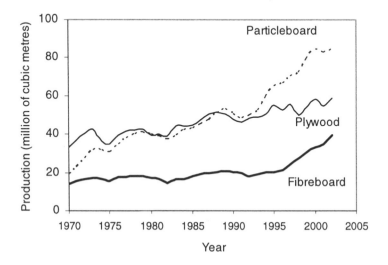

Fig. 4.10. World production of fibreboard, plywood and particleboard (compiled from FAO statistics on-line).

Wood Composites

Wood composites are increasing in importance, mainly wood with plastic and wood with cement. Composites of wood with thermosetting plastics are a hybrid having some of the advantages of both wood and plastic. The wood fibre filler provides strength along the direction of the fibres and reduces overall cost compared to plastic alone. The plastic provides lightness, durability and the ability to extrude profiles, inject mould components and heat-weld joints. The major market in wood plastic composites has been in extruded profiles for applications such as decking, railings and window and door components. More recently, injection moulding is producing trim components such as caps, finials and skirts for decks and posts. Wood with plastic and carbon is also promising for external applications requiring lightness and strength. Wood and cement composites are manufactured by mixing cement with either long wood shavings, wood particles or wood fibres. These products have taken over from asbestos cement composites which are no longer used because of health issues. Cement composites with long wood shavings are used in thermal and acoustic insulation as well as for partitioning in low cost housing. Cement composites with wood particles or fibres are very water resistant and are suitable for both internal and external cladding and panelling applications.

Pulp and Paper

Paper is made from fibres that may be of animal, vegetable or synthetic origin, but mostly come from wood. Wood fibres from trees (anatomically tracheids in coniferous species and fibres in non-coniferous species) provide about 50% of the global material used to produce paper. Recycled paper makes up a further 20% and the remainder is made up of additives to improve paper characteristics and alternative fibres such as rice straw, wheat straw, rye straw, kenaf, bagasse, bamboo, jute, sisal, esparto and hemp. The manufacture of paper is a two stage process. The first is turning wood into wood pulp. This is done by first chipping the wood and then mechanically or chemically breaking the chips down into individual fibres. The second stage (which may take place at a different location) is when the suspended fibres are deposited as an intermeshing mat which is pressed and dried into a sheet of paper. Wood is made up of cellulose, hemicellulose, lignin and extractives. Lignin is important in standing trees to impart strength and extractives to impart colour and durability. However, both lignin and extractives are a nuisance in paper making and they must be removed during the chemical pulping process. Typically wood has about 40 to 45% cellulose, 30% hemicellulose and 25 to 30% lignin. The best raw material for paper making will have high holocellulose (cellulose + hemicellulose), low lignin, and fibres with a high ratio of length to diameter. Coniferous species have relatively long fibres (about 3-4mm) and non-coniferous species relatively short fibres (about 1.2mm) (Bhojvaid and Rai, 2004). Important non-coniferous species include eucalypts, aspen, oak and maple and important coniferous species include pine, spruce, fir and larch (Doshi and Dyer, 2004). The raw material can come from thinnings, small trees and processing residues. Plantation grown trees are becoming increasingly important, and non-coniferous species, especially eucalypts, are becoming more prominent in the mix. Southern hemisphere sources, especially South America, are becoming more important as suppliers of wood for pulp (AbuBakr, 2004).

The amount of virgin wood pulp has been increasing steadily over the last 40 years but paper and paperboard production have been increasing at a faster rate because of the increasing contribution of recycled fibre (Fig. 4.9). There is no evidence that recent advances in information technology and electronic communication have dampened paper consumption. Consumption is closely related to *per capita* income and emerging economies, particularly China, will continue to increase demand. Paper manufacturing is a very large and capital intensive industry whose profitability depends on the base price paid for a raw material that is towards the bottom of the quality range in wood-based industry. Currently wood waste is an important source of pulp. The relative importance of plantations will increase in the future in order to meet increased demand. Commodity outputs from the pulp and paper industry are second only to oil in the value of production (Table 4.6).

The pulp and paper industry, by the very nature of the technologies involved, has been one of the dirtiest and smelliest of all industries in terms of water and air pollution. It has had to make considerable investment in recent years in cleaning up its act. The industry, perhaps because of its size, has also been singled out as a major deforesting agent and as an environmental villain. There have been calls for using alternative fibre materials such as hemp. There is no doubt that the pulp and paper industry (and other wood-based industries) has, to a greater or lesser extent, been responsible for poor and non-sustainable forest practices in the past and present. The solution, however, is not to dampen demand for wood fibres but rather to sustainably manage existing forests and to establish more forests. The trend towards more pulp wood coming from plantations is a step in the right direction.

Ultimately the package of sustainable management of forests, the establishment of plantations and the efficient recycling of paper is preferable to promoting any alternative source of fibre (although this could easily provoke an argument).

Table 4.6. Commodity outputs in 1992 (derived from Bolton, 1998).

Commodity	Production (millions of tonnes)	Value in 1992 (billions of USD)
Crude oil	3,000	411
Pulp and paper	412	395
Steel	735	350
Tobacco	8	32
Iron ore	950	31
Aluminium	19	20
Bauxite	110	16
Wool	3	7
Rubber	6	6
Cocoa beans	2	3

There is a range of chemicals which are produced as a byproduct of the pulping process. Terpenes, the major constituents of turpentine, are produced from the sulphate pulping process. Resin acids, the main components of rosin, are produced from the tall oil obtained as a byproduct of the kraft black liquor recovery process. Fatty acids are also recovered from the tall oil and these have applications in ore separation, metal working, detergents, drying agents and finishes. Rayon is a manufactured fibre made from specially treated wood pulp. It is used as a fabric for clothes and soft furnishings, in sanitary hygiene fabrics and as a strengthening agent in tyres.

Non-wood Forest Products

Non-wood forest products (NWFP) can sometimes be as economically important as or more important than traditional wood products. NWFP can come from woodlands and isolated trees as well as from trees in the forest. Sometimes when a forest tree species is established as a plantation to provide NWFP, the boundary between forestry and horticulture becomes blurred. An example would be a plantation of walnuts that are planted initially for the nuts but eventually are utilized for the timber. FAO considers that rubber plantations are part of the forest plantation estate but that palm oil plantations are not. Rubber trees are tapped for latex when young and utilized for timber when they are older and past their efficient latex producing years. The variety of NWFP is very great but reliable global data on their production and trade are lacking. There is clear evidence of over-exploitation of some NWFP but sustainable management of NWFP is virtually non-existent globally. This is of great concern not only because of loss of biodiversity, but also loss of livelihood to those communities who depend on NWFP for their very survival. FRA 2000 undertook a global survey of NWFP (FAO, 2001) and much of what follows draws on their data. China and India are the world's largest producers and consumers of NWFP.

Food and Fodder

Food products from plants include fruits, seeds, flowers, bark, roots and mushrooms, and food products from animals include meat (from both vertebrates and invertebrates), honey, bird nests and eggs. The range of food coming from 'wild' plants and animals in the forest is so diverse that only isolated examples can be given here. In Africa, edible NWFP are important for sustenance during periods of food shortage and honey and bees wax are also important. Southern and eastern Asia are the major producers of wild culinary herbs and wild spices (nutmeg, mace, cinnamon and *Cassia*) and of the wild edible mushrooms *Morchella* spp. (morels), *Auricularia auricula-judae* (jews ear), *Tremella fuciformis, Lentinus edodes* (shiitake), *Trichloma matsutake* (matsutake) and *Pleurotus ostreatus* (hiritake). Edible pine nuts are produced from *Pinus gerardiana* in India, Afghanistan, Pakistan and China, *Pinus koraiensis* in Russia, Manchuria, China and Korea, *Pinus pinea* (stone pine) from around the Mediterranean and *Pinus cembroides* (Mexican pinyon) from south-western USA and Mexico. Brazil nuts and palm hearts are important wild food products from South American forests as well as *Araucaria* seeds, mushrooms (*Boletus luteus, Lactarius deliciosus*) and maté, an addictive but harmless stimulant obtained from leaves of *Ilex paraguariensis* which is extremely popular as a beverage in Argentina, Uruguay, Paraguay and southern Brazil. The forests of Central America contain more than 100 tree or palm species with edible fruits. The latex (chicle) from sapodilla (*Manilkara zapota*) is used for making chewing gum. Hearts from the manicole palm (*Euterpe oleracea*) are the principal source of income for Amerindian communities in the coastal wetlands of Guyana. Berries, honey, mushrooms (for example the truffle *Tuber melanosporum*) and nuts are important in Europe and mushrooms, berries and maple syrup in North America. Many forest trees for food have become 'domesticated' and are grown in plantations (pecans, walnuts, beech nuts, pine nuts, acorns, cashews, maple syrup, dates, coconuts, *Macadamia* nuts, avocados, citrus, stone fruits, bananas and almonds). Forest trees produce a range of oil seeds, although oil from olives (*Olea europaea*) and oil palm (*Elaeis guineensis*) is now obtained from plantations and would not be considered to be a NWFP. Illipe nuts, the fruits of some *Shorea* trees, contain an oil with properties similar to cocoa butter. In Morocco, edible oils are obtained from the argan tree (*Argania spinosa*).

The consumption of bushmeat (antelopes, gazelles, monkeys, wild boar and porcupines) in Africa has raised legitimate concerns about wildlife conservation. For example, bushmeat consumption in the Congo basin is estimated at 5 million tonnes per annum which is double the production rate (Fa *et al.,* 2002). Meat is a significant NWFP in Europe and North America, but here it is called game meat rather than bushmeat and is obtained more for sport and recreation rather than as a staple for basic sustenance. For the most part, hunting and fishing in European and North American forests are under some form of management and regulation.

Forests can provide fodder for domestic animals and flowers for bees. Fodder from trees is very important in the drier parts of Africa (*Acacia, Khaya, Faidherbia* and *Balanites*), South America (*Prosopis* spp) and in continental and southern Asia.

Medicinal Plants and Animals

Forests are a very rich source of plants and animal parts used in traditional medicine and also in western medicine. Forests are also a rich genetic resource for future medicines yet to be discovered and developed. More than 80% of the world's population uses traditional medicines from plants for their primary health needs. This is mostly concentrated in the developing world but the developed world is steadily increasing its use of medicinal plants. In the developed world, 25% of all medical drugs are plant-based and in the developing world it is near to 75%. Over 25% of all prescriptions in Western Europe and North America and near to 60% of those in Eastern Europe are unmodified or slightly altered plant products. In 1996, the USA had a turnover of 4 billion dollars in herbal drugs (Tiwari *et al.*, 2004). Traditional medicine based on medicinal plants is very well developed in Asia (particularly China and India), Latin America and Africa. One-third of the plants listed by the ancient Egyptians as being of pharmaceutical value are still used for this purpose today. The ancient Greeks used extracts from willow (*Salix*) to relieve pain. The active ingredient in willow led to the development of aspirin, one of the most widely used drugs in the world. The anti-malarial quinine is obtained from the *Cinchona* tree and this is still used, often in combination with other drugs, to treat malaria today. Many plants have proven medicinal properties. There are so many species that no attempt will be made here to single out any more. In Ethiopia alone there are 600 plant species listed as being important in traditional medicine. About 4000 species of medicinal plants are recognized in Asia.

The growing global demand for medicinal plants has led to over-exploitation and species endangerment. Conservation issues are discussed by Bhattarai and Karki (2004). The Convention on International Trade in Endangered Species of Wild Fauna and Flora (CITES) is an international convention aimed at conserving threatened and endangered species. Signatories to the convention have agreed not to trade in species listed as endangered. Also attempts are now being made to certify NWFP (Shanley *et al.*, 2002). China is the world's largest producer and exporter of medicinal plants followed by India. Egypt and Morocco are the main African exporters of wild medicinal plants, while Chile and Mexico are the major producers and exporters in Latin America. Germany, France and Bulgaria are significant European exporters and the USA is also an exporter. The proportion of medicinal plants that are cultivated compared to those obtained from the wild is increasing. Conflicts between western medicine and traditional medicine have been intense in the past and continue to be so, but there has been a growing trend towards co-existence and learning from each other.

Aromatics for Perfumes and Cosmetics

Essential oils are volatiles that are extracted from the plant by distilling, compressing, steaming or dissolving. They are called essential from the word 'essence.' Essential oils from aromatic plants (*Thymus, Rosmarinus, Acacia* and *Eucalyptus*) are important products of Egypt, Morocco and Tunisia. Essential oils derived from the seeds of the nutmeg tree (*Myristica fragrans*) are a major export of Grenada. Southeast Asia and eastern Asia are major producers of essential oils from sandalwood (*Santalum* spp), agarwood (*Aquilaria* spp.), snowbell (*Styrax* spp.), and blue gum (*Eucalyptus globulus*). In South America, essential oils are distilled from rosewood (*Aniba rosaeodora*), andiroba (*Carapa* spp.) and sassafras (*Ocotea pretiosa*). Tea tree oil (*Melaleuca alternifolia*) is produced in Australia. Essential oils traditionally have been used in cosmetics, perfumes and as aromatic additives to otherwise non-aromatic or unpleasant substances such as in disinfectants and cleaners. Recently, however, they have become very popular in aromatherapy where they are vigorously promoted, often without any scientific evidence, to cure many ills and to provide good health and well-being.

Frankincense or olibanum (*Boswellia* spp.), myrrh (*Commiphora myrrha*) or opopanax (*Commiphora* spp.) are aromatic resins from Somalia, Ethiopia and Sudan that have been prized since ancient times. The first record of plant hunting was in 1495 BC when Queen Hatshepsut of Egypt commissioned an expedition to Somalia to collect myrrh trees which she subsequently planted in her garden (Musgrave *et al.*, 1998). Camphor from *Cinnamomum camphora* is a product of Southeast Asia.

Fibres for Construction, Craft and Utensils

Rattan (*Calamus, Daemonorops, Plectocomia* and *Korthalsia*) is the most important internationally traded NWFP. There are about 600 species of which ten are important commercially as canes used principally in furniture making. Production is centred in Southeast Asia and particularly Indonesia. However rattans have been depleted through overexploitation and through loss of forest habitat. Bamboo (*Phyllostachys* and *Dendrocalamus*) has very wide use in Asia for both construction and handicrafts and is now being extensively grown as a crop by farmers. Kapok (*Ceiba pentandra*) is an insulating material and comes mainly from Thailand and Indonesia. In Ecuador, fibres from the palm *Carludovica palmata* are used for making panama hats. In Chile, young branches of *Salix viminalis* are woven to produce handicrafts and furniture. In Central America, various palms (*Desmoncus* spp., *Sabal* spp., *Astrocaryum* spp. and *Carludovica palmata*) provide leaves, fibres and canes suitable for construction and handicraft. In the Caribbean there are a wide variety of NWFP used for making utensils as well as for handicrafts (see FAO, 2001).

Materials for Commerce and Industry

Resins are found in special ducts in the wood and bark called resin canals. Resins exude if the tree is wounded, their function being to heal the wounds and impart resistance to fungal infection and insect attack. Resins may be classified as hard resins or oleoresins. Hard resins are solid, transparent, colourless, odourless and brittle substances that are low in oil content. They are the best material for manufacturing varnishes. Oleoresins have high amounts of oil as well as resins and have a distinct aroma and flavour. The source and use of the main hard and oleoresins are given in Table 4.7.

The main oleoresins are from *Pinus*. Pine oleoresins are highly water resistant. The ancient Greeks used them to caulk the hulls and decks of their boats and they remained important as a source of pitch throughout the history of wooden ships. Because of their traditional use as a caulking agent in boats, they became known as naval stores. The resins were originally obtained by destructive distillation of wood, but for the last 300 years they have been obtained by tapping live trees in the same way as rubber trees are tapped for latex. The main products derived from pine resin are rosin and turpentine (Table 4.7). Pine resin is produced from about 10 main pine species and production is dominated by China, Russia, Scandinavia, Italy, France, Portugal, Spain and the USA. Today, however, terpenes (the main constituents of turpentine) and resin acids (the main constituents of rosin) are mainly obtained as by-products of the pulp and paper industry (see earlier in this chapter).

Table 4.7. Hard resins and oleoresins (adapted from Bhojvaid and Chaudri, 2004).

Resin	Tree	Use
Hard resins		
Copal	*Agathis*	manufacture of linoleum
Dammar	Dipterocarps	paper, varnish, lacquer, paint, inks, polishes, water-resistant coatings, incense, medicine
Mastic	*Pistacia*	chewing gum, flavouring, varnish, medicine
Dragon's blood	*Daemonorops*	violin varnish
Oleoresins		
Rosin	*Pinus*	paper sizing, printing ink, paint, varnish, leather, soap, batteries, synthetic rubber, perfumes, fireworks, adhesives
Turpentine	*Pinus*	paint, varnish, waxes, insecticides, germicides, perfumes, pharmaceuticals, soap, aromatics
Benzoin	*Styrax*	aromatics, pharmaceuticals, medicine
Styrax	*Liquidamber*	perfumes, pharmaceuticals
Balsam	*Myroxylon*	perfumes and medicine
Copaiba	*Copaifera*	shampoos, soaps and cosmetics, perfumes,
Elemi	*Canarium*	lacquers, varnishes

Gums are also exudates from trees but are carbohydrates for the most part and are used as additives in food and cosmetics. Perhaps the best known is gum arabic, an exudate from *Acacia senegal* from sub-Saharan Africa. Gum arabic has an osmotic pressure and colloidal content similar to blood and has been used as a surgical adhesive. It was also used as an adhesive on stamps. Today it is used in the pharmaceutical industry as a thickening and suspending agent. *Acacia senegal* is a multi-purpose tree. It not only produces gum arabic, but also is a useful fodder tree and a nitrogen fixer. India is the main producer of karaya gum, an exudate from *Sterculia urens*, which is used as a fibre deflocculating agent in paper production, as a thickener for foodstuffs and cosmetics and for finishing textiles. Argentina produces gum brea from *Cercidium australe*, which can be used as a substitute for gum arabic. (Frankincense, myrrh and opopanax are also gums.)

Tannins are carbohydrates found in wood, bark, roots, leaves and seeds of plants that traditionally have been used in curing (tanning) animal skins to make leather. The process of tanning is very old, at least dating back to 10,000 BC, and oak and chestnut were historically important species. Tannins are extracted by boiling or soaking in water and then concentrating the extract. Most tannins of tree origin come from the fruit of tara (*Caesalpinia spinosa*) and Peru is the major producer. The quebracho tree (*Schinopsis* spp.) is an important source of tannins from Argentina and Paraguay. *Acacia* and mangroves are other sources. Synthetic tannins are also used in modern leather tanning and tannins are also used in the manufacture of adhesives.

Dyes were used by the earliest civilizations and trees of historical importance were alder, ash, walnut, oak and birch. Today most dyes and colourants are synthetically produced from coal tar derivatives. Vegetable dyes, however, are still used in developing countries and their use is increasing in developed countries in art and craft. Peru is the major producer of the rich yellow orange colourant annatto (from the seeds of *Bixa orellana*) and of a carmine colourant from cochineal (from the insect *Dactilopius coccus*).

Brazilwood (*Caesalpinia sappan*) from Southeast Asia and mainly India, is valued for red dyes and *Caesalpinia echinata* from South America for its red-yellow dies. *Pterocarpus* from western Africa produces a bright red dye and *Chlorophora* from South America produces a yellow dye.

Latex is a milky emulsion that exudes from cells under the outer bark called laticifers. Some contain long-chain hydrocarbons that are elastic and the most commercially significant elastic latex is that produced by the rubber tree (*Hevea brasiliensis*), a native of South America that is extensively planted in Southeast Asia, particularly Malaysia. Over 90% of the world's supply of natural rubber comes from Southeast Asia with the residual production coming from equatorial Africa and South America. The latex comes from tapping the tree and the latex is coagulated to form natural rubber. The tree is first tapped at about age 5 years and tapping continues to about age 20. After tapping, the plantation can be left to about 30 years at which time it provides a timber crop that is often used to make furniture. Natural rubber is considered to be the best rubber but two-thirds of world supply is now manufactured from petrochemicals. Natural rubber has some unique properties that synthetics have not been able to reproduce. It has high resilience, low heat build-up in tyres, excellent strength at high temperatures and excellent fatigue resistance (Allen, 2004).

Cork is produced from the bark of many species but the best quality comes from the bark of the cork oak, *Quercus suber*, grown in Portugal, Spain and North Africa (Algeria, Morocco and Tunisia). Cork is a good insulator, is light, and is a viscoelastic material that can withstand large deformations without fracture and with good dimensional recovery. This has led to its use in flotation devices, as an insulation material, as a floor covering, in antivibrational joints in structures and as sealants in engines (Pereira and Tomé, 2004). Lower quality cork can be high temperature bonded with natural resins to make agglomerates that can be used as sheets for floor coverings and insulation. However, the economic viability of the cork industry is determined by the use of cork for stoppers in the wine industry. Currently cork stoppers are under serious competition from screw top seals which many wine aficionados consider to be at least as good as, if not better than, the best quality cork. The future of cork as a stopper for wine depends on marketing it on life style values rather than on its technical merits.

Shellac is a traditional and contemporary finishing agent for fine furniture. It is produced from lac, the protective coating of the scale insect *Coccus lacca* from India and Southeast Asia. Tung oil from the Chinese trees *Aleurites fordii* and *Aleurites molucanna* is another fine protective finish for furniture and wood work. In Brazil, a hard wax, carnuba wax, is obtained from the seeds of the carnaúba palm (*Copernicia prunifera*) and is used as a wax finish on cars. Citronella (*Cymbopogon* spp.) from Southeast Asia and the Caribbean and neem (*Azadirachta indica*) from India have insect repellant properties. Indeed neem has an almost cult-like following as an eco-friendly tree with a myriad of uses. Sphagnum moss (*Sphagnum cymbifolium*) is the live moss that grows on top of the dead peat moss in a bog in cool moist climates and is most often found in the open but also occurs in forests. Both mosses are used in horticulture and plant propagation.

Ornamentals, Trophies, Live Plants and Animals

Ferns, *Ficus* spp., orchids, and aquatic plants are among the wide range of plants that come from the forest and are traded. Animals traded include amphibians, birds, reptiles, frogs, primates and lemurs. Some species traded are included on the CITES endangered species lists. Christmas trees are produced commercially in Europe, North America, Canada, Japan, Australia and New Zealand. Decorative foliage from the forest is produced in both Europe and the USA, often as by-products of the pruning necessary to shape the Christmas trees (Hammett and Murphy, 2004). Decorative NWFP include vines, floral arrangements, dried decorations, wreaths and religious ornaments. Hunters in Europe, North America, New Zealand and elsewhere take pelts and wall trophies from their kills. There are no reliable data on the amount of pelts and trophies (or of game meat) taken by the hunters.

References and Further Reading

AbuBakr, S. (2004) World paper industry overview. In: Burley, J., Evans, J. and Younquist, J.A. (eds) *Encyclopedia of Forest Sciences*. Elsevier Academic Press, Amsterdam, pp. 694-700.

Allen, P.W. (2004) Rubber trees. In: Burley, J., Evans, J. and Younquist, J.A. (eds) *Encyclopedia of Forest Sciences*. Elsevier Academic Press, Amsterdam, pp. 627-633.

Andersson, G., Asikainen, A., Björheden, R., Hall, P.W., Hudson, J.B. Jirjis, R., Mead, D.J., Nurmi, J. and Weetman, G.F. (2002) Production of forest energy. In: Richardson, J., Björheden, R., Hakkila, P., Lowe, A.T. and Smith, C.T. (eds) *Bioenergy from Sustainable Forestry*. Kluwer Academic Publishers, Dordrecht, pp. 49-123.

Arnold, M., Köhlin, G., Persson, R. and Shepherd, G. (2003) *Fuelwood Revisited, What has Changed in the Last Decade?* CIFOR Occasional Paper No. 39, Center for International Forestry Research, Bogor, Indonesia.

Bhattarai, N. and Karki, M. (2004) Medical and aromatic plants: ethnobotany and conservation status. In: Burley, J., Evans, J. and Younquist, J.A. (eds) *Encyclopedia of Forest Sciences*. Elsevier Academic Press, Amsterdam, pp. 523-532.

Bhojvaid, P.P. and Chaudhari, DC. (2004) Resins, latex and palm oil. In: Burley, J., Evans, J. and Younquist, J.A. (eds) *Encyclopedia of Forest Sciences*. Elsevier Academic Press, Amsterdam, pp. 620-627.

Bhojvaid, P.P. and Rai, A.K. (2004) Paper raw materials and technology. In: Burley, J., Evans, J. and Younquist, J.A. (eds) *Encyclopedia of Forest Sciences*. Elsevier Academic Press, Amsterdam, pp. 701-707.

Biermann, C.J. (1996) *Handbook of Pulping and Papermaking*. 2nd Edition. Academic Press, San Diego.

Boberg, J. (2000) Woodfuel Markets in Developing Countries – A Case Study of Tanzania. Ashgate, Aldershot and Vermont.

Bolton, T. (1998) *The International Paper Trade*. Woodhead, Cambridge.

British Petroleum (2002) *BP Statistical Review of World Energy, June 2002*. British Petroleum, London.

Broadhead, J., Bahdon, J. and Whiteman, A. (2001) Woodfuel consumption modelling and results. Annex 2. In: *Past Trends and Future Prospects for the Utilization of Wood for Energy*. Working Paper No. GFPOS/WP/05, Global Forest Outlook Study, FAO, Rome.

Buchanan, A. and Levine, S.B. (1999) Wood-based building materials and atmospheric carbon emissions. *Environmental Science and Policy* 2(6), 427-437.

Cook, E. (1971) The flow of energy in an industrial society. *Scientific American* 225,135-144.

Dengg, J., Hillring, B., Ilavsky, J., Ince, P., Stolp, J. and Perez-Latorre, M. (2000) *Geneva Timber and Forest Discussion Papers - Recycling, Energy and Market Interactions*. United Nations Economic Commission for Europe and Food and Agriculture Organization of the United Nations, Geneva.

Doshi, M.R. and Dyer, J.M. (2004) Paper recycling science and technology. In: Burley, J., Evans, J. and Younquist, J.A. (eds) *Encyclopedia of Forest Sciences*. Elsevier Academic Press, Amsterdam, pp. 667-678.

Eckholm, E.P. (1975) *The Other Energy Crisis, Firewood*. Worldwatch Institute, Washington.

Elder, T. (2004) Chemicals from wood. In: Burley, J., Evans, J. and Younquist, J.A. (eds) *Encyclopedia of Forest Sciences*. Elsevier Academic Press, Amsterdam, pp. 607-612.

Emrich, W. (1985) *Handbook of Charcoal Making*. Reidel, Dordrecht.

Eschete, G. (1999) Assessment of fuelwood resources in *Acacia* woodlands in the Rift Valley of Ethiopia. *Sylvestria* 104, Swedish University of Agricultural Sciences Doctoral Thesis, Umeå, Sweden.

Fa, J.E., Peres, C.A. and Meeuwig, J. (2002) Bushmeat exploitation in tropical forests: an intercontinental comparison. *Conservation Biology* 16(1), 232-237.

FAO (1983) *Simple Technologies for Charcoal Making*. Food and Agriculture Organization of the United Nations, Forestry Paper 41, Rome.

FAO (1985) *Industrial Charcoal Making*. Food and Agriculture Organization of the United Nations, Forestry Paper 63, Rome.

FAO (1993) *A Decade of Wood Energy Activities within the Nairobi Programme of Action*. Food and Agriculture Organization of the United Nations, Forestry Paper 108, Rome.

FAO (2001) *Global Forest Resources Assessment 2000 - Main Report*. FOA Forestry Paper 140, Food and Agriculture Organization of the United Nations, Rome.

FAO (2003) *State of the World's Forests*. Food and Agriculture Organization of the United Nations, Rome.

Goodell, B., Nicholas, D.D. and Schultz, T.P. (1993) *Wood Deterioration and Preservation – Advances in our Changing World*. ACS Symposium Series 845, American Chemical Society, Washington, DC.

Gundimeda, H. and Köhlin, G. (2003) *Fuel Demand Elasticities for Energy and Environment Studies: Indian Sample Survey Evidence*. Environmental Economics Unit, Department of Economics, Göteborg University, Sweden.

Hakkila, P. and Parikka, M. (2002) Fuel resources from the forest. In: Richardson, J., Björheden, R., Hakkila, P., Lowe, A.T. and Smith, C.T. (eds) *Bioenergy from Sustainable Forestry*. Kluwer Academic Publishers, Dordrecht, pp. 19-48.

Hall, P.J. (2002) Sustainable production of forest biomass for energy. *Forestry Chronicle* 78(3), 391-396.

Hammett, A.L. and Murphy, M.A. (2004) Seasonal greenery. In: Burley, J., Evans, J. and Younquist, J.A. (eds) *Encyclopedia of Forest Sciences*. Elsevier Academic Press, Amsterdam, pp. 633-638.

Hollingdale, A.C, Krishnan, R. and Robinson, A.P. (1999) *A Charcoal Production Handbook*. Natural Resources Institute, Chatham Maritime, UK.

Hummel, F.C., Palz, W. and Grassi, G. (1988) *Biomass Forestry in Europe: A Strategy for the Future*. Elsevier Applied Science, London and New York.

IEA (2002) *Sustainable Production of Woody Biomass for Energy*. A position paper prepared by IEA Energy, International Energy Agency, Paris.

Koch, P. (1992) Wood versus nonwood materials in U.S. residential construction: some energy-related global implications. *Forest Products Journal* 42(5), 31–42.

Lee, S. (1996) *Alternative Fuels*. Taylor and Francis, Washington, DC.

Lewin, M. and Goldstein, I.S. (1991) *Wood Structure and Composition*. International Fiber Science and Technology Series, Volume 11, Dekker, New York.

Meil, J., Lippke, B., Perez-Garcia, J., Bowyer, J. and Wilson, J. (2004) *Environmental Impacts of a Single Family Building Shell - From Harvest to Construction*. Corrim Phase 1 Final Report, August 23, 2004, www.corrim.org visited 14 September 2004.

Moss, R.P. and Morgan, W.P. (1981) *Fuelwood and Rural Energy*. Tycooly International Publishing, Dublin.

Musgrave, T., Gardner, C. and Musgrave, W. (1998) The Plant Hunters – Two Hundred Years of Adventure and Discovery around the World. Ward Lock, London.

National Academy of Sciences (1980) *Firewood Crops – Shrub and Tree Species for Energy Consumption*. National Academy Press, Washington, DC.

National Academy of Sciences (1983) *Firewood Crops – Shrub and Tree Species for Energy Production Volume 2*. National Academy Press, Washington, DC.

Newcombe, K. (1976) *A Brief History of Concepts of Energy and the Use of Energy by Humankind*. Centre for Resource and Environmental Studies Publication HEG 1-76, Australian National University, Canberra.

Ninnin, B. (1994) *Principles of Regional Economy Applied to Traditional Fuel Resources. The Example of Five Sahelian Countries: Burkina Faso, Gambia, Mali, Niger, and Senegal.* RPTES, World Bank, Washington, DC.

NZFOA, NZFIC, MAF (2004) *New Zealand Forest Industry Facts & Figures 2003/2004.* NZ Forest Owners Association Inc, NZ Forest Industries Council, Ministry of Agriculture and Forestry, Wellington.

Pereira, H. and Tomé, M. (2004) Cork oak. In: Burley, J., Evans, J. and Younquist, J.A. (eds) *Encyclopedia of Forest Sciences.* Elsevier Academic Press, Amsterdam, pp. 613-620.

Reed, T.B. and Gaus, S. (2001) *A survey of Biomass Gasification 2001*, 2nd edition. The National Renewable Energy Laboratory and the Biomass Energy Foundation, Inc. Golden, Colorado.

Richardson, J., Björheden, R., Hakkila, P., Lowe, A.T. and Smith, C.T. (eds) (2002) *Bioenergy from Sustainable Forestry.* Kluwer Academic Publishers, Dordrecht.

Saddler, J.N. (1993) *Bioconversion of Forest and Agricultural Plant Residues.* CAB International, Wallingford.

Saxena, N.C. (1997) *The Saga of Participatory Forest Management in India.* Center for International Forestry Research, Bogor, Indonesia.

Shanley, P., Pierce, A.R., Laird, S.A. and Guillén, A. (eds) (2002) *Tapping the Green Market.* Earthscan, London.

The Economist (2004a) The future's a gas. *The Economist*, August 28, 2004, 53-54.

The Economist (2004b). The future is clean. *The Economist*, September 4, 2004, 65.

Tiwari, B.K., Tynsong, H. and Rani, S. (2004) Medicinal, food and aromatic plants. In: Burley, J., Evans, J. and Younquist, J.A. (eds) *Encyclopedia of Forest Sciences.* Elsevier Academic Press, Amsterdam, pp. 515-523.

Tsoumis, G. (1991) *Science and Technology of Wood – Structure, Properties, Utilization.* Van Nostrand Reinhold, New York.

United Nations (2001) *World Population Prospects: The 2000 Revision, Vol.1, Comprehensive Tables.* United Nations Publishing, New York.

Walker, J.C.F. (1993) *Primary Wood Processing – Principles and Practice.* Chapman and Hall, London.

Wayman, M. and Parekh, S.R. (1990) *Biotechnology of Biomass Conversion – Fuels and Chemicals from Renewable Resources.* Prentice Hall, Englewood Cliffs, New Jersey.

Zerbe, J.I. (2004) Energy from wood. In: Burley, J., Evans, J. and Younquist, J.A. (eds) *Encyclopedia of Forest Sciences.* Elsevier Academic Press, Amsterdam, pp. 601-607.

Zsuffa, L. (1982) The production of wood for energy. In: W.R. Smith (ed) *Energy from Forest Biomass.* Academic Press, London, pp. 5-17.

Chapter 5

Deforestation and Forest Degradation in the Tropics

Some Definitions

'Deforestation' is the conversion of forest to an alternative permanent non-forested land use such as agriculture, grazing or urban development. FAO considers a plantation of trees established primarily for timber production to be forest and therefore does not classify natural forest conversion to plantations as deforestation (but still records it as a loss of natural forest). FAO does not consider tree plantations that provide non-timber products such as food (e.g. oil palm) to be forest, although they do classify rubber plantations as forest. 'Reforestation' is where a recently deforested area is regenerated either naturally or by planting, and the term is usually, though not exclusively, applied to the re-establishment of plantations. 'Afforestation' occurs when tree plantations are established on non-forested land. 'Natural forest expansion' is where forest establishes on non-forested (often abandoned) land and this is sometimes called 'forest reversion' when forest re-establishes on cleared land that carried forest in the recent past. Forest that re-establishes after deforestation or heavy logging is sometimes called 'secondary' forest, but the distinction between 'primary' (virgin) forest and 'secondary' forest is confusing because it assumes that primary forest is pristine and untouched, which is usually not the case. Forest degradation occurs when the ecosystem functions of the forest are degraded but where the area remains forested rather than cleared. Providing ecological damage is not severe, degraded forest has the capacity to recover.

The boundary between deforestation and degradation is blurred and this complicates the reporting of forested and cleared areas. For example, areas cleared by slash and burn farmers that have been left to fallow may either be abandoned to revert to secondary forest or reused for agriculture at varying frequencies. Harvesting trees for timber production (logging) is not deforestation unless the forest is degraded to such an extent that it re-establishes as grassland or shrub. However, such areas eventually may revert to forest. Logging and fuelwood gathering can accompany deforestation for agriculture and grazing when the trees from clearing operations are utilized rather than burnt or discarded. Logging does not necessarily lead to forest degradation but it often does so in tropical forests. The challenge for sustainable forest management is to harvest the forests without degrading them. The science and technology to do this are available but their application is often superseded by social, political and economic factors as well as land ownership distortions. Poore (1989) estimated that, in 1985, less than one million hectares of tropical forests out of an estimated area of 828 million hectares were demonstrably sustainably managed and there is no reason to believe that this has substantially improved since then. A more subtle form of forest degradation occurs when forests are high-graded (selectively removing boutique species or the better specimens only of desirable species), when fuelwood, litter and other materials are removed from the forest and if invasive pests are not controlled.

Tropical forests are dynamic ecosystems that have advanced and receded with changes in climate over geological time. Indigenous peoples prior to European colonization also modified the forests through slash and burn farming and hunting and gathering. It is misleading to consider the tropical forests as static entities that are here today and may be gone tomorrow (Anderson *et al.*, 2002).

Where and How Much Deforestation

Deforestation at the hands of humans has been occurring since the hunter-gathers but has accelerated since the age of agriculture (Chapter 1). Historically and over extended time-scales, deforestation has not always caused a permanent loss of forest: rather there have been cycles of depletion and recovery. Over the last 10,000 years, humans have reduced the global forest cover from about one half of global land surface area to about one quarter. Most deforestation has occurred in the temperate and sub-tropical domains. Most of the world's remaining forests are located in the tropical and boreal domains (see Fig. 2.4). The temperate world has been through a period of wide-scale deforestation. There was extensive deforestation and forest degradation in Europe in the 17th and 18th centuries and in North America in the 18th and 19th centuries, mainly to provide agricultural land for an increasing population (Chapter 1). Deforestation is no longer significant in the temperate developed world and, indeed, many countries in the temperate world now are recording increases in forest area. By contrast, extensive tropical deforestation is a relatively modern event that gained momentum in the 20th century and particularly in the last half of the 20th century. Today the developed temperate world has low population growth and productive agricultural systems and has no need to clear more forest for agriculture. In many ways the tropical world of today is similar to the temperate world of the 17th to 19th centuries. Forest is being cleared to provide land for agriculture to feed an increasing population and to raise their standard of living and *per capita* consumption. However, tropical forests are also being cleared to provide food and fibre to the developed world.

FAO (2001a) has made a good attempt at estimating recent deforestation and reforestation. It is considerably more difficult to estimate forest degradation and the boundary between deforestation and forest degradation is subject to dispute. A summary of deforestation during the decade 1990–2000 is given in Table 5.1. This table shows there

Table 5.1. Annual gross and net changes in forest area, 1990–2000 (million ha/year). Increase in forest area is the sum of natural expansion of forest area and afforestation (abridged from FAO, 2001b).

Domain	Deforestation	Increase in forest area	Net change in forest area
Tropics	-14.2	+1.9	-12.3
Non-tropics	- 0.4	+3.3	+ 2.9
World	-14.6	+5.2	- 9.4

was considerable deforestation in the world during 1990-2000 but this was almost entirely confined to the tropical regions. Non-tropical regions recorded a net increase in forest area but not enough to compensate for the forest loss in tropical regions. In most instances developed nations are located in temperate domains and developing nations in tropical domains.

Table 5.2 shows the eight worst deforesting countries and the four worst deforesting regions over the period 1990-2000. This table confirms that deforestation is mainly occurring in tropical developing countries and in both moist and dry forest types. Brazil and Indonesia dominate, between them accounting for almost 40% of net forest loss over the decade. However, Africa overall has more deforestation than all other regions, a considerable proportion of this being in the drier forest types. Even though Brazil is the top deforesting country by area, the forests in Brazil are so extensive that this represents a loss of 0.4% per year. By contrast some smaller countries have very high percentage losses per year and risk losing virtually all their forests within the next few decades if current rates are maintained (Table 5.2). Indeed some countries do not even make the list in Table 5.2 because they have already removed most of their forest. Bryant *et al.* (1997) listed 31 countries that have lost all of their frontier forests with those that remain being seriously fragmented and degraded.

Table 5.2. Annual deforestation of the eight worst deforesting countries and the four worst deforesting regions over the period 1990–2000 (compiled from data given in FAO, 2001b).

Net Forest Loss (million ha/y)		Net Forest Loss (% /y)	
Brazil	2,309	Haiti	5.7
Indonesia	1,312	Saint Lucia	4.9
Sudan	959	El Salvador	4.6
Zambia	851	Micronesia	4.5
Mexico	631	Comoros	4.3
Dem. Republic of the Congo	532	Rwanda	3.9
Myanmar	517	Niger	3.7
Zimbabwe	320	Togo	3.4
Africa	5,262		
South America	3,711		
Southeast Asia	2,386		
Central America	958		

Care is needed when looking at trends over time based on published figures because of differences in the standard and nature of reporting. FRA 2000 (see FAO, 2001a) now has a consistent methodology and published trends in the future should be safer. Recognizing the problem of inconsistent reporting, FRA 2000 undertook a study using consistent methodology over an area representing 87% of the world's tropical forests to compare land cover changes between the decades 1980-1990 and 1990-2000. They found that deforestation over this area occurred at the rate of 9.2 million hectares per annum for 1980-1990 and at 8.6 million hectares per annum for 1990-2000 but that the difference was not statistically significant. However, deforestation was significantly less in tropical moist deciduous forest in 1990-2000 than 1980-1990. Achard *et al.* (2002) using satellite imagery concluded that FAO overestimated deforestation of tropical rainforests by 23%. However the definition of what is and what is not forest remains controversial. The tropical rainforests capture most attention but 60% of the deforestation that occurred in tropical forests during 1990-2000 was in moist deciduous and dry forests (FAO, 2001a). It is comforting to realize that many of the dire predictions of future tropical deforestation that were made before 1990 have been shown to be overestimates. The tendency towards hyperbole in predictions has not helped the debate, but there is still no reason for complacency.

Some Differences of Opinion

The deforestation of tropical forests and of tropical rainforests in particular, is an issue that arouses great passion and considerable conflict. To some extent discussion of tropical deforestation has become polarized between those who promote development and those who promote conservation as solutions.

Broadly speaking, those promoting development consider that overpopulation and poverty are the driving forces in tropical deforestation in developing countries. They note that tropical countries of today are in a similar situation to Europe in the 17th and 18th centuries and North America in the 18th and 19th centuries, where and when deforestation occurred to provide agricultural and grazing land to support a rapidly increasing population. They note that today these regions are no longer clearing forest for agriculture, that forest areas are stable or increasing, that standards of living have been raised and that population growth has decelerated. They argue that the way forward for developing nations is to follow the same path and that some deforestation is inevitable in order to achieve a stable agricultural land base. They recognize that people from developing countries resent the suggestion that they should deny themselves the very resources that the developed countries utilized in order to become developed. They do not believe that the developing countries should pay the cost of conserving their forests where the benefits of doing so accrue to the developed world. In summary, they argue that policies, programmes and incentives to promote development are the way to proceed.

On the other hand, those promoting conservation see the conservation of the tropical forest as being the foremost consideration. They believe that the consequences of clearing too much tropical forest would be catastrophic to global life support systems and that the fate of the planet is at stake. They argue that all biodiversity has an intrinsic right to exist and that humans are the guardians of that right. They argue that the extent of clearing is greatly contributing to global climate change and that biodiversity is being recklessly reduced. They argue that overpopulation and poverty are not the main drivers of tropical deforestation but that inefficient use of already cleared land, inequitable distribution of land, unjust systems of land tenure and the disintegration of traditional agricultural systems are the main causes. They say that, in the long-term, the value of keeping the forests is greater than any alternative land uses. They argue that developed countries, power brokers in developing countries, multinational corporations and donor agencies are conspiring to ensure that the developed world can maintain its current levels of consumption.

Regrettably the debate between these extremes has become somewhat entrenched. Clearly both have valid points but neither can claim to own the whole truth. There are no generic off-the-shelf solutions and, as in most social issues, some compromise is necessary to achieve useful outcomes. It is exceedingly unlikely that the tropical forests will be conserved by persuading the developing countries that they have an obligation to do so, at least not until the bulk of the population of these countries are well-fed, well-clothed and well-housed. The issue of tropical deforestation is rich in rhetoric but poor in knowledge.

How Much (if Any) Deforestation is Acceptable

The environmental, social and economic significance of the tropical forests was established in Chapters 3 and 4. There can be no doubt that tropical forests are a resource to be cherished and conserved. Deforestation results in a loss of environmental services, as well as forest products. Also about 300 million people, many of them indigenous, live in or near the edge of the tropical forests and rely on the forests for their livelihood and survival.

Table 5.3. Estimated rates of extinctions resulting from tropical deforestation (reprinted from Reid, 1992, copyright 1992, with kind permission of Springer Science and Business Media).

Estimate	% Global Loss per Decade[a]	Method of Estimation	Reference
One million species between 1975 and 2000	4	Extrapolation of past exponentially increasing trend	Myers, 1979
15-20% of species between 1980 and 2000	8-11	Estimated species-area curve; forest loss based on Global 2000 projections	Lovejoy, 1980
12% of plant species in neo-tropics; 15% of bird species in Amazon basin[c]		Species-area curve $(z = 0.25)$[b]	Simberloff, 1986
2000 plant species per year in tropics and sub-tropics	8	Loss of half the species in area likely to be deforested by 2015	Raven, 1987
25% of species between 1985 and 2015	9	As above	Raven, 1988a; b
At least 7% of plant species	7[de]	Half of species lost over the next decade in 10 'hot spots' covering 3.5% of forest area	Myers, 1988
0.2-0.3% per year	2-3[d]	Half of rainforest species assumed lost in tropical rainforests to be local endemics and becoming extinct with forest loss	Wilson, 1988; 1990
2-13% loss between 1990 and 2015	1-5[d]	Species-area curve $(0.15<z<0.35)$[b]: range includes current rate of forest loss and 50% increase	Reid, 1992

[a]   Based on total species number of ten million. Estimates in bold face indicate the actual loss of species over that time period (or shortly thereafter). Estimates in standard type refer to the number of species that will be committed to extinction during that time period as a new equilibrium is attained.

[b]   See Reid, 1992 for definition of z.

[c]   Extinction estimates apply to the number of species committed to extinction by the year 2000 at the current rates of forest loss. How long it will take for the new equilibrium to be obtained is not known.

[d]   Estimate refers to number of species committed to eventual extinction when species numbers reach equilibrium following forest loss.

[e]   This estimate applies only to hot spot regions, thus the global extrapolation is conservative.

The tropical forests are rich in biodiversity. Table 5.3 (from Reid, 1992) provides a range of estimates of species extinctions made by various authorities over the period 1979 to 1992. These may well be overestimates because they are based on rates of deforestation that we now know to have been overestimates. Also the species area curve method of estimation has a tendency to overestimate (Brown and Pearce, 1994). The number of known extinctions is much less than shown here. However, even if the actual rate is one half of that shown in Table 5.3, it is still alarming. The International Union for the Conservation of Nature maintains a Red List of threatened species on their web-site. Table 5.4 summarizes parts of this. This list shows the number of threatened species both as a percentage of the number of species described and as a percentage of the number of species evaluated. There is a tendency to preferentially evaluate those species thought to be threatened and so the actual percentage will be somewhere in-between the two, and probably nearer the lower. Clearly species, some as yet unknown and undescribed, have already become extinct and will become extinct in the future as a result of tropical deforestation. The number of species that have become extinct as a result of timber harvesting alone rather than clearing for agriculture and grazing is comparatively very small.

Table 5.4. Numbers of threatened species by major groups and organisms in 2004 (from the Red List of IUCN at www.redlist.org accessed on 8 January, 2005).

	Number of described species	Number of species evaluated	Number of threatened species	% threatened of described species	% threatened of evaluated species
Vertebrates					
Mammals	5,416	4,853	1,101	20	23
Birds	9,917	9,917	1,213	12	12
Reptiles	8,163	499	304	4	61
Amphibians	5,743	5,743	1,770	31	31
Fishes	28,500	1,721	800	3	46
Invertebrates					
Insects	950,000	771	559	0.06	73
Molluscs	70,000	2,163	974	1	45
Crustaceans	40,000	498	429	1	86
Others	130,200	55	30	0.02	55
Plants					
Mosses	15, 000	93	80	0.5	86
Ferns	13,025	210	140	1	67
Gymnosperms	890	907	305	31	34
Dicotyledons	199,350	9,473	7,025	4	74
Monocotyledons	59,300	1,141	771	1	68
Others					
Lichens	10,000	2	2	0.02	100

Do we have to conserve all tropical forest and, if not, how much deforestation is too much? Is it inevitable that deforestation must occur as tropical peoples move from poverty to a better quality of life? Does it follow that tropical peoples must have the same experience as the temperate peoples and go through a phase of deforestation before they can stabilize their areas of agricultural land? Should developing nations be denied the opportunity to develop so that the tropical forests can be conserved? Are there socio-political models that will allow developing countries to develop without the necessity for further deforestation? To what extent are the developed nations responsible for tropical deforestation? What responsibility do the developed nations have in assisting developing nations in reducing deforestation? How can this be achieved? Is the urbanized industrial model of the developed nations the most appropriate model anyway? There is no consensus to the answers to these questions but some of the issues surrounding these questions will be discussed in this chapter. The issues surrounding deforestation are a complex of social, economic and political interactions at local, national and international levels. These interactions vary with region and change with time. There are no simple answers.

Forestry and Agriculture

Forestry and agriculture are inextricably linked and any discussion of deforestation cannot be considered in isolation from agriculture. The main reason for clearing forests is to provide land for agropastoral activities. One of the main strategies suggested to reduce deforestation is to promote sustainable forest management. It is probably more important, however, to improve the efficiency and productivity of agriculture. This was one of the main reasons for the stabilizing of deforestation in Europe in the 19th century (Mather, 2001) and the USA in the 20th century (Williams, 1989). Many authorities consider that increasing crop productivity will eventually decrease the need to clear more forest. The green revolution has greatly improved the productivity of cereals by providing improved breeds that respond to fertilization. FAO (2003) is convinced that the green revolution and technological advancements following this have already reduced deforestation. However, the impact of the green revolution has often been remote from the forest frontier where deforestation is occurring and the beneficiaries have usually been the large land owners rather than poor farmers. Also, under some circumstances, increasing crop productivity could have the opposite effect. Increased profitability of agriculture could result in a greater rate of deforestation if it feeds export markets at a constant price or if it displaces labour to the forest/agriculture frontier (Angelsen and Kaimowitz, 2001). Perhaps there are some circumstances where a transitional period of increased deforestation is a necessary precursor to stabilization of agricultural areas.

Globally there is a clear relationship between the clearing of tropical forests and the expansion of tropical agriculture (crops and pastures). FAO (2001a) found that most of the conversion in the area of tropical forests during the period 1990-2000 was to provide land for permanent crops and pastures (Fig. 5.1) but that considerable forest conversion also was to a category called 'other'. Barraclough and Ghimire (2000) discussed the importance of 'other' in assessing the relationship between agriculture and forestry. Other land presumably represents abandoned land, vacant land, waste land, forest degraded to the extent that it has degenerated to shrub or non-pasture grasses, forest that has been cleared and abandoned but will revert to secondary forest, land taken up in urbanization and infrastructure development as well as any land without a designated use for the purpose of compiling statistics. They looked at the period 1950-1992 and listed 14 countries where agricultural area increased and forest and other land decreased, 33 countries where agricultural land and other land increased and forest land decreased, 19 countries where agricultural and forestry areas increased and other land decreased, 21 countries where agriculture and forest areas decreased while other land increased and 7 countries where agricultural area decreased and both forest area and other uses increased.

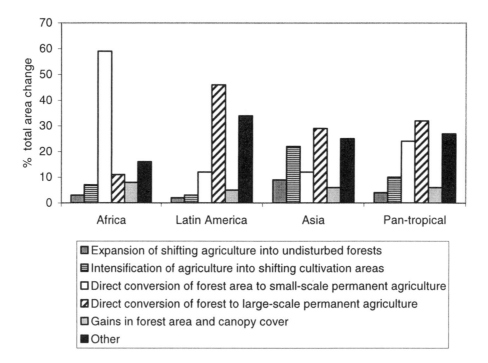

Fig. 5.1. Percentage of total area change of forests by individual change processes at regional and pan-tropical levels, 1990-2000 (from FAO, 2001a).

There is more forest being cleared than land taken up for shifting agriculture, permanent cropping and the establishment of pastures. Regrettably, comprehensive details on what the category 'other' includes are not available. Obviously the pressure on the forests will be reduced if agricultural expansion comes from other land rather than clearing more forest. However, it is not unusual to find that when tropical forests are cleared, the soils are marginal and not suited to agriculture. Cleared forest not taken up by agriculture may return to forest (secondary forest) or may be locked into shrub land or non-forage grasses. Often this is fire-induced. For example, there are about 45 million hectares of cleared tropical forest in Southeast Asia that has regenerated to fire-induced tropical grasses, mainly *Imperata cylindrica* (alang-alang) (Myers, 1994). Perhaps also one of the reasons for the wide disparity in estimates of the relative contribution of shifting agriculture to deforestation is that it is partly embedded and undefined within the 'other' category. The large area of 'other' suggests there is considerable failure and wastage in forest conversion to agropastoral uses.

What are the Causes of Deforestation

The people actually doing the deforesting are often responding in a rational manner to other forces that compel them to do so. They are not necessarily to blame and not necessarily the point at which to address the problem. Direct agents and causes of deforestation are relatively easy to identify but the indirect causes, which are usually the main drivers of deforestation, are the ones that cause most disagreement and the ones that are the hardest to quantify. Table 5.5 shows direct and indirect causes of deforestation.

The causes listed in Table 5.5 will be discussed individually for convenience. However, none of them act in isolation and the degree of interaction between them is large. Causes vary regionally: in general the main causes of deforestation are cattle ranching in Latin America, fuelwood collecting and small-scale agriculture in Africa and permanent crop production in Southeast Asia. In all regions these causes are preceded by or occur in parallel with shifting agriculture.

Table 5.5. Direct and indirect causes of deforestation and degradation of tropical forests.

Direct causes	Indirect causes
Shifting agriculture	Population pressures
Permanent crops	Poverty
Permanent pasture	Transmigration
Logging	Inequitable land and resource distribution
Forest plantations	Development and fiscal policies
Fuelwood gathering	Markets and consumerism
Mining	Undervaluing the forest
Transport	
Dams and reservoirs	
Urbanization	
Fire	
Overgrazing	

Direct Agents and Causes of Deforestation and Degradation

Shifting Agriculture

Shifting agriculture, also called slash and burn agriculture, is the clearing of forested land for raising crops, growing the crops until the soil is exhausted of nutrients and/or the site is overtaken by weeds, and then moving on to clear more forest. Soils in tropical forests are not inherently fertile: there are a relatively higher proportion of nutrients in the standing crop and less in the soil in tropical forests than in temperate forests. Consequently if the standing crop is slashed and then burnt, a high proportion of the nutrients is lost from the site. The burning will give a sudden but short burst of available nutrients and the crop for the first year is likely to be good but then will be reduced in subsequent years. A tropical rainforest site usually cannot be cultivated for cropping without fertilization after one to three years. If the cultivated site is abandoned within the first year or two and allowed to rest for a sufficient length of time (10+ years) to re-establish nutrient cycling and replenish nutrient stocks in the soil, then the damage caused is not great. The abandoned site will re-establish with secondary forest and this is called 'long fallow.' 'Short fallow' is when the abandoned site is re-cultivated after a shorter time. This is insufficient time for the site to regain its fertility and the damage to the site is more serious. If the site is cleared and cultivated for an extended period without fallow, the damage to the site can be very serious unless some form of nutritional management such as fertilization is practised. As the frequency of clearing and burning increases, so to does the risk of soil erosion, pest infestations, disease outbreaks and the encroachment of fire-induced weeds. Shifting agriculture occurs over the whole tropical world from the wet humid tropics to dry areas. In the drier parts of the tropics when crop yields become reduced through declining fertility, the exposed soil becomes vulnerable to wind erosion. Commercial farmers and grazing interests often come in behind the shifting agriculturalists and use the already cleared land for permanent agriculture or grazing.

Shifting agriculture has often been reported as the main agent of deforestation. Most estimates have shifting agriculture being responsible for about one half of tropical deforestation although there are estimates of up to two-thirds. It is not easy to estimate the extent of shifting cultivation because of its dynamic nature and difficulty in defining where the boundary is drawn between areas under shifting cultivation, under fallow and under permanent cropping. However, recent estimates suggest it is now less than it has been (or that it has been overestimated in the past). Figure 5.1 estimates the proportion of different types of forest conversion from 1990-2000. Shifting agriculture was greatest in Asia (about 30%) but only about 15% over the whole tropical world. It appears that the proportion of direct conversion of forest to agriculture is increasing and the proportion of shifting agriculture is decreasing with time.

Shifting agriculture occurs at the forest frontier. About 300 million people live in or around the tropical forests and most of these would be shifting agriculturalists who depend on clearing forests. Shifting agriculturalists represent the poorest, most marginalized and powerless sections of the population. They are predominantly subsistence farmers who must clear forest in order to eat and survive. They may have no land rights at all. They squat on land for which they have no legal tenure. If somebody comes along and claims a right to the land, they have no option but to move on to clear more forest. The result is that the forest frontier retreats further. As shifting cultivators are displaced, they often are pushed towards poorer soils on steeper slopes with higher rainfall. The consequences are high rates of soil erosion, excessive leaching of soil nutrients and the need for longer fallow periods. An integral part of any scheme to arrest tropical deforestation must start with bringing some equity to these victims of the system. Shifting cultivation is the only rational choice for these frontier farmers. Shifting cultivators may sell a small proportion of their crop but they are family concerns where forested land is freely available at no cost, family labour is available but capital is severely restricted. Subsistence is more important to them than profits and consequently they will adopt risk minimization rather than profit maximization. Rather than have a crop failure, which would be catastrophic, they adopt practices like mixed cropping, phased planting, intercropping, crop rotations and alternating crop mixtures (Jepma, 1995). For the shifting agriculturalist, returns per unit of labour rather than per hectare of land are more important. The transition from shifting to settled agriculture requires capital.

Shifting agriculture has been practised all over the world for thousands of years, which suggests that it is a sustainable form of agriculture. However crop yields are very low and the quantity and range of food is restricted. The long-term sustainability of shifting agriculture depends on low land-use pressure. If a traditional shifting cultivation regime comes under pressure, such as from increased population, the alternatives are either to clear ever more forest or to intensify the shifting agricultural system. Intensification means longer periods under cultivation and shorter under fallow until ultimately permanent agriculture is achieved. Intensification cannot be achieved without injection of capital, security of land tenure and the adoption of sustainable nutrient management regimes. The risk in intensification is that sustainable soil management regimes are not implemented. As a consequence, soil nutrients will be depleted and the enterprise will collapse causing the forest frontier to be at risk once more. Also, increasing agricultural productivity does not necessarily mean that deforestation will be reduced. Depending on circumstances, it may have the opposite effect where increased profits in the agricultural sector will stimulate more clearing of forest. Not everybody agrees that the best long-term future for shifting agriculturalists is to make them settled agriculturalists. However, their current state is fairly grim and history would suggest that an agricultural system that delivers higher productivity, increased on-farm income, reduced soil degradation and reduced deforestation is the preferred objective. To achieve all four of these at once is a very difficult proposition. The pathway from shifting agriculture to more advanced forms of permanent agriculture is not smooth and may result in accelerated deforestation prior to stabilization of forested areas.

Permanent Crops

This comprises small to large landowners who either clear forest or take over land already cleared by shifting cultivators. The emphasis is on commercial cash cropping rather than subsistence farming. The relative importance of cash crop production, particularly by large commercial interests, is increasing as a cause of tropical deforestation. Staples are rice in Asia, cassava in Africa and maize in Latin America. Typical cash crops in the tropical world are sugar, coffee, bananas, cocoa, coca, cotton, soybeans, manioc, oil palm, coconut, and rubber. Cash cropping is focused towards fewer crops (including non-food crops) and may be oriented towards an export market. This can have an adverse impact on both rural and urban populations when it replaces a more comprehensive self-provisioning agriculture. Ironically, the clearing of forests to provide more land for agriculture can produce local food shortages. Commercial farmers acquire land tenure and the landless subsistence farmers are forced to relocate to clear more forest. Commercial farmers occupy the better soils and the subsistence farmers relocate on poorer soils. The relocated farmers may have no experience of poorer soils and may be unaware that they need longer fallow periods to restore fertility. Also, commercial farming requires less local labour which further marginalizes and displaces the rural poor.

Forest conversion to support large scale permanent crop production is most pronounced in Southeast Asia and small scale permanent crop production is most pronounced in Africa. There was acceleration in the establishment of oil palm (*Elaeis guineensis*) in the 1990s, particularly in Malaysia and Indonesia which between them had 82% of the world production in 2002 (Fig. 5.2). Malaysia is a major exporter and Indonesia is rapidly becoming so (Fig. 5.3). Up until 2000, more than 50% of Indonesia's palm oil was consumed domestically. Only a small proportion of tropical soils (alluvial and volcanic) are suitable for continuous cropping and one of the tragedies of the tropics has been that large areas of forest have been cleared only to find that they are unsuitable for permanent agriculture. Even so, progress has been made in improving the quality of less fertile soils and the challenge is to ensure that this can be sustained. Conversion of undisturbed forest to crop land increases the likelihood of soil erosion but some cropping systems in the humid tropics are quite successful at controlling soil erosion (Critchley and Bruijnzeel, 1996).

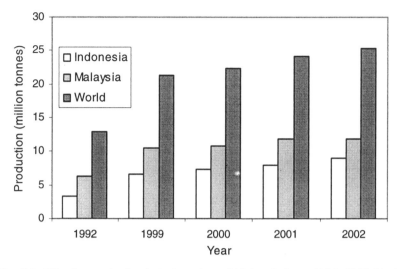

Fig. 5.2. Oil palm production in Indonesia and Malaysia from 1992–2002 (derived from FAO statistics on-line).

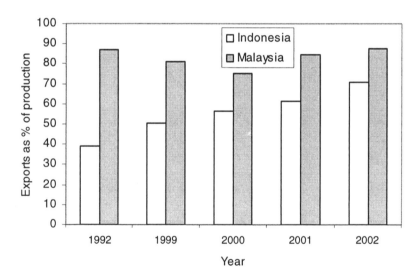

Fig. 5.3. Oil palm exports as a % of production in Indonesia and Malaysia from 1992–2002 (derived from FAO statistics on-line).

Permanent Pasture

The conversion of tropical forest to permanent pasture for mainly cattle grazing predominates in South America and in Central America. For example FAO 2004 statistics on-line show that Brazil increased its area of permanent pasture by 50 million hectares in the period 1961 to 1980 and by a further 25 million hectares from 1980 to 2002. The cattle herd size and cattle stocking intensities increased greatly over the same period (Table 5.6). The rapid expansion from 1961 to 1980 was a result of subsidies offered by the Brazilian government. Forest clearing, however, still persisted at a lower rate following the scaling down of the subsidies in the 1980s. Central America increased its area of permanent pasture by 7.5 million hectares between 1980 and 2002.

Table 5.6. Cattle herd sizes and stocking intensities in the Brazilian Amazon from 1970 to 1995 (extracted from Anderson *et al.*, 2002).

	1970	1975	1980	1985	1995
Herd size	6,461	9,390	15,220	18,999	36,045
Cattle per km^2 of private pasture	27	31	39	44	70

Cattle ranchers either directly cut and burned the forest or purchased land from small-scale farmers or land speculators at the frontier. Cutting and burning of rainforest to produce pastures (and the repeated re-burning) is similar to slash and burn agriculture. The initial burst of nutrients provided good pasture, but fertility fell away on the poorer soils unless some form of remedial treatment including fertilization was applied. Overgrazing was also common. Consequently large areas of so-called permanent pasture became degraded to shrubs and non-forage grasses. The rehabilitation of degraded lands is one important way of arresting the rate of deforestation of tropical forests.

Cattle ranching has often been claimed to be uneconomical unless propped up by subsidies. However, in the Amazon, subsidies have been removed and forests continue to be cleared to expand cattle ranching. Apparently cattle ranching has developed into an economical form of land use. Ranching is a relatively flexible and low risk investment and more profitable than any current alternative investment, such as timber production and extractives, that keeps the forest intact. Timber production is a less attractive investment because it is long-term, inflexible and does not return cash to the owner for the non-timber values. By contrast, cattle ranching requires little labour and has well established markets without the volatile price fluctuations of cash crops. There is also a cultural dimension to cattle ranching in Latin America. The cowboy is often looked upon with respect, almost as a cult hero. Cattle ranching, however, has the socially undesirable consequence of large areas of land being held in the hands of few people with limited prospects of offering employment to the rural and displaced poor. Also there has been a considerable amount of degraded land created from unsuccessful conversion of forest to pasture.

Logging

Fuelwood is the main wood product extracted from tropical forests and this will be discussed under its own heading later in this chapter. World production of tropical logs and export of tropical logs and processed products has been falling over the last five years (ITTO, 2001, 2002, 2003). Most log production is in the Asia-Pacific region but production is falling. Production is relatively small and decreasing in Africa but is increasing in Latin America (Table 5.7). In 2003, 55% of log production in the tropics came from the Asia-Pacific with Latin America and Africa contributing 30% and 15% respectively. More than one half of the tropical log production for timber products is consumed domestically. In 2003, the Asia-Pacific exported 56% of its production of logs and processed products (sawnwood, plywood and veneer) while Latin America and Africa exported 12% and 46% respectively. The major importers are countries from eastern Asia (China, Japan and Taiwan) and the European Union (Fig. 5.4). Japan's imports appear to be decreasing while China's are increasing. Japan is importing increasing amounts of conifer logs and sawnwood from Europe and Russia. In 2003, China was the largest importer of tropical logs and sawnwood and Japan the largest of tropical plywood. Most industrial roundwood production and trade is in the non-tropical world. In 2003, 88% of production, 93% of imports and 77% of exports were from non-tropical sources (ITTO, 2003).

Table 5.7. Composition of exports from tropical regions (million m^3 round wood equivalent; from ITTO, 2003; data from ITTO producer countries; totals may not sum exactly due to rounding).

Region	Log Production		Log Exports		Processed Exports		Total Exports	
	2001	2003	2001	2003	2001	2003	2001	2003
Africa	20.1	19.0	4.7	4.7	4.0	4.1	8.7	8.8
Asia-Pacific	86.0	74.0	11.5	8.6	36.0	32.9	47.5	41.5
Latin America	37.7	40.2	0.2	0.1	3.6	4.8	3.8	4.9
Total	143.8	133.2	16.4	13.4	43.5	41.8	59.9	55.1

Logging usually does not cause deforestation but can cause forest degradation. Logging in Latin America is often of low intensity, taking only selected commercial species. Logging in Southeast Asia is more intensive and can be quite destructive. Ironically it is not what is removed from the forest that causes the damage but what is left. Considerable damage can be done to the trees remaining after logging and soil disturbance and erosion can be extensive. Road construction is a major cause of soil erosion and also opens up the frontier to agents of deforestation, particularly shifting cultivators. Logging can modify the composition of the forest as can silvicultural intervention. From an ecological perspective this may or may not be a bad thing. Indeed some silvicultural techniques are specifically designed to modify species composition and to promote the regeneration of desired species. There are two indices of sustainability that are used in forest management. The first of these is 'sustained yield' where log removals should not be at a greater rate than the capacity of the growing forest to replace it. The second is 'ecologically sustainable forest management' where management should not impair the ecological functions of the forest. The second is more difficult to achieve and to monitor than the first. Regrettably most tropical forests are not being sustainably managed under either definition. Some conservationists believe that the tropical rainforests are so complex and fragile that ecologically sustainable management is impossible and that all tropical forests should be reserved. However, this is an unsustainable argument. Developing countries with large areas of forest will use their forests to develop and increase their standard of living. Owners of the forest cannot be expected to forego the benefits of clearing the forest unless they are adequately compensated for doing so. The better way to conserve the forest is to provide an option that

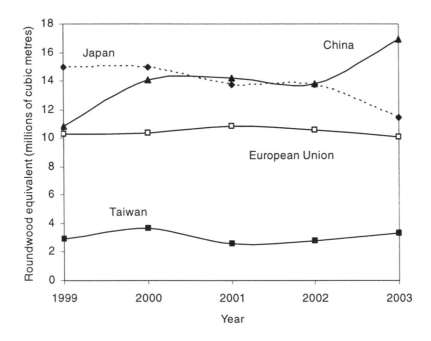

Fig. 5.4. Imports of tropical timbers by selected countries (unpublished data supplied by ITTO Economic Information and Market Intelligence).

is more attractive than clearing. One, and perhaps the only option is ecologically, economically and socially sustainable forest management where the real values of the forest are recognized and where the full range of beneficiaries meet the costs. This is a daunting but worthy aim by no means impossible. There has been a long history of research into the management and silviculture of tropical forests (Dawkins and Philip, 1998). There is little doubt that ecologically sustainable forest management is technically achievable and that it has been successfully practised in tropical forests in isolated cases. Also, there are very competent and well trained foresters in the tropical world. The reason ecologically sustainable forest management is rarely practised is that currently it is not economically and socially sustainable. Land use options to deforest are often more profitable. Logging is usually done by agents with a short time horizon and no vested interests in the long-term future of the site. Unsustainable logging propped up by weak government, powerful minorities and corruption gives greater short term profits. The way forward to arrest tropical deforestation and forest degradation is to promote ecologically, economically and socially sustainable forest (and agriculture) management. Ways of doing this will be discussed later in this chapter.

Forest Plantations

Plantations will be discussed in greater length in Chapter 7. On balance, plantations are a positive benefit and should assist in reducing the rate of deforestation. Plantations are often intensely managed and more productive than native forest. Consequently they can take the pressure off the native forest for producing wood for fuel, pulp, logs, sawnwood and other processed wood products. Plantation products could supply the volume commodity market while the native forest could supply the lower volume high quality end of the market. About one-half of the plantations in the tropics are established on native forest cleared for the purpose. This is unfortunate. Fortunately, however, plantation establishment is often first directed at unproductive forest fallow and forest that has been seriously degraded by irresponsible logging. Even so, every encouragement should be given to plantations being established on already cleared land. Plantation establishment is an obvious choice for rehabilitating degraded and waste lands wherever this is possible. Plantations are sometimes criticized as being a less biodiverse alternative to native forest, which is of course true. However, plantation establishment on a smaller land base reduces the need to degrade native forests over a larger land base. Plantation establishment can promote deforestation by constructing roads that improve access of shifting cultivators and others to the forest frontier. The proposition that sufficient plantations should be established to meet 100% of wood requirements is arguably flawed because this will not stop deforestation. The conservation of the tropical forests requires that they be given a value that is equal to or exceeds the alternative of clearing it. Sustainable forest management for a range of values including timber production and extraction of non-wood forest products is the way to proceed.

Fuelwood

Fuelwood is discussed at greater length in Chapter 4. Fuelwood (including charcoal production) comprises over 80% of wood removed from tropical forests and in some regions (e.g. sub-Saharan Africa) fuelwood can comprise more than 80% of energy consumption. Fuelwood is not usually the major cause of deforestation in the humid tropics although it can be in some populated regions with reduced forest area such as in the Philippines, Thailand and parts of Central America. For example, fuelwood gathering was considered to be the main cause of deforestation and forest degradation in El Salvador where three-quarters of the population used fuelwood at a national rate of 3.1 kg per person per day, the highest in Central America (Heckadon, 1989).

In the drier areas of the tropics, fuelwood gathering can be a major cause of deforestation and degradation. In the 1980s there were forecasts of dire fuelwood shortages and these have not eventuated. However, there is no reason for complacency. Fuelwood gathering, along with small scale agriculture, is the main cause of deforestation in tropical Africa. Fuelwood shortages are increasing in Africa, particularly in sub-Saharan Africa, owing to increased consumption (Fig. 5.5) and decline in forest area.

Considerable amounts of fuelwood are gathered outside the forest and therefore do not contribute to deforestation. Fuelwood can be obtained from isolated trees and as residue from wood processing plants. It is also obtained as a by-product of forest clearing. The impact of fuel gatherers on the forests can range from negligible to drastic as gatherers

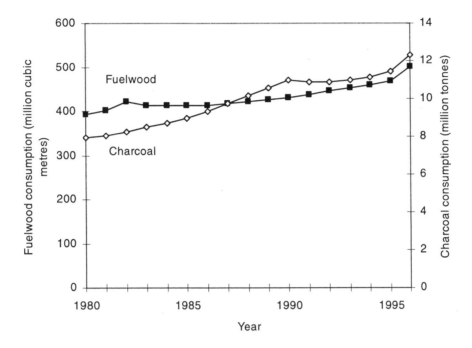

Fig. 5.5. Consumption of fuelwood and charcoal in tropical Africa from 1980-1996 (derived from FAO statistics on-line).

graduate from gathering woody litter, to removing live branches, to stripping bark to felling whole trees. Prolonged fuelwood gathering from the forest can alter the composition of the forest and reduce its regenerative capacity. The intensity of fuelwood collection increases with population and is most pronounced in forests that are unregulated and used as communal property. Usually *per capita* consumption in urban areas is less than in rural areas but the pressure of population in urban areas can result in serious deforestation in the proximity of cities. Urban populations have the tendency to remove whole trees while rural populations, more attuned to the necessity to keep the growing stock, take branches.

Mining

Mining is very intensive and very destructive. The area of land involved is quite small and it is not seen as a major cause of primary deforestation. Rather the environmental problems caused by mining are reductions in water quality and adverse effects on indigenous peoples. Mining is a lucrative activity, promoting development booms which may attract population growth with consequent deforestation. Roads constructed to support the mining operations will open up the area to shifting agriculturalists, permanent farmers, ranchers, land speculators and infrastructure developers. If wood is used as a fuel in mining operations, and unless it is sourced from plantations established for the purpose, it can cause serious deforestation in the region. On the other hand, mining can be labour intensive and take labour away from forest clearing. Wealth created by minerals and oil can encourage rural to urban migration, thereby decreasing pressure at the forest frontier and reducing deforestation (Wunder, 2003).

Transport

The construction of roads, railways, bridges and airports opens up the land to development and brings increasing numbers of people to the forest frontier. A prime example is the Trans-Amazon highway which, more than any other factor, was responsible for the deforestation in Amazonia. Cattle ranching is often blamed as the primary cause but deforestation from whatever cause was inevitable after the road was built and the government instituted aggressive expansionary policies to encourage the movement of people into the area. Eighty-five per cent of deforestation has occurred within 50 km of the highway. Currently the Brazilian government has plans to complete the paving of the highway (The Economist, 2004) which will greatly increase access. There is the fear of another rush to deforest and displace indigenous peoples and small farmers. It is widely recognized that the building of the highway had disastrous environmental consequences. There is a meeting of the minds at all levels of Brazilian government, multinationals and environmental NGOs, to ensure that, with careful planning and cooperation, this will not happen again with the paving of the last section of the highway. Discussions are still short of consensus but they are encouraging. The rate of deforestation in Costa Rica was estimated to have increased 5 times along the Pacific slope following the construction of the Inter-American highway (Silliman, 1981). Utting (1993) considered that one of the reasons why El Salvador lost most of its forest resources before other countries in the region was that it was more advanced than the other countries in road and bridge construction.

Dams and Reservoirs

The creation of reservoirs for water supply and for the generation of hydroelectricity can flood forested land. Often dam construction meets considerable resistance from those wishing to arrest deforestation and ecological damage as well as those who have land uses that will be affected by changes in water flow. However, the generation of hydroelectricity is a relatively clean and sustainable form of generating energy. Any deforestation resulting from it should be considered alongside the environmental consequences of having alternative forms of energy production. The construction of rights of way for power transmission lines also causes deforestation.

Urbanization

Expanding cities and towns require land to establish the infrastructures (housing, roads, utilities, golf courses) necessary to support growing populations. The influence of expanding populations on deforestation and degradation extends further than the residential fringe because of pressure on the forests for fuelwood and readily accessible agricultural land.

Fires

Fires are a major tool used in clearing the forest for shifting and permanent agriculture and for developing pastures. Fire provides an initial flush of nutrients but the net effect is a loss of nutrients from the site. Also burning of forest vegetation increases carbon dioxide concentrations in the atmosphere and this has implications for climate change which will be discussed in Chapter 6. Fire is a good servant but a poor master. Fire used responsibly can be a valuable tool in agricultural and forest management but if abused it can be a significant cause of deforestation. Continual burning will promote pasture over forest and repeated burning of pastures can degrade them and reduce their productivity. Consequently more deforestation is required to replace lost productivity. Also degraded pastures easily burn and encroach into the forests. Uncontrolled fires initially set by humans can create havoc. For example the extensive and damaging fires in Sumatra and Kalimantan in 1997 and 1998 started as fires escaping from burning operations associated with land clearing. Fires are responsible for the regeneration of very large areas of cleared or degraded forest in Southeast Asia to non-forage grasses such as *Imperata cylindrica*.

Overgrazing

Overgrazing occurs when the carrying capacity of the pasture is exceeded and pastures lose their productivity, degrading to non-palatable grasses and shrubs. The degraded pasture may be abandoned and may or may not regenerate to forest. It is commonly claimed that overgrazing has contributed to pasture degradation in Brazil. Table 5.6 shows that both cattle numbers and stocking intensities in the Brazilian Amazon greatly increased from 1970 to 1995. However Anderson *et al.* (2002) noted that the stocking intensities are still low compared with other parts of Brazil and internationally. This suggests that land is not scarce, and that overgrazing, though acute in specific areas and circumstances, is not an overall major problem in the Brazilian Amazon. Long-term sustainability requires that pastures be cared for and not overgrazed. Overgrazing is more common in drier areas of the tropics where pastures degraded by overgrazing are subject to soil erosion. Stripping trees to provide fodder for grazing animals can also be a problem in some dry areas of the tropics but is probably not a major cause of deforestation.

Indirect causes

Population

The role of population in deforestation is a contentious issue. Some consider that blaming population and poverty for deforestation is an excuse not to tackle the root causes such as inequality of land distribution and non-use of already available cleared land (Colchester and Lohmann, 1993). Also there are many examples of where deforestation is occurring in areas where population is low (Westoby, 1989). However, there is good evidence that rapid population growth is a major indirect and over-arching cause of deforestation. More people require more food which requires more land for agriculture. This, in turn, results in more clearing of forest. There are thousands of years of history to support this connection (Chapter 1). In earlier times the impact of increase in population on deforestation was felt near to the location where the increase in population occurred. This was because of poorly developed transport and communication. In modern times the impact of increased population may or may not be felt at the site of deforestation. For example, Palo *et al.* (2000) in a study of 67 tropical countries spread across the tropical world found that population and income at the national level were the most important factors explaining deforestation at the sub-national (regional) level. In Africa, and particularly sub-Saharan Africa, deforestation is mainly caused by small-scale cultivation and cattle herding and by fuelwood gathering. Deforestation in sub-Saharan Africa is at its most intense where increasing populations have operated in the area over an extended period and crop productivity has steadily declined. El Salvador is cited as a classical case study where population pressure is related to deforestation. The population of El Salvador increased from less than one million to more than 5 million during the 20th century and El Salvador became the most densely populated nation in Latin America (Utting, 1993). Population growth, however, does not act as an isolated cause. Rather, it operates in conjunction with other indirect causes, such as transmigration, inequitable land distribution and uneven distribution of wealth.

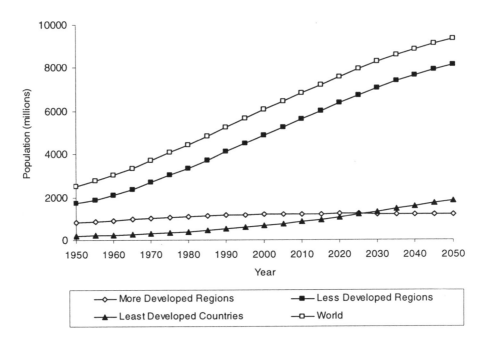

Fig. 5.6. Estimates of world population from 1950-2000 and predictions from 2000-2050. Drawn from data supplied by United Nations (2001). Predictions based on median variant. Most developed regions are Europe, North America, Australia/New Zealand and Japan. Less developed regions are Africa, Asia except Japan, Latin America and the Caribbean, Melanesia, Micronesia and Polynesia. Least developed countries are as defined by the United Nations General Assembly in 1998 and comprise 48 countries of which 33 are in Africa, 9 in Asia, 1 in Latin America and the Caribbean and 5 in Oceania.

Malthusians argue that if population increases exponentially, food production increases only linearly and that the inevitable consequences are misery, starvation and a savage population correction. The extension of this argument is that accelerated deforestation must occur in an attempt to maintain the food supply to an ever increasing population. However, because of increased agricultural productivity and increased real incomes, these dire consequences have not eventuated. Neoclassical economists argue that population growth need not hamper development but indeed could enhance it by providing more demand, more labour, better skills and technology, economies of scale and lower production and distribution costs (Palo, 1994). However, some ecologists and conservationists point out that many of the natural resources consumed to support the non-Malthusian model are finite and that, despite lip-service to the contrary, sustainable management of natural resources is not occurring now and may not do so in the future. The relationship between development and deforestation is complex and dynamic and can go either way depending on circumstances (see later).

During the second half of the 20th century it appeared as though the world population was increasing exponentially. Current predictions, however, suggest that the relationship extrapolated to year 2050 is sigmoidal and that population growth rates are falling (Fig. 5.6). Even so, the world population is expected to increase by about 50% over the next 50 years. Population growth is dominated by the less developed regions: nearly 90% of the population in year 2050 is expected to come from what are now the less developed regions. The rate of population growth in the least developed countries (which are almost all tropical countries) is predicted to continue to rise to 2050 and beyond. Population growth in the developed regions is much less than in the developing regions and is predicted to be negative after 2040. Arguably increasing population is the biggest challenge of all to sustainable management of human life support systems and controlling population growth is perhaps the best single thing that can be done to promote sustainability. There is a strong negative relationship between stage of development (as measured by affluence and literacy) and population growth. This supports the notion that programmes to support development ultimately will control populations and reduce deforestation of tropical forests. The history of temperate Europe, North America, Australia and New Zealand supports this. This suggestion, however, is strongly contested by some.

Poverty

Poverty and overpopulation are inextricably linked. Deforestation is affected mainly by the uneven distribution of wealth. Shifting cultivators at the forest frontier are among the poorest and most marginalized sections of the population. They usually own no land and have little capital. Consequently they have no option but to clear virgin forest, which comes at no cost, in order to eke out a fairly miserable existence. The shifting cultivators cannot escape this fate unless they can find the capital to sustainably intensify their agriculture and increase land productivity. There is a circular argument about poverty and its relationship to deforestation: development will/may ultimately reduce deforestation as well as poverty, but national poverty restricts the capacity of a country to develop. Also there is the argument that increased development resulting from increased income can increase deforestation under certain conditions. Palo *et al.* (2000) found that increasing *per capita* income increased deforestation up to a threshold of about USD 1500 after which deforestation decreased.

Transmigration

Transmigration of people to the forest frontier, whether forced or voluntary, the result of development policy or dislocation from war, is a major indirect cause of deforestation. Dispossessed and landless people bring increased population pressure to the forest frontier. The migrants may not be farmers or may come from an area where the farming conditions are quite different. They may be unfamiliar with the traditional shifting agricultural practices of the new area. They may further accelerate deforestation by clearing more marginal land and shortening or eliminating the fallow periods thereby promoting soil infertility, soil erosion and land degradation. Relocated peasants are sometimes called 'shifted' rather than 'shifting' agriculturalists. Another unfortunate outcome of transmigration is that indigenous peoples may be forced off their land (Downing *et al.,* 1992).

In southern and eastern Asia and Latin America there have been official programmes where millions of families have been relocated into forested areas for relieving population pressure in congested areas and to open up new areas for development. Planned settlement programmes have been promoted in Paraguay, Bolivia, Peru, Ecuador, Colombia, Venezuela, Brazil, Panama, Costa Rica, Nicaragua, Honduras, Guatemala, Mexico, Ethiopia, Indonesia, Malaysia, Philippines, Bangladesh and Vietnam (Colchester and Lohmann, 1993). For example, the aggressive expansionary policy of the Brazilian government in the 1960s resulted in the non-indigenous population of the Brazilian Amazon increasing from 2 to 18 million during which time there was an unprecedented rate of deforestation, greater than in any other country (Anderson *et al.,* 2002). In Indonesia, an official transmigration programme from the late 1960s to the mid 1980s sponsored the resettlement of over 500,000 families from Java to mainly Sumatra, Kalimantan and Irian Jaya, which resulted in an estimated one million hectares of forests being converted to agriculture (Pearce *et al.,* 1990). Other families voluntarily followed and their lack of sponsorship combined with unfamiliarity led to less sustainable and more damaging land use (Jepma, 1995). There have been several wars in Central America and in Africa since 1980. Refugees from war search out new agrarian frontiers and may cross national borders to find land to grow their crops. For example, during 1966-1980 settlers from Mali and Burkina Faso moved along logging roads into Côte d'Ivoire and increased the population of these forested areas by 600%, doubling the area of forest damaged by slash and burn agriculture. The migrants were unfamiliar with forest agriculture and caused massive and irreversible deforestation (Martin, 1991).

Just to confuse the issue, the move in the opposite direction from rural to urban areas is also occurring in tropical countries. Indeed the rural to urban migration in Latin America increased from 26% in urban areas in 1950 to 70% in the 1990s, the highest in the developing world. Rural to urban migration occurred in the developed countries in the 18th and 19th centuries and came at the cost of considerable social and environmental upheaval as these countries moved from a rural agrarian culture to an urban industrialized culture. However, deforestation has now ceased in these countries. There is the possibility that a similar transition will have the same outcome in the tropical countries but there can be no guarantees.

Inequitable Distribution of Land and Resources

Cultivators at the forest frontier often do not hold title to land and are displaced by others who gain tenure over the land they occupy. This means they have to clear more forest to survive and the cycle continues. Shifting agriculturalists on untitled land cannot get bank loans and therefore are unable to raise the capital to establish permanent agriculture. Also, without tenure, the land occupiers have no incentive to engage in long-term forest protection and improvement practices. Sometimes land tenure will only be granted to untenured land if it has been 'improved'. Unfortunately deforestation is often considered to be an improvement for this purpose. Land speculators take advantage of this and clear forests to obtain title which they sell on. In the tropical developing world large tracts of land are owned and controlled by the privileged few who have strong political connections and who are very efficient at resisting any attempts at land reform to remove these inequities. Another problem is that land owned by the poorer sections of the community is often too small in area to become an economic production unit. An example is given in Table 5.8 for the distribution of rural families by size of land holding in Central America in 1970 following the export cotton, beef and sugar booms of 1950 to 1970.

Table 5.8. Distribution of rural families by size of holding in Guatemala and El Salvador in 1970 (derived from Utting, 1993).

Size of holding (ha)	Guatemala (%)	El Salvador (%)
Landless	26.6	26.1
Less than 0.7	15.0	24.4
0.7-4	42.3	36.2
4-350	15.7	13.1
More than 350	0.4	0.2

Those with the power to reform the land either own the land or have the patronage of those who do. It is difficult to correct this and usually attempts at land reform fail. Governments frequently address the problem of inequitable land distribution by sponsoring resettlement schemes which just adds to the problem and promotes further deforestation. There is no doubt that social injustice has a large part to play in deforestation. Inequitable distribution of capital keeps the peasant farmer locked into subsistence farming and forest clearing. Peasant farmers at the forest frontier may be living in abject poverty and struggling to grow enough food to support themselves in a nation that has well developed and efficient food production in developed agricultural areas. Some argue that population pressure in itself is less important in explaining deforestation in the tropics than is the marginalization of small farmers caused by inequitable distribution of land and resources, structural changes in agriculture, excessively small holdings, and contraction of rural labour markets (Downing *et al.*, 1992).

Development Policies

The relationship between development and deforestation is complex and dynamic. One point of view is that development will increase land productivity and thereby reduce the need to clear forests to meet food requirements. Another is that development will provide further capital and incentive to expand and clear more forest. Both are probably true depending on circumstances. The former may be the case when constrained by a fixed food demand. The latter may be the case when food demand cannot be satisfied owing to a continuing export market and rising internal population with rising levels of consumption. The transition argument is that the latter will precede the former and therefore some degree of deforestation is first necessary to ultimately reach a non-deforesting equilibrium. In any case it is unreasonable to expect the rural poor in developing nations to remain poor in order to save the forests. The challenge is to develop policies that both assist the rural poor and save the forest. Successful policies are most likely to be those that directly assist the rural poor at the forest frontier. Many development policies have failed because they have supported, either wittingly or unwittingly, the development of those who already have land, power, influence and political clout. This further alienates the rural poor and puts the pressure back on the forests.

The Brazilian government made the conscious decision to colonize the Brazilian Amazon and they did so by constructing highways and railroads, by encouraging transmigration into the area and by providing subsidies for large-scale agro-pastoral activities, mainly cattle ranching. These subsidies were in the form of grants, tax holidays, subsidized credits, exemption from import taxes and land concessions. The government was successful in meeting its development objective but at the expense of 15% of the Brazilian Amazonian rainforest being cleared by 1995. Large scale agro-pastoral activities were the main causes of deforestation in the Amazon and small-scale farmers and shifting agriculturalists had a lesser role than in other areas of the tropics. The Amazon has now developed to the extent that it has a life of its own and no longer needs nor receives subsidy to support it. The Amazon continues to be the world's most deforesting region (Table 5.2).

Subsidies have been used to improve agricultural productivity, such as subsidies for fertilizer to aid the development of oil palm plantations in Indonesia. Generally over the tropical world, subsidies to improve agricultural activity have gone to the richer portion of the population and to large scale corporate agriculture rather than to the peasant farmers who need it. Subsidies for fertilizer may contribute to land degradation if they reduce the farmer's incentives to adopt sustainable soil management practices. There has been considerable subsidy to improve irrigation in developing countries and this has resulted in increased agricultural productivity albeit sometimes at considerable environmental cost. Price support and price stabilization schemes can assist in developing agriculture but care needs to be taken that they are not propping up an inherently unprofitable venture. Ironically, world recession can aid in combating deforestation if it slows down export oriented trade in food and forest products. Another form of subsidy is the injection of foreign aid and investment. The Central American cattle boom of the 1960s and 1970s was strongly supported by aid from the World Bank and the Inter-American Development Bank. Subsidies tend to concentrate agro-pastoral activities into large focused operations at the expense of smaller more diversified operations. The external debt in the developing world is currently about 2 trillion US dollars. Brazil (USD 235,000 million) and Mexico (USD 140,000 million) are among the most indebted of nations. Brazil is the most deforesting nation on earth and Mexico is the fifth (Table 5.2). External indebtedness diverts financial resources from initiatives that could conserve the forests and tempts countries with extensive forest cover to service their debts by accelerating deforestation.

In many tropical countries concessions are given for logging tropical forests with little consideration being given to regeneration and long-term forest management. Often the number of years of the concession is less than the rotation length for sustained yield. Consequently the concessionaire has no vested interest in the long-term benefit of the forest. Under these circumstances it is inevitable that the concessionaire will adopt a get in and cut out policy, doing as little as possible to meet the mandated environmental standards.

There has been an increased globalization of forestry over the last decade and there are a range of global instruments and conventions arising mainly out of the United Nations Conference on Environment and Development (UNCED) in 1992 and the United Nations Forum of Forests (UNFF) established in 2000. The numerous conventions and international organizations comprise a bewildering variety of acronyms but demonstrate the growing trend of international collaboration. Immediately the battle lines are drawn between the proponents of globalization who see it as a rational response to efficiently manage natural resources and to those who disagree, seeing globalization as yet another means for developed nations to subjugate developing nations and for the developed nations to maintain inequitable and unfair levels of consumption.

There are of course policies that have achieved their aim and resulted in benefiting the rural poor and reducing the rate of deforestation. Policies aimed at improving access to better quality land, providing cheap credit, guaranteeing land tenure and providing advice and assistance on more productive farming methods sometimes work, and work well. There are many success stories. For example, rubber was introduced into western Kalimantan without clearing primary forest and forest cover increased in areas where land use became less intensive as a result (de Jong, 2001). Also, lowland irrigation projects in the Philippines provided employment to upland residents and improved their welfare. Less time was

Table 5.9. Changes in human welfare and forest cover in the Brazilian Amazon (derived from Anderson *et al.*, 2002).

Changes in human welfare	Changes in forest condition
Life expectancy increased from 50 to 61 (1970-1991)	335 000 km^2 of forest converted to mainly pasture (1970-1991)
Literacy rates increased from 56% to 72% (1970-1991)	Biodiversity irreversibly reduced by an unknown amount
Infant mortality fell from 124 deaths per 1000 to 57 per 1000 (1970-1991)	3.3 billion tonnes of carbon dioxide poured into the atmosphere (1970-1995)
Rural GDP *per capita* increased at 5.8% per year in real terms (1970-1995)	15% of forest in Brazilian Amazon cleared overall
Rural GDP per hectare increased by 2.6% per year from USD 106 to USD 203 (1970-1995, expressed in 1995 dollars)	

spent clearing upland forests to grow cash crops (Shively and Martinez, 2001). It would be a mistake to have the impression that all development policies are perverse and that all interventions and subsidies are failures. Indeed a lack of policy is a greater threat. Weak government institutions with no coherent policies and government officials who lack the will and the means to combat land rich elites also aid in promoting deforestation. Much can be learned from failed policies. Policies that focus on the least privileged and the rural poor are the most likely to benefit the forests.

Achieving the appropriate balance between human welfare and conservation is difficult with a range of opinion. The Brazilian Amazon serves as an example (Table 5.9) where increases in human welfare have been achieved at the expense of the forest. However it should not automatically follow that that human welfare can only be increased at the expense of the forest. The challenge is to find a win-win situation. The strict anti-development stance taken in much of the literature appears to offer no constructive solutions. Arguably the best way forward is to promote sustainable development which by its very definition incorporates environmental protection.

Markets and Consumerism

Rampant consumerism by the developed countries frequently has been claimed as a major reason for tropical deforestation. The opening of tropical countries to the world commodity markets accelerated deforestation. The products include coffee, sugar, bananas, cotton and beef in Central America and oil palm, rubber and timber in Southeast Asia. The relative importance of export markets, however, is decreasing with time. There is a general trend for an increasing proportion of food and fibre to be consumed domestically or exported within the tropical world to meet increased population and increasing *per capita* consumption. For example, Central America is an importer of maize which is the main crop planted by the slash and burn farmers who clear the forest. Indonesia (the number two deforesting nation) is a net importer of rice and Brazil (the number one deforesting nation) consumes more than 80% of its beef production. There has been extensive deforestation in Central America to support cattle ranching (and cotton, sugar and banana production) but Central America consumes almost all of its own beef and is a net importer (Table 5.10).

Table 5.10. Production, imports and exports of beef and veal (Mt) in Central America from 1980 to 2002 (derived from FAO statistics on line).

	Production	Exports	Imports
1980	1,061,096	29,452	1,809
1985	1,213,576	28,934	10,675
1990	1,464,220	22,555	30,552
1995	1,764,645	5,691	9,674
2000	1,755,306	3,607	34,732
2002	1,806,818	9,291	34,963

The USA is the major producer of beef in the world, the largest exporter and a net exporter by value. There may have been a hamburger connection between the USA and Latin America prior to 1980 but now it is disconnected. Exports of timber from the tropical world continue but these are dominated by exports to eastern Asia (Fig. 5.4). There is considerably more timber traded within the temperate developed world than traded from the tropical to the temperate world. The largest producers and exporters of timber products are in the developed cool temperate/boreal world. Sweden, Finland and Canada are the largest net exporters. The industrialized North America and Europe have economies that have been founded on agriculture and they have a strong impetus towards self-sufficiency and export. Developing countries have to work hard to penetrate their markets. The global commodity market has undoubtedly caused deforestation and the extent to which it may or may not continue to do so will depend on whether agro-pastoral activities intensify (increase land productivity) or extensify (increase area of forest cleared). There is no longer a substantive case that northern consumerism is the major culprit in causing continued deforestation.

Undervaluing the Forest

Forests have traditionally been treated like the oceans, as a limitless resource belonging to nobody in particular and which can be accessed on a come as you please basis. Indeed sometimes forests gain value only after they are cleared to obtain legal title through 'improvement'. Large-scale agro-pastoralists and corporations operating at the forest frontier will choose the best short-term profit option and on this basis the better option is almost always to clear the forest or to mine the timber. Small-scale farmers and shifting agriculturalists will choose the option that provides them with their basic needs, that has the opportunity to generate a little cash and that minimizes their risks. Again their only rational option is to clear the forest. The extraction of non-wood forest products has been suggested as a way to add value to the forest but it is not economical when compared to clearing options. The economics of sustainable forest management are also questionable. Sustainable forest management is a long-term investment into the whole suite of forest values including environmental services and is unattractive to short-term investors. The problem is that the ecological, amenity and environmental values of the forest (soil and water conservation, conservation of biodiversity, carbon storage) cannot be realized by the agent at the forest frontier. If some means could be devised where those who benefit from the environmental values could pay the forest owners or agents of deforestation for them, then the option to not clear would become more competitive. Alternatively, if the national government values the environmental benefits, it could apply a tax or disincentives to clear. However, even though maintenance of the environmental services is essential for sustained economic development, deforesting nations usually have more immediate goals and are unprepared to take this step. Those most concerned with the environmental services are usually outside of the deforesting country. The solution is for the international community to pay but no reliable mechanism for doing so currently exists.

Corruption

Corruption and bribery are considered to be normal and acceptable behaviour in some tropical countries and the cultural and political structures that support it are not easily broken. Corruption and weak government usually go hand in hand if not hand in pocket. It is not unusual for donor funds for development to get diverted into private income. Corruption keeps poor people in poverty which places pressure at the forest frontier. Illegal logging is considered to be the major single cause preventing sustainable forest management in Southeast Asia. It is not unusual for the amount of illegal logging to greatly exceed the amount of officially sanctioned and regulated logging. Huge wood processing plants have been and continue to be established with their capacities based on the illegal plus legal cut. The necessity to keep these plants supplied means that sustainable forestry cannot be achieved.

Strategies to Reduce Tropical Deforestation

Ways of reducing tropical deforestation must go hand in hand with improving the welfare of cultivators at the forest frontier. Any policy that does one without the other is unacceptable. Outcomes cannot be guaranteed. A considerable amount has been learnt from mistakes and successes of the past and by systematic research and analysis. There are no general solutions and strategies will vary with region and will change over time. All strategies require cooperation and goodwill. The main strategies are: reduce population growth and increase *per capita* incomes; increase the area and standard of management of protected areas; increase the area of forest permanently reserved for timber production; increase the perceived and actual value of the forest; promote sustainable forest management; make better use of already cleared land; sustainably increase agricultural productivity; increase the area of forest plantations; strengthen government institutions and policies; support environmental NGOs; encourage participatory forest management; increase investment in research, education and extension; and improve the information base (see also Roper and Roberts, 2003).

Reduce Population Growth and Increase *Per Capita* Incomes

Reduction of population growth is pivotal in reducing tropical deforestation. Reduction in population growth and increase in *per capita* income will occur as a consequence of increased real incomes and literacy rates. This is a somewhat simplistic analysis and not applicable to all situations and not all would agree. However, in general it has proved to be true. Anti-development strategies are unlikely to work in the long-term. The challenge is to promote sustainable development. History has shown that a transitional period is required and that deforestation initially may increase before it stabilizes.

Increase the Area and Standard of Management of Protected Areas

The provision of protected areas is fundamental in any attempt to conserve biodiversity. Protected areas alone, however, are not sufficient to conserve biodiversity. They should be considered alongside, and as part of, a wider strategy to conserve biodiversity. The minimum area of forests to be protected is generally considered to be 10% of total forest area. FRA 2000 estimates that 12.4% of the world's forests are located within protected areas. Not all of this has been set aside for conservation of biodiversity and some of it is too small and fragmented to be optimal for conservation. Many conservationists argue that 10% is not enough and they stress the need for representation of complete ecosystems, particularly targeting those 'hot spots' known to be rich in biodiversity and known to be threatened. Tropical forests are very rich in biodiversity (Chapter 2). The protection of a sufficient area of representative tropical ecosystems, clustered if possible to provide connecting corridors, is an essential component of conserving biodiversity.

Tropical and temperate forests have the highest proportions of their forests in protected areas and boreal forests have the least: the Americas (North, Central and South) have the greatest proportions and Europe the least (Table 5.11).

Protected areas are at risk from inadequate long-term tenure and various forms of encroachment. Strong government and enforcement of regulations are necessary to make

Table 5.11. Forests in protected areas (from FAO, 2001a).

	Forest area 2000 (million ha)	Forest in protected areas 2000 (million ha)	Proportion of forest in protected area (%)
Region			
Africa	650	76	11.7
Asia	547	50	9.1
Oceania	187	23	11.7
Europe	1,039	51	5.0
Nth. and Cent. America	549	111	20.2
South America	886	168	19.0
Ecological domain			
Tropical	1,997	304	15.2
Subtropical	370	42	11.3
Temperate	507	83	16.3
Boreal	995	49	5.0
Total	3,869	479	12.4

sure that the protected status of an area remains permanent and that protected areas are not degraded by fishing, hunting, fuelwood gathering, grazing, road construction, fires, invasive pests, shifting cultivators, and illegal loggers (and this is not a comprehensive list). Protected areas should not be 'preserved' and left alone for nature to take its course. Rather they need to be actively managed to achieve conservation of biodiversity and other environmental values. The provision of protected areas is just one part of an overall strategy for sustainable forest management.

Increase the Area of Forest Permanently Reserved for Timber Production

Arguably, the most serious impediment to sustainable forest management is the lack of dedicated forests specifically set aside for timber production. If the forest does not have a dedicated long-term tenure for timber production then there is no incentive to care for the long-term interests of the forest. Under these circumstances, agents of deforestation (shifting cultivators, land speculators and agropastoralists) will take advantage of the roads and infrastructure provided by a logging operation and they will occupy the land, which then alienates it from any possibility of a sustainable future as forest. Reservation for timber production requires commitment and vision from strong government because currently options to deforest or unsustainably log the forest are easier and more profitable than the sustainable forest management option. FAO (2001a) found that 89% of forests in industrialized countries were under some form of management but that probably only about 6% were in developing countries. Just 2% of the Brazilian Amazon is currently set aside as National Forest. If 20% could be set aside, not only could timber demand be sustainably met but buffer zones could be established to consolidate protected areas (Veríssimo *et al.*, 2002). This would form a conservation estate that would be one of the largest and most important in the world.

Increase the Perceived and Actual Value of the Forest

There are several ways of achieving this: governments can impose realistic prices on stumpage and forest rent; governments can invest in improving the sustainable productivity of the forest; and national and international beneficiaries of the environmental services of the forest could be required to pay for the services. There has been some success in devising schemes to collect payments for environmental services (carbon sequestration, conservation of biodiversity, catchment protection and ecotourism) but the rural poor still seem to be missing out (Landell-Mills and Porras, 2002; Angelsen and Wunder, 2003). Even so, progress is encouraging and things are moving in the right direction. The policies determining the value of the forest need to be controlled and spread in such a way that it is at least as attractive to sustainably manage the forests as it is to mine the timber.

Promote Sustainable Forest Management

In order to promote sustainable forest management, it must be sustainable ecologically, economically and socially. To achieve ecological sustainability means that the ecological values of the forest must not be degraded and if possible they should be improved. This means that silviculture and management should not reduce biodiversity, that soil erosion should be controlled, that soil fertility should not be lost, that water quality on and off site should be maintained and that forest health and vitality should be safeguarded. Some argue that ecological sustainability is impossible to achieve and in any case that techniques to monitor whether or not ecological sustainability is being achieved are unavailable. They promote the precautionary principle which says that if you are not sure whether an intervention will have a deleterious effect, then don't intervene at all.

However, management for environmental services alone is not economically and socially sustainable. It will not happen until or unless the developing nations have reached a stage of development and affluence that they can accommodate the costs of doing so. Alternatively, the developed world must be prepared to meet all the costs. It is more realistic to compare alternatives, and it needs to be understood that all alternatives to sustainable forest management are demonstrably more damaging. (The definition of sustainable forest management in this context includes setting aside the maximum area of protected forest that can be economically and socially sustained.) This is still putting a negative spin on the issue: that choosing sustainable forest management is a matter of choosing the least of evils. Sustainable forest management is better considered in a positive light where products of the forest are appreciated as renewable resources constructed by solar energy and which are environmentally superior to their competitors.

There is a considerable literature on silviculture of tropical rainforests. In its simplest terms the silviculture of tropical rainforests is concerned with manipulating canopy dynamics to promote regeneration and controlling harvesting activities to minimize damage. One of the outcomes of the recent globalization of forestry has been the agreement by 149 countries to work towards agreed environmental standards as defined by 'criteria' of sustainable forest management and 'indicators' of whether this is being achieved. There are nine processes of criteria and indicators in all and the degree of implementation of these processes varies widely between countries (see Chapter 6). However the processes are still in their infancy. FAO has had great difficulty in obtaining data from the developing tropics on the areas of forest under management and the nature of that management. This suggests that forest management in tropical countries is still rare.

Economically sustainable forest management requires realistic valuation of the forest and increased forest productivity. There are many examples to demonstrate that increased productivity can be achieved through improved silviculture and management. Sustainable forest management is a long-term investment and therefore requires a different way of thinking among investors. Socially sustainable forest management requires equitable distribution of costs and benefits and the participation and collaboration of all parties. The marginalization of peasant farmers has been a major impediment to socially sustainable forest management. Socially sustainable forest management also includes sensitivity to cultural issues and respect for indigenous rights and values.

The mean annual increment (MAI) of trees in native forests is on average about 2 m^3/ha/y. The MAI of industrial plantation forests are of the order of 10 to 40 m^3/ha/y. If the MAI of native forests could, through improved management, be increased to just 3 m^3/ha/y, this would be an increase of 50% and would make an astonishing difference to wood availability (and to the amount of carbon sequestered, see Chapter 6).

Make Better Use of Already Cleared Land

Reference has already been made to the vast areas of unused land. Some of this is degraded and of low fertility. Technological advances are being made to bring this land back into production. This should be a major priority since a significant proportion of cleared tropical forest will eventually end up as degraded land of low fertility anyway.

Sustainably Increase Agricultural Productivity

There has been considerable success in promoting increased agricultural activity in the tropics, including on marginal soils. Much of this has been through developing improved breeds and by the use of fertilizers and pesticides. However, along with the developed world, the challenge is to ensure that increases in productivity can be sustained. Ultimately this requires more conservative systems of nutrient and pest management such as: minimum burning, mulching, conservation of soil organic matter, minimum tillage, contour planting, integrated pest management, crop rotation, use of legumes and cover crops, deferred grazing and agroforestry (see Chapter 8 for discussion of agroforestry). An example is soybean in Brazil. Research has aided a spectacular increase in the productivity of soybean on acidic low nutrient marginal soils (Anderson *et al.,* 2002). However, the rapid development of soy has increased deforestation by displacing ranchers into the forest who in turn displace small farmers. Also the large amount of fertilizers and pesticides used in soybean production has contaminated ground waters and streams. Increasing land productivity can both increase and decrease deforestation. If sustainable increases in land productivity are combined with other policies to deal with marginalization and displaced labour, deforestation should be reduced, although not necessarily immediately. Under some circumstances shifting agriculture should not necessarily be discouraged. Rather it should be done with sufficient fallow time for complete recovery of soil fertility.

Increase the Area of Forest Plantations

Increasing the area of forest plantations should have a net positive benefit. Even though plantation establishment provides access to agents of deforestation and native forest is sometimes cleared to establish plantations, the high productivity of plantations permits exploitation for timber at lower intensities in native forests. Plantations will be discussed at greater length in Chapter 7.

Strengthen Government Institutions and Policies

Strong and stable government is essential to slow down the rate of deforestation. FAO (1999) considered that half of the current tropical deforestation could be stopped if the governments of the deforesting countries were determined to do so. Palo and Uusivuori (1999) give the example of boreal Finland where deforestation and degradation was replaced by expansion of forest resources through balanced use of both markets and policies by a stable government with a strong political will. The process, however, was transitional and took one hundred years. Governments in the tropical world have sometimes promoted development policies that they know will cause deforestation. Governments need to be convinced of the importance and value of forests. Many of the environmental services provided by forests are of direct economic benefit to a developing country. If a government is keen to reduce deforestation, it has to reduce the benefits accruing to the deforesting agents and increase the benefits to those involved in keeping the forest intact.

A critical component to reducing deforestation is for governments to promote and enforce policies that provide more equitable distribution of land. Governments are slow to do this because vested interests in or close to the government often hold the land. Again strong government is required. Government policies that have failed are those that have caused marginalization and displacement of the poor at the forest frontier. It is the style of development (inequitable policies and social injustice) rather than development *per se* that has been largely responsible for deforestation.

Corruption and illegal logging are scourges. The problem is that illegal logging in a region or country can completely distort the market for other regions or countries and make it impossible for them to consider sustainable forest management as an economic option. Eliminating corruption requires structural and cultural change. International donors and lenders are increasingly requiring safeguards against corruption before releasing their funds.

Support Environmental NGOs

Environmental NGOs have matured considerably over the last two decades. Their contribution towards conservation management has been enormous. Most (not all) have shifted their ground from an extreme preservation ethic to one where they are actively assisting those at the agricultural/forest frontier to improve their welfare in a way that reduces deforestation. Environmental NGOs have the advantage over government organizations and large international organizations that they are not constrained by government to government bureaucracy and inertia. They are better equipped to bypass corruption and they are very effective at getting to the people at the frontier who are in most need. Environmental NGOs are also effective in lobbying consumers and particularly retailers to promote stock that has been produced from sustainable agriculture and forestry.

They have been at the forefront of establishing processes for certification to ensure that agricultural and forest products have been produced and processed sustainably. Forest certification will be discussed in Chapter 6.

Participatory Forest Management

In order for forest management to succeed at the forest frontier, all parties with an interest in the fate of the forest should be communally involved in planning, management and profit sharing. Parties may have competing or conflicting interests. Community forestry will be discussed in Chapter 8.

Increase Investment in Research, Education and Extension

There is a lack of knowledge and information in the general community about forests and forestry. More worrying is the ignorance of those making the policy decisions that influence deforestation. Much of the information in the community about deforestation has been generated in the popular media by single-issue interests. The deforestation of tropical forests is an emotive issue and the temptation to sensationalize often overcomes the necessity to search for the facts. There is no substitute for research and critical analysis. Forest managers and those developing forest policies need to be comprehensively educated and need to appreciate the complexity of the interacting ecological, economical, social, cultural and political factors involved. Many universities provide programmes in forestry and produce foresters and specialist scientists to meet this need. However the numbers of students undertaking such education have been lower in recent years due in part to false perceptions of forestry generated in the community.

Improve the Information Base

Knowledge of how much forest, where it is and what it is composed of would seem to be straightforward. Surprisingly this most basic of information is not always available. Modern remote sensing technology is now assisting FAO in obtaining basic global data but it is still the responsibility of individual countries to monitor their forests. Information on the ecological and environmental values of forests is very scarce in almost all instances. It is not possible to properly manage a forest ecosystem without first understanding it.

A Final Word

This chapter has concentrated on tropical deforestation because gross estimates of forest change suggest that deforestation is not occurring in the developed world. However these estimates do not account for forest degradation and there is good evidence that significant areas of forest in the developing world are continuing to be degraded particularly in the boreal forests of the northern hemisphere.

References and Further Reading

Achard, F., Eva, H.D., Stibig, H-J., Mayaux, P., Gallego, J., Richards, T. and Malingreau, J-P. (2002) Determination of deforestation rates of the world's humid tropical forests. *Science* 297, 999-1002.

Anderson, L.E., Granger, C.W.J., Eustáquio, J.R., Weinhold, D. and Wunder, S. (2002) *The Dynamics of Deforestation and Economic Growth in the Brazilian Amazon.* Cambridge University Press, Cambridge.

Angelson, A. and Kaimowitz, D. (2001) *Agricultural Technologies and Tropical Deforestation.* CAB International, Wallingford.

Angelsen, A. and Wunder, S. (2003) *Exploring the Forest-Poverty Link: Key Concepts, Issues and Research Implications.* CIFOR Occasional Paper No.40, Center for International Forestry Research, Bogor, Indonesia.

Barraclough, S.L. and Ghimire, K.B. (1995) *Forests and Livelihoods – The Social Dynamics of Deforestation in Developing Countries.* Macmillan, London.

Barraclough, S.L. and Ghimire, K.B. (2000) *Agricultural Expansion and Tropical Deforestation.* Earthscan Publications Ltd, London.

Bee, O.J. (1993) *Tropical Deforestation – The Tyranny of Time.* Singapore University Press, Singapore.

Brookfield, H., Potter L., Byron, Y. (1995) *In Place of the Forest - Environmental and Socio-economic Transformation in Borneo and the Eastern Malay Peninsula.* United Nations University Press, Tokyo.

Brown, D. and Schrekenberg, K. (1998) *Shifting Cultivators as Aagents of Ddeforestation: Assessing the Evidence.* Natural Resource Perspective Number 29, April 1998, Overseas Development Institute, London.

Brown. K. and Pearce, D.W. (1994) *The Causes of Tropical Deforestation – The Economic and Statistical Analysis of Factors Giving Rise to the Loss of the Tropical Forests.* UBC Press, Vancouver.

Bryant, D., Nielsen, D. and Tangley, L. (1997) *The Last Frontier Forests – Ecosytems and Economies on the Edge.* World Resources Institute, Washington, DC.

Cattaneo, A. (2002) *Balancing Agricultural Development and Deforestation in the Brazilian Amazon.* International Food Policy Research Institute, Washington, DC.

Colchester, M. and Lohmann, L. (eds) (1993) *The Struggle for Land and the Fate of the Forests.* World Rainforest Movement, Penang.

Critchley, W.R.S. and Bruijnzeel, L.A. (1996) *Environmental Impacts of Converting Moist Tropical Forest to Agriculture and Plantations.* UNESCO International Hydrological Programme, Paris.

Dawkins, H.C. and Philip, M.S. (1998) *Tropical Moist Forest Silviculture and Management.* CAB International, Wallingford.

de Jong, W. (2001) The impact of rubber on the forest landscape in Borneo. In: Angelson, A. and Kaimowitz, D (eds) *Agricultural Technologies and Tropical Deforestation.* CAB International, Wallingford, pp. 367-381.

Downing, T.E., Hecht, S.B., Pearson, H.A. and Garcia-Downing, C. (eds) (1992) *Development or Destruction – The Conversion of Tropical Forest to Pasture in Latin America.* Westview Press, Boulder, San Francisco and Oxford.

FAO (1999) *State of the World's Forests.* Food and Agriculture Organization of the United Nations, Rome.

FAO (2001a) *Global Forest Resources Assessment 2000 - Main Report.* FAO Forestry Paper 140, Food and Agriculture Organization of the United Nations, Rome.

FAO (2001b) *State of the World's Forests.* Food and Agriculture Organization of the United Nations, Rome.

FAO (2003) *State of the World's Forests.* Food and Agriculture Organization of the United Nations, Rome.

Fairhead, J and Leach, M. (1998) *Reframing Deforestation – Global Analysis and Local Realities: Studies in West Africa.* Routledge, London and New York.

Faminow, M.D. (1998) *Cattle, Deforestation and Development in the Amazon – An Economic, Agronomic and Environmental Perspective.* CAB International, Wallingford.

Fisher, G., Shah, M., van Velthuizen, H. and Nachtergaele, F.O. (2002) *Global Agro-ecological Assessment for Agriculture in the 21st Century.* Report for IIASA and FAO, Laxenburg and Rome.

Grainger, A. (1993) *Controlling Tropical Deforestation.* Earthscan Publications Ltd, London.

Heckadon, M.S. (1989) Los Viveros Comunales en El Salvador. *El Chasqui* 20, 3-24.

ITTO (2001, 2002, 2003) *Annual Review and Assessment of the World Timber Situation.* International Tropical Timber Organization, Yokohama, Japan.

Jepma, C.J. (1995) *Tropical Deforestation – A Socio-economic Approach.* Earthscan Publications Ltd, London.

Kaimowitz, D. (1996) *Livestock and Deforestation : Central America in the 1980s and 1990s : a Policy Perspective.* Center for International Forestry Research, Jakarta.

Kaimowitz, D. and Angelsen, A. (1998) *Economic Models of Tropical Deforestation : a Review.* CIFOR, Bogor, Indonesia.

Landell-Mills, N. and Porras, I.T. (2002) *Silver Bullet or Fools' Gold.* Research Report prepared by the International Institute for Environment and Development (IIED), London.

Lovejoy, T.E. (1980) A projection of species extinctions. In: *Council on Environmental Quality (CEQ) The Global 2000 Report to the President. Vol. 2*, CEQ, Washington, DC, pp. 328-331.

Martin, C. (1991) *The Rainforests of West Africa: Ecology, Threats, Conservation.* Birkhäuser Verlag, Basel, Boston, Berlin.

Mather, A. (2001) Transition from deforestation to reforestation in Europe. In: Angelson, A. and Kaimowitz, D (eds) *Agricultural Technologies and Tropical Deforestation.* CAB International, Wallingford, pp. 35-52.

Myers, N. (1979) *The Sinking Ark: a New Look at the Problem of Disappearing Species.* Pergamon Press, Oxford.

Myers, N. (1988) Threatened biotas: 'hot spots' in tropical forests. *Environmentalist* 8(3), 1-20.

Myers, N. (1994) Tropical deforestation: rates and patterns. In: Brown. K. and Pearce, D.W. (eds) *The Causes of Tropical Deforestation – The Economic and Statistical Analysis of Factors Giving Rise to the Loss of the Tropical Forests.* UBC Press, Vancouver, pp. 27-40.

Palo, M. (1994) Population and deforestation. In: Brown. K. and Pearce, D.W. (eds) *The Causes of Tropical Deforestation – The Economic and Statistical Analysis of Factors Giving Rise to the Loss of the Tropical Forests.* UBC Press, Vancouver, pp. 42-56.

Palo, M. and Uusivuori, J. (1999) Forest-based development in Finland – a unique success? In: Palo, M. and Uusivuori, J. (eds) *World Forests, Society and Environment.* Kluwer Academic Publishers, Dordrecht, pp. 300-318.

Palo, M. and Vahanen, H. (eds) (2000) *World Forests from Deforestation to Transition.* Kluwer Academic Publishers, Dordrecht, London, Boston.

Palo, M., Lehto, E. and Uusivuori, J. (2000) Modelling Causes of Deforestation with 477 subnational units. In: Palo, M. and Vahanen, H. (eds) *World Forests from Deforestation to Transition.* Kluwer Academic Publishers, Dordrecht, London, Boston, pp. 101-124.

Pearce, P., Barbier, E and Markandya, A. (1990) *Sustainable Development, Economics and Environment in the Third World.* Edgar Elwood Publishing Limited, Aldershot.

Poore, M.E.D. (ed) (1989) *No Timber without Trees - Sustainability in the Tropical Forests.* Earthscan Publications, London.

Raven, P.H. (1987) The scope of the plant conservation problem world-wide. In: Bramwell, D., Hamann, O., Heywood, V. and Synge, H. (eds) *Botanic Gardens and the World Conservation Strategy.* Academic Press, London, pp. 19-29.

Raven, P.H. (1988a) Biological resources and global stability. In: Kawano, S., Connell, J.H. and Hidaka, T. (eds) *Evolution and Coadaptation in Biotic Communities.* University of Tokyo Press, Tokyo, pp. 3-27.

Raven, P.H. (1988b) Our diminishing forests. In: Wilson, E.O. and Peter, F.M. (eds) *Biodiversity.* National Academy Press, Washington, DC, pp. 119-122.

Reid, W.V. (1992) How many species will there be? In: T.C. Whitmore and J.A. Sayer (eds) *Tropical Deforestation and Species Extinction.* Chapman and Hall, London, pp. 55-73.

Richards, J.F. and Tucker, R.P. (eds) (1988) *World Deforestation in the Twentieth Century.* Duke University Press, Durham, London.

Roper, J. and Roberts, R.W. (2003) *Deforestation: Tropical Forests in Decline.* CFAN CIDA Forestry Advisers Network, www.rcfa-cfan.org (accessed 2 August 2004).

Sharma, N.P. (ed) (1992) *Managing the World's Forests: Looking for the Balance between Conservation and Development.* Kendall Hunt, Iowa.

Shively, G. and Martinez, E. (2001) Deforestation, irrigation, employment and cautious optimism in Southern Palawan, the Philippines. In: Angelson, A. and Kaimowitz, D. (eds*) Agricultural Technologies and Tropical Deforestation.* CAB International, Wallingford, pp. 335-346.

Silliman, J. (ed) (1981*) Draft Environmental Profile on the Republic of Costa Rica.* Arid Lands Information Center Office of Arid Lands Studies, University of Arizona, Tucson, Arizona.

Simberloff, D. (1986) Are we on the verge of a mass extinction in tropical rain forests? In: Elliott, D.K. (ed*) Dynamics of Extinction.* Wiley, New York, pp. 165-180.

Sponsel, L.E., Headland, T.N. and Bailey, R.C. (1996) *Tropical Deforestation – The Human Dimension.* Columbia University Press, New York.

The Economist (2004) Asphalt and the jungle. *The Economist* July 24th, 2004, pp. 35-37.

United Nations (2001) *Forest Products Annual Markets Review 2000-2001.* Timber Bulletin – Volume LIV No.3, United Nations Economic Commission for Europe and Food and Agriculture Organization of the United Nations, New York and Geneva.

Utting, P. (1993) *Trees, People and Power - Social Dimensions of Deforestation and Forest Protection in Central America.* Earthscan Publications Ltd, London.

Veríssimo, A., Cochrane, M.A. Jr., Souza, G. Jr. and Salomão, R. (2002) Priority areas for establishing national forests in the Brazilian Amazon. *Conservation Ecology* 6, 4.

Westoby, J. (1989) *Introduction to World Forestry: People and their Trees.* Blackwell, Oxford and New York.

Whitmore, T.C. and Sayer, J.A., (eds) (1992*) Tropical Deforestation and Species Extinction.* Chapman & Hall, New York.

Wickramasinghe, A. (1994) *Deforestation, Women and Forestry – The Case of Sri Lanka.* International Books, Utrecht.

Williams, M. (1989) *Americans & their Forests – A Historical Geography.* Cambridge University Press, Cambridge.

Williams, M. (2003) *Deforesting the Earth – From Prehistory to Global Crisis.* The University of Chicago Press, Chicago.

Wilson, E.O. (1988) The current state of biological diversity. In: Wilson, E.O. and Peter, F.M. (eds) *Biodiversity.* National Academy Press, Washington, DC, pp. 108-116.

Wilson, E.O. (1990) Threats to biodiversity. *Scientific American*, September 1990, 108-116.

Wunder, S. (2003) *Oil Wealth and the Fate of the Forest – A Comparative Study of Eight Tropical Countries.* Routledge, London and New York.

Key on-line resources
www.fao.org
www.iucn.org
www.wwf.org
www.cifor.cgiar.org
www.eldis.org
www.worldbank.org

Chapter 6

Sustainable Forest Management

What is Natural

The word 'natural' is used in a confusing and inconsistent manner. Usually natural forests are considered to be those in a so-called pristine condition that have had little human impact. This use of the word natural is so endemic in the literature that there is no alternative but to continue to use it in this way in this chapter and throughout this book. However, it suggests that somehow humans are outside of nature which is ecological nonsense. Consider on the one hand a pristine forested landscape that has had no human impact and on the other hand a busy city office enclosed by concrete and glass with never a tree, plant or animal in sight. Which is natural and which is not? They both are natural and this is an important point to understand when considering the relationship between humans and the rest of everything. Humans are completely contained within and constrained by nature and its laws. Humans can only exist by consuming natural resources and all resources are natural. Humans are a part of nature and cannot escape this. By the same argument the use of the word 'exotics' is questionable when it suggests that introduced species are somehow 'unnatural'.

Ecology is the relationship between all living things (including humans) and of all living things with their environment. Humans are components of ecosystems and their activities, for better or for worse, are subject to the same ecological principles as for all other living things. This being the case, it is an ecologically flawed argument that humans can choose to stand outside of nature and leave nature to take its course. Such thinking has led to the idea that human impacts on forests must always be negative and that forests should be left alone in their own best interests. Humans have the capacity to degrade ecosystems and they have demonstrated that they are very proficient at this. However, from an ecosystem perspective, they also have the capacity to enhance ecosystems. As such, humans should actively manage their ecosystems in their own interests and that of other living things. This may mean low or no impact interventions under certain circumstances, but it should be an informed management decision.

What is Sustainable Management

Sustainability is a current catchword that has become somewhat degraded through overuse and it is often used glibly as a pretence to best practice. Many articles and indeed books (Aplet *et al.,* 1993) have been written about defining just what sustainability is and, although there is a general consensus about what it means in general terms, its precise definition usually reflects the self interests of the proposer. The three pillars of sustainable management are economic, environmental and social values. One definition of sustainable management from the New Zealand Resource Management Act is given here as an example:

'Managing the use, development and protection of natural and physical resources in a way and at a rate which enables people and communities to provide for their social, economic and cultural well-being, and for their health and safety, while:
(a) sustaining the potential of natural and physical resources (excluding minerals) to meet the reasonable foreseeable needs of future generations;
(b) safeguarding the life-supporting capacity of air, water, soil and ecosystems; and
(c) avoiding, remedying or mitigating any adverse effects on the environment.'

Most disagreement about defining sustainability is in the relative significance of ecology, economy and society in the definition. Sustainability is about making sure that our resources are not run down and that they are available for future generations. This entails promoting the use and development of renewable resources, avoiding the degradation and dilution of natural resources, reducing waste, encouraging recycling, improving efficiency of production and energy use, modifying over-consumptive life styles, and protecting the biotic and abiotic components of ecosystems. Implicitly, sustainability means that the world should be handed over in good condition to future generations.

The economic and social dimensions of sustainability are equally important to the ecological ones. These include population control, poverty alleviation, equality of opportunity and access to resources, fairness in trade, freedom from oppression and sensitivity to spiritual and cultural values. Probably a better and more equitable way of thinking about sustainability is to consider 'sustainable development' and the World Commission on Environment and Development defines this as:

'Development that meets the needs of the present without compromising the ability of future generations to meet their own needs.'

The sustainable management of forests cannot be considered in isolation because impacts within the forest are felt outside of the forest and *vice versa*. Management of forests should be considered as part of overall land management at catchment, regional and national levels. Achieving sustainable agricultural management is a fundamental precursor to achieving sustainable forest management, particularly in the tropical developing world where the main driver of deforestation is the clearing of land to support agriculture and grazing (Chapter 5). The quest to achieve sustainable forest management is a journey that is by no means complete. There are numerous examples of unsustainable forest management and it would be easy to become depressed and negative. However, there also are some great success stories and the movement is towards rather than away from sustainability. The ideas around sustainable forest management commenced with sustained yield, then progressed to management for multiple use and from there to whole ecosystem management.

Sustained Yield

The beginning of forest management in Europe was to protect wildlife and keep poachers out. However, the emphasis later shifted to management for timber production largely in response to unregulated cutting and consequent deforestation and degradation. The modern concept of sustained yield was first developed in Germany in the 18th century (Chapter 1). In its broadest terms, management for sustained yield is where the frequency and amount of logging is such that the forest can maintain a sustained yield in perpetuity. A main objective was to develop forests with about equal areas of each age size or class in order to have more or less constant supply of the largest trees for harvest. In practice, managers chose to cut to maximum sustained yield rather than to be conservative. At about the same time, the Germans developed the basic silvicultural systems (single tree selection, group selection, shelterwood and clearcut) which remain the basis of silvicultural practices today. The emphasis in management for sustained yield was (and still is in some quarters) ensuring that timber supplies were continuously available and less attention was given to other values of the forest.

Managing for sustained yield was a step towards sustainable forest management. Forests managed for sustained yield were, by definition, not over-cut and for the most part were in good condition and free of disease. Emphasis was on the crop tree and the impact of management on other parts of the ecosystem was usually not monitored. Often the composition of the forests was altered by promoting the growth of the commercially favoured trees and suppressing the growth of less desirable species. There was no knowledge of, or interest in the effect of this on ecosystem function. There was the tacit assumption that, if the forest was not over-cut, then it was probably in reasonably good condition and that management for sustained yield was a responsible and conservative management option. Over time the preoccupation with timber production became relatively less and interest in other values of the forest became relatively more. In this way management for sustained yield became embedded in a more comprehensive management strategy called management for multiple use.

Management for Multiple Use

Management for multiple use recognizes that forests have many values of which timber production is just one. The various values (conservation of soil, water and biodiversity; recreation and amenity; aesthetic, cultural and religious values; climate amelioration; protection against natural hazards; timber production; non-wood forest products) have been discussed in Chapters 3 and 4. The concept that forests can be managed for multiple use has been a cornerstone of the forestry profession for the last 50 years at least and is held as almost sacred by the profession. In its purest and most idealistic form it is the management of a forest for all of its values, for all of the people for all of the time. This is, of course, impossible and compromises need to be made. Management for multiple use is easier in countries where a significant proportion of the forests are owned by the government, in which case the requirement to manage for multiple use may be embodied in legislation. For example the USA Multiple Use, Sustained Yield Act of 1960 requires:

'......*the management of all the various renewable resources of the national forests so they are utilized in the combination that will best meet the needs of the American people......harmonious and coordinated management of the various resources each with the other, without impairment of the productivity of the land....... such that management will not necessarily offer the combination of uses that will give the greatest return....'*

For privately owned forest, use can still be regulated to some extent, but the owners will favour their special interests as much as possible.

The relative importance of forest values and uses differs between people. In some cases the uses will be compatible between different groups of people and in others the uses will not be compatible and will create conflict. For example, conservationists who oppose logging will not be satisfied with any management option that includes logging; bushwalkers will not happily share their track with motor bikes and four wheel drive vehicles; and fishermen and fisherwomen are unlikely to be impressed with any activity that changes the nature of their streams. The concept of multiple use is one of harmony but it has caused enormous disharmony. Single-interest groups will always promote their interest if it is threatened by a competing interest. Under management for multiple use, the returns for any single use will be sub-optimal. Therefore management for multiple use completely satisfies nobody except the professional foresters who attempt to carry it out. Even so, the concept of meeting the reasonable expectations of all parties is a noble ideal and the greatest challenge to the forestry profession is how to do this in a sustainable manner. The ideal model of multiple use may need to be modified in order to achieve a practical outcome. One way of doing this is to separate spatially the various multiple uses into single use units at the regional or national level. An extreme example of this is in New Zealand where timber production is almost entirely confined to intensively managed plantations and all of the government owned native forest (83.5% of the total) is reserved as a conservation estate. Another modification of multiple use is where a forest is managed primarily for a single-use but where the other values of the forest are protected by compulsory or voluntary codes of practice.

Forest Ecosystem Management

Over the last few decades there has been the increasing realization that management for multiple use will not necessarily achieve sustainability. It is generally accepted today that a better way to approach sustainability is to manage the forest as an ecosystem and for silviculture and management to be sympathetic with those ecological processes that enhance ecological function, that maintain biodiversity and that conserve water and soil. Clearly this requires an in-depth understanding of ecosystem dynamics. Kaufmann *et al.* (1994) provided six principles as an ecological basis for ecosystem management:

'Humans are an integral part of today's ecosystems and depend on natural ecosystems for survival and welfare; ecosystems must be sustained for the long-term well being of humans and other forms of life

In ecosystems, the potential exists for all biotic and abiotic elements to be present with sufficient redundancy at appropriate spatial and temporal scales across the landscape

Across adequately large areas, ecosystem processes (such as disturbance, succession, evolution, natural extinction, recolonization, fluxes of materials, and other stochastic, deterministic, and chaotic events) that characterize the variability found in natural ecosystems should be present and functioning

Human intervention should not impact ecosystem sustainability by destroying or significantly degrading components that affect ecosystem capabilities

The cumulative effect of human influences, including the production of commodities and services, should maintain resilient ecosystems capable of returning to the natural range of variability if left alone, and

Management activities should conserve or restore natural ecosystem disturbance patterns.'

Implicit in ecological management is the recognition that an ecosystem is a dynamic system in which disturbances such as fires, floods, insect outbreaks, fungal infestations and storms are a normal part of ecosystem function.

Sustainable Forest Management

Within the context of forest ecosystem management, sustainable forest management may or may not include management for timber production. Forests may be managed for conservation and environmental services alone or in conjunction with timber production. Ecological aspects of sustainable management are just one element of sustainable forest management and not the whole, which also includes economic, political and social aspects.

Conservation Management

A major component of conservation management is the establishment of reserves or protected areas. Currently 12.5% of the world's forests are in protected areas but conservationists argue that this is not enough (Chapter 5). There is a bias towards reserving forests of grandeur and scenic value and to conserving the larger animals to which humans have an affinity. Reservation often followed agricultural, forestry and urban development and therefore over-represents areas of lesser economic interest such as mountains. In developed regions reserves usually are made for a variety of reasons (local community interest, unused land, roadside corridors) other than ecological considerations. It is important to redress all these imbalances so that reserves cover the largest possible range in biodiversity and in ecosystems. An ecosystem is scaleless and it is difficult to define its boundary. The boundary of a reserve is an artificial construct. The land mosaics adjacent to the reserve need to be managed in sympathy with the reserve, as most reserves are neither large enough to protect all species, nor to allow for migrations to replace populations that become locally extinct (Norton, 1999). Animals and plants will migrate across reserve boundaries. A landscape boundary such as a catchment is likely to be a good reserve boundary. However, reserve boundaries usually are set by political and economic realities or by land tenure rather than by ecological considerations.

The immediate challenge is to ensure that reserves are comprehensive (sampling the full range of biodiversity), adequate (viable populations and areas of habitat) and representative (covering the geographic range of species and ecological communities) (Burgman and Lindenmayer, 1998). There needs to be some sort of ecological classification scheme to achieve this. The measurement of the full extent of biodiversity is not possible but surrogates (such as species richness) are useful. Also, ecosystems are dynamic successional entities, strongly influenced by disturbance. Consequently, it is necessary to have a historical perspective to understand where the ecosystem has come from and where it is going. These constraints may be overwhelming but it is possible to make some generalizations about reserve design (Diamond, 1975). Generally speaking, reserves should be as large as possible and there should be as many of them as possible. There is considerable unresolved argument about whether a single large reserve is better or worse than several small reserves of the same area (Simberloff and Abele, 1982). As far as possible, the length of edges should be kept to a minimum and this means that a circular reserve is better than a long thin rectangle, for example. Also, clusters of reserves in close proximity are probably better than reserves in isolation and connectivity corridors may aid in movement between reserves (although not all agree). Reserves close together are preferred for species dispersal and richness but disadvantageous for representing the wider geographical range of biodiversity, for transmitting diseases and invasive species and for mitigating the effect of disturbance. The use of buffer zones around reserves, as a gradual rather than a sudden change to an alternative land-use, is also strongly recommended (Hunter, 2002).

Some argue that human impact in reserves should be kept to a minimum. This is not always possible or practical in much of the developing world where reserves have large human populations in and around them. Also this is at odds with the argument that humans are part of the natural environment. The main argument for keeping human impact to a minimum is the reasonable assumption that no impact will have the best chance of conserving biodiversity. Another reason is that minimum impact reserves can then be used as a reference to determine, for example, the effects of presumed sustainable forest management on ecosystem processes (Norton, 1999). However, management for minimum impact does not mean that reserves should be abandoned to let nature take its course. Reserves still need to be managed and some of the management decisions are difficult to make in the context of minimizing human impacts. For example, there is a range of natural disturbances that are an integral part of ecosystem function. A wild fire can reduce a forest reserve to ashes. To some this would be a catastrophe that should be avoided at all costs and they would promote a fire prevention plan. However, others would argue that the reserve should not be managed for fire protection because this would interfere with natural succession and change the forest into something that it otherwise would not have been. What should the manager do if an exotic plant or animal species migrates into the reserve and creates havoc? For example, stoats and weasels introduced into New Zealand's native forests are seriously threatening the native bird-life and introduced possums are threatening the native trees and birds. The management decision here is obvious and non-controversial: attempt to control the stoat, weasel and possum populations. However, not all decisions are this obvious. What about the situation where a native plant or animal threatens the viability of other species. Should this be seen as just another example of natural disturbance? In the 1980s, the kangaroo populations in the Little Desert National Park in Victoria, Australia, increased to the stage that the kangaroos were threatening the survival of rare plant species. This created a dilemma for both the managers and conservationists. The kangaroo is not only an Australian native animal but is a national icon. There was a clash of opinion between those conservationists with an ecosystem view and those with a biocentric view and the conservation community was divided about the appropriate action. The managers decided to cull the kangaroo herd but this had to be done as a covert operation over one night in order to avoid confrontation with those conservationists who disagreed with killing native animals in a national park.

People visit national parks for recreation and the potential impacts from this can be enormous. A prime example is the Smoky Mountains National Park in the USA which has more than 6 million visitor days per year (Hunter, 2002). The park has to be managed to minimize these impacts without unduly restricting access. There are very few forests in the world that have not had any human impacts and forests from regions having a long history of human population have all been impacted by humans to a greater or lesser degree. The forests of Europe for example have been impacted by humans for thousands of years and they are not pristine wild places. What is the manager's responsibility here: to set the benchmark at the time of reservation or to attempt to create a forest that might have occurred if humans had not intervened? Also some forests are appealing because they have had a long period of management, but public pressure to reserve these forests does not always recognize that the very features that they wish to protect have been created by humans. Another area of uncertainty is how long an introduced species needs to be established in an environment before it is considered to be a native. Humans have been moving plants around for 10,000 years or more and plants also migrate without any human influence. Australia's forests, for example, are a legacy of 40,000 years of aboriginal land management. It is conceivable that an introduced species that is initially considered to be a weed will after (say) 200 years be considered to be a native species to be valued and protected. This may well happen to *Eucalyptus*, an Australian native, which is becoming widely distributed on all continents. Douglas fir (*Pseudotsuga menziesii*) is another species that may be considered to be a native species in due course in Europe and in New Zealand. It is surprising how many people expect a protected area to be like a photograph, frozen in time. Part of management responsibility is to communicate to the public the dynamic nature of forest ecosystems.

Conservation of biodiversity can be aided by *ex situ* initiatives. Botanical and zoological gardens have an important role to play although there is the tendency for these gardens to feature plants and animals that humans are interested in. Seed banks can assist in preserving the gene pool of a plant species but emphasis here is on plants of commercial significance.

Protected areas alone are insufficient to conserve biodiversity. They should be just one part of an overall conservation strategy which, in turn, should be part of an overall landscape approach to natural resources management. Boyle (2001) has comprehensively reviewed the impact of human interventions on forest biodiversity and the capacity of management to enhance forest biodiversity. The best way to conserve forests is to make sure that they are kept as fully functioning ecosystems and not cleared for alternative land uses. It is not feasible for all forest to be reserved as protected areas and, in order to protect the remaining forest, keeping the forest intact has to be competitive against alternative land uses and sustainable management for timber production has the capacity to do this. However, ecosystem integrity and environmental services should not be compromised (and not all agree that this is possible).

Sustainable Forest Management for Timber Production and Environmental Services

Forest ecosystem management in its purest form is management that mimics ecosystem processes. An idealized forest is one where there are mixed species having a large range of size/age classes. The older trees die and create gaps in the canopy and new trees regenerate in the gap from seed shed from surrounding trees. In this way the size/species structure of the forest is maintained. No forests are like this but some forests, for example climax tropical rainforests and the mixed angiosperm/conifer forests of the northern hemisphere temperate zone, approach it. The appropriate timber management regime for this forest would be to carefully log the large trees and the poorest stems to create a regeneration gap beneath. The emphasis should be on which trees to retain rather than which to remove. This is the basis of the silvicultural system called single tree selection. Central European mixed species forests were over-exploited in the 17th and 18th centuries (Chapter 1). In response to their poor condition, there was a concerted move at the beginning of the 19th century to replace them with conifer forests managed as monocultures by area and age class regimes. The monocultures were clearly more productive. However, by the middle of the 19th century there was a general recognition that this had strayed too far from the norm and there has been a steady move back to mixed species forests. This has progressed to the extent that there is now a steady transition towards 'near natural' selection forestry with the emphasis on continuous canopy cover and selecting trees for retention rather than for removal (Benecke, 1996). Every indication is that this has been a success.

The other extreme is a forest that periodically undergoes a disturbance such as wildfire which creates a very large gap in the forest and even-aged regeneration of a single species fills the gap. By so doing, areas of even-aged monocultures are created. No forests are exactly like this either but some, for example the *Eucalyptus regnans* forests of southeast Australia and the *Pinus contorta* forests of the USA, closely approach this. This is the basis of the silvicultural system called clear-cutting. Clear-cutting creates a visual eye-sore and is opposed by many. However, it can be the most appropriate silvicultural treatment for certain forests and any attempt to impose selection silviculture on these forests will change the structure and species composition of the forest. There are many instances where the imposition of single tree selection on a forest unsuited to it has failed dismally. Species that colonize areas that have been completely deforested by disturbance such as wildfire or glaciation are called pioneer species. Such species are very well suited to grow in plantations, which is not all that different from how they occur in 'nature'.

Forest ecosystems lie on a continuum between these two extremes and their composition and ecological function depends on the relationship between gap size and gap-creating frequency. Small gap size and high frequency of gap formation favours the single tree selection model and large gap size and low frequency of gap formation favours the clear-cutting model. The appropriate silviculture depends on where the forest ecosystem lies on the continuum between these two extremes. For example, a wildfire may not be extreme enough to kill all the trees and some larger trees will remain. Regeneration will occur in the residual gaps. This will result in a stand that has two cohorts. The word cohort is used here rather than age class because regeneration in gaps may occur over several years. When gaps are larger and less frequent than in the single tree selection model, there may be a number of identifiable cohorts (multi cohorts). Usually the most appropriate silvicultural treatment will lie somewhere on the continuum and not at the extremes. The best silviculture is all about choosing what and when to cut and more importantly what to retain in order to mimic ecosystem processes. This is extensively reviewed in Smith *et al.* (1997) and Franklin *et al.* (1997).

The ideals of forest ecosystem management are often compromised for expediency and for economic advantage. Clear-cutting at a frequency greater than the natural disturbance that created the gap does not mimic the ecosystem processes and should be avoided. The uniform cutting of trees above a size class does not take into account that different species grow and mature at different rates. Rotation length (or cutting cycle) is a manager's tool of convenience and has no ecological basis in a mixed species forests. Cutting preferred commercial species and leaving non-commercial species will change the ecosystem into something else. Focusing on what to cut rather than what to retain is an incorrect ecological approach. This is not to say that management should always aim to maintain the *status quo*. Indeed there are examples where management, informed or by default, has apparently increased biodiversity, enhanced site productivity, reduced soil erosion and increased water quality. What it does mean is that, if managers wittingly choose to alter an ecosystem, they should do so in full recognition of the ecological consequences. It is important to leave some old, dying and dead trees scattered around because these have important roles in conserving biodiversity, both as standing trees and also as coarse woody debris when they eventually fall to the ground.

Sustainable management implies maintenance of site productivity in perpetuity and this strongly depends on soil management (Nambiar, 1996; Powers; 2001). In native forests this generally means ensuring that cycling of carbon and nutrients is favoured; that nutrient rich litter and plant residues are not removed from the site; that soil biota are encouraged; that wood is not removed from the site at too great a frequency; that soils are not contaminated by pesticides, salts, heavy metals or acids; and that erosion, soil disturbance and compaction are avoided during management operations, particularly during harvesting. Harvesting can be particularly damaging to soil unless conscious effort is made to reduce adverse impacts on the soil. Ground harvesting on wet erodible soils on steep slopes can be very damaging, whereas single tree logging using helicopters has scarcely any impact. Technologies are available to reduce harvesting impacts to acceptable limits. Site productivity and soil management can be more aggressively managed in plantations and this will be discussed in Chapter 7.

Increasing forest productivity means that less forest needs to be harvested at a given intensity to achieve the same yield. Alternatively the same area of forest can be managed less intensively to achieve the same yield. Either way, increasing productivity is a better sustainable option. There has been an understandable and entirely appropriate pre-occupation among agronomists and silviculturalists to improve plant productivity, but this does not necessarily improve ecosystem stability, resilience or richness. Some very rich ecosystems occur on infertile soils. Maintenance of productivity alone is not sufficient to guarantee sustainability. Management should not degrade the physical, chemical and biological properties of soil that support plant growth. In the final analysis, all life depends on ecosystem productivity which is driven by intercepted radiation, atmospheric carbon dioxide, and water, nutrients and oxygen from the soil. Soil management is arguably the most important factor in the sustainable management of natural renewable resources. Maintenance of site productivity also depends on controlling vertebrate and invertebrate pests and fungal diseases. Pesticides are often ineffective over the long-term and have the potential to contaminate the environment. The future of pest management and disease control must be environmentally friendly and multi-disciplinary.

Ecological forest management may prove to be a wise economical choice over time because it safeguards environmental values. Forest owners, however, are keen to maximize their short term profits and ecological forest management usually will not achieve this. Ecological forest management therefore needs to be supported by government policy and regulation, by compliance with international conventions and by compulsory or voluntary codes of practice. It is debatable whether ecological forest management alone could supply world wood requirements. Some are quite convinced that it will but others not so. The argument is centred on the acceptable intensity of human impact. This raises another contentious issue - the role of plantations. Plantations are highly productive and can efficiently produce wood on a small land area. This should permit less intensive (more extensive) management of 'native' forests. The optimum solution to forest management could be to have a mixed bag of management intensities from extensive to intensive. Figure 6.1 shows two quite radically different ways that Germany and New Zealand achieve this.

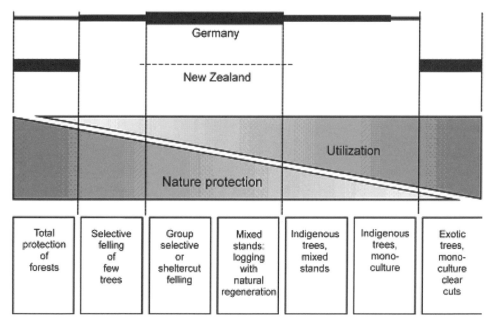

Fig.6.1. The range of forest management in Germany and New Zealand (data of Volz from Sands, 2003). The thickness of the bars represents the relative proportion of activities within the zones.

Germany has a range of forest management from completely protected areas through to plantation monocultures. Most activity is concentrated near the middle of the range, with group selection, shelterwood felling and natural regeneration in mixed stands. New Zealand, on the other hand, only operates at the extremes of the range. Timber production in New Zealand is almost totally confined to plantations of monocultures of exotic species, mainly *Pinus radiata* and all of the native forest that is publicly owned (83.5%) is in protected areas. Both of these are examples of multiple-use or of 'balanced forestry' (Kimmins, 1992). The development of plantations in New Zealand has allowed the reservation of one of the highest percentages of protected areas in the world and this should be recognized as an excellent example of conservation management. It is criticized, however, on the basis that insufficient funding is available to guard the protected areas from invasive pests. The argument is that a small amount of low impact logging in these forests would provide the funds to do so. The appropriate balance between intensive versus extensive management systems remains controversial.

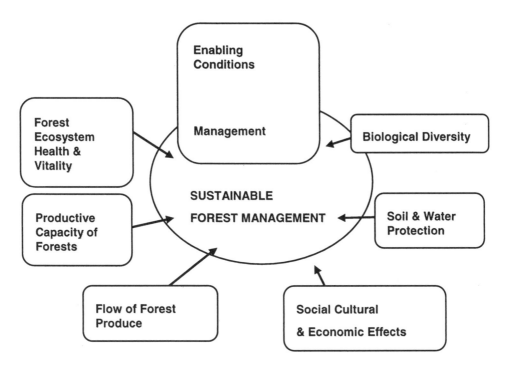

Fig. 6.2. The main components of sustainable forest management (from Wijewardana, 2005).

In many ways, forest management is a process of learning by doing. Rigorous monitoring should highlight discrepancies and foreshadow improvements. Management will then become progressively refined and this is called adaptive management. Some argue that sustainable forest management cannot be realized because it is impossible to reconcile the ecological, social and economic components. Even so, this cannot be used as an argument to not manage. Ultimately forest management is a balancing act to achieve the best outcome for humans and all other living things amidst an environment of single interest groups, each vying for supremacy and each of which on their own advocate a point of view that is patently unsustainable. The best available evidence strongly supports the proposition that timber harvesting appropriately managed need not compromise ecosystem values (Powers, 2001).

The Broader Aspects of Sustainable Forest Management

There has been a tendency to become preoccupied with the ecological basis of forest management, perhaps because this is the aspect that is most contentious. Figure 6.2 shows a broader vision of sustainable forest management which forms the basis of the Criteria and Indicators for Sustainable Forest Management which will be discussed in more detail in the next section. All of the components need to be met in order to achieve sustainable forest management.

Monitoring Sustainable Forest Management

FAO (2001) reported that 89% of forests in developed countries (45% of total world forest area) were being managed according to a formal or informal management plan. This does not mean, however, that they were being sustainably managed. Data supplied to FAO from developing countries was variable in quality, but preliminary estimates showed that at least 123 million hectares or about 6% of the total forest area was being managed under a nationally approved forest management plan covering a period of at least five years. Again, there is no suggestion that these forests were being sustainably managed. This sounds fairly depressing but FAO reported that there were significant improvements in most regions in the period 1990-2000. These data are not very revealing. They are obtained by asking the various countries what areas of forest they are managing. Many do not respond at all and others provide information that suggests that there is a lack of uniformity in the way that the definition of forest area managed has been interpreted and applied (FAO, 2001). For example, at least one country assumes that, if the forests are state-owned, they must therefore be managed. At least one other country only includes areas managed for timber production. There are no reliable figures on areas of forest that are presumed to be sustainably managed and this problem is exacerbated by the fact that there is considerable disagreement and confusion about what sustainable forest management is anyway. One way of getting a feel for what progress there has been towards sustainable forest management is to assess the degree of commitment that nations have to international initiatives to promote sustainable forest management.

There are many international bodies that are interested in forest policy and promoting sustainable forest management. Indeed it is an acronymic nightmare. Various conventions have arisen from the United Nations Convention on Environment and Development (UNCED) in Rio de Janeiro in 1992: the Intergovernmental Panel on Forests (IPF), the Intergovernmental Forum on Forests (IFF), the United Nations Forum on Forests (UNFF), and the Collaborative Partnership on Forests (CPF). International organizations with a stake in sustainable forest management include: the Food and Agriculture Organization of the United Nations (FAO), the United Nations Development Programme (UNDP), the United Nations Environment Programme (UNEP), the World Bank, the World Agroforestry Centre (ICRAF), the Center for International Forestry Research (CIFOR), the International Tropical Timber Association (ITTO), the Economic and Social Council of the United Nations (ECOSOC), the United Nations Department of Economic and Social Affairs (DESA), the Global Environment Facility (GEF), the International Union of Forest Research Organizations (IUFRO), and the Organization for Economic Cooperation and Development (OECD). Other related international conventions include the Convention on Biological Diversity (CBD), the United Nations Convention to Combat Desertification (UNCCD), the Convention on International Trade in Endangered Species of Wild Fauna and Flora (CITES), the United Nations Commission on Sustainable Development (UNCSD), the United Nations Framework Convention on Climate Change (UNFCC), and the Kyoto Protocol. International non-government organizations (NGOs) include the World Conservation Union (IUCN), the World Wide Fund for Nature (WWF), Greenpeace International, and the World Resources Institute (WRI). National governments have, to varying degrees, been active in supporting these international initiatives.

Criteria and Indicators for Sustainable Forest Management

UNCED proposed 'Forest Principles' which led to a number of national, regional and international initiatives to develop criteria and indicators for sustainable forest management. IPF/IFF proposed and UNFF endorsed the role of criteria and indicators in monitoring, assessing and reporting progress towards sustainable forest management by countries (Rametsteiner and Wijewardana, 2003). By the end of 2001, there were 149 out of 213 countries involved in one or more of nine initiatives (or processes) to develop and implement criteria and indicators for sustainable forest management: 56 countries from Africa, 49 from Asia, 20 from Oceania, 41 from Europe, 34 from North and Central America and 14 from South America. The nine different processes are: the Dry Zone Africa Process, the Pan European Forest Process, the Montreal Process, the Tarapoto Proposal, the Near East Process, the Lepaterique Process of Central America, the African Timber Organization Initiative, the Regional Initiative for Forests in Asia, and ITTO. The geographical distribution of these processes is shown in Fig. 6.3.

A criterion represents conditions or processes by which sustainable forest management can be assessed and there are seven criteria (thematic areas) that are common to all nine processes: the extent of forest resources; biological diversity; forest health and vitality; productive functions of forest resources; protective functions of forest resources; socio-economic benefits and needs; and legal, policy and institutional framework. The indicators are measurements of aspects of criteria and can be monitored over time to detect trends. Indicators of sustainable forest management within criteria vary between processes because of differences in climate, forest types, social environment, cultural environment, legal environment, political environment and economic environment.

The quality and degree of development and implementation has differed between countries and between processes. The processes are still evolving. The development of criteria and indicators has largely been a national government-led process and monitoring and reporting is done at the national level. The extent to which national reporting reflects regions or forest management units varies between countries. The processes should provide valuable tools in the development of national forest policy. The use of criteria and indicators is still in its infancy and progress has been far from smooth. However, it does indicate a willingness among much of the international community to embrace the concept of sustainable forest management. The most recent information on the state of progress is found in the papers resulting from conferences in Guatemala (FAO, 2003) and the Philippines (FAO, 2004).

Criteria and indicators offer a great opportunity to promote sustainable forest management because they are internationally recognized and government sanctioned and they encompass all of the ideals of triple bottom line (ecological, social and economic) accountability. They may be the main tool for promoting sustainable forest management in the future. ITTO is running workshops to train countries to use them and UNFF has given it a high priority. There is a likelihood that aid and private investment could flow towards implementing criteria and indicators when countries seek funding.

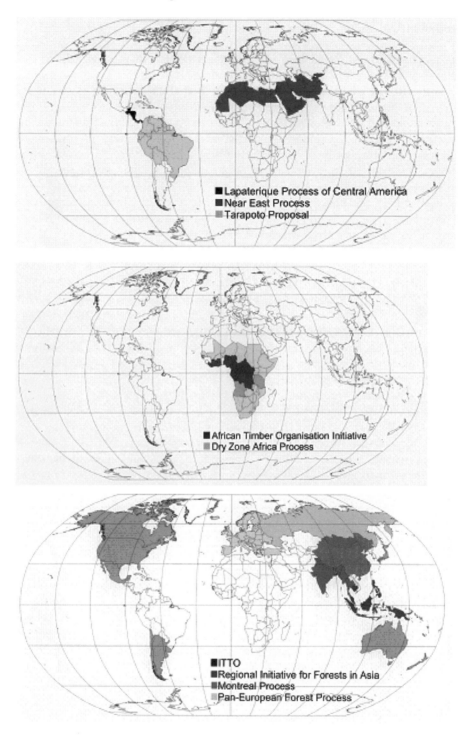

Fig 6.3. The geographical location of the Criteria and Indicator processes (from FAO, 2001).

Forest Certification

Another instrument to promote sustainable forest management is forest certification. This arose out of concern by mainly environmental NGOs about the role of logging in tropical deforestation and degradation. Initially the objectives of the NGOs were to reduce demand for tropical timber or stop it altogether. They promoted bans and boycotts of tropical timber with mixed success. Certainly this brought the issue of tropical deforestation to the consciousness (and conscience) of the international community. However, it became clear that boycotts were not very effective and indeed were unfair to those whose livelihood depended on logging and boycotts also disadvantaged indigenous peoples. It is unreasonable, impractical and unsustainable to reserve all tropical forests as protected areas. The best way to protect tropical forest from clearing for agriculture or grazing is to keep it as forest and manage it under secure tenure for timber production (see Chapter 5). Although forest areas in the developed world were stable or even increasing, there was considerable concern, again mainly among NGOs, that management (or non-management) was degrading the forest, that biodiversity was being lost and that other environmental values of the forest were being threatened. The most contentious issue was the fate of 'old growth' forests, forests with large old trees that were being felled and replaced with a management system that would never again allow trees to grow again to this age and size. Irrespective of any technical argument for and against logging in old growth forests, these old trees are both awesome and grand and, once removed, are never likely to be seen again in the life-time of our grandchildren or perhaps ever again, except in reserves. Surprisingly, some claim that old growth forests are being felled to create plantations. This is substantially not true. Rather, the establishment of plantations could assist in protecting the old growth forests.

The essence of certification is that it is a voluntary incentive-based process whereby forest managers agree to adopt documented and verifiable ecological, social and economic standards. A disinterested third party assesses the forest management against the agreed standards and certifies accordingly. There are two approaches to setting the standards: performance-based and systems-based. Performance-based approaches monitor whether the manager is actually doing what they contract to do and the Forestry Stewardship Council (FSC) certification is an example of this. Systems-based approaches are based on whether there are appropriate management systems in place to achieve the desired environmental performance and the International Standards Association's ISO 14000 series (14001, 14004, 14024) is an example of this. Implicit in forest certification is having a 'chain of custody', to ensure that each step in the path from standing tree to processed product in the marketplace is scrutinized by the assessor. Once the chain is broken, subsequent products can no longer be certified and cannot carry the certification logo. Developing robust systems for chain of custody is enormously difficult because of the large number of steps in the trail and the great variety of operators involved in placing a product in the marketplace. The most difficult part of the chain is assessing the use of certified timber in secondary processing, for example in the manufacture of kitchen cabinets, furniture, windows and doors. Under these circumstances the certifier might accept that a certain proportion of the components should come from certified wood.

Forest certification is a going concern but it is still in its infancy and still evolving. The degree of success of certification depends on the consumers, whether they be domestic or industrial. They need to be aware of certification, care enough about the welfare of forests to take an interest, and then be prepared to pay a premium for a certified product. Very few consumers are so environmentally motivated that they are prepared to pay a significant premium. Probably the success of certification depends on placing a certified product in the marketplace at a price equal to or not that much greater than an uncertified competitor. Unless the merchant can sell certified stock, she/he is unlikely to be excited about certification. Another way of promoting the use of certified products has been for protagonists to vigorously lobby large retailers to preferentially stock certified products. This has been successful in some cases, for example B&Q in the United Kingdom and Home Depot in the USA. The number of different environmental labels found in the marketplace can be confusing to the consumer. Outdoor furniture, for example, may be branded as 'made from timber from sustainably managed forests' where there has been no form of certification at all. In order for certification to be effective it has to be supported by strong certification labels that are recognized and respected by consumers, and retailers need to have systems to expose false labels. There are many certification systems, national and international, that are in various stages of development and implementation. Some have been championed by environmental NGOs, others by the timber industry organizations, and others by national governments. Inevitably, the proponents favour their own prejudices and the whole process of certification involves the bringing together of quite disparate groups and working towards a consensus solution. When it works, it is a major step forward and a triumph for common sense.

Perhaps the best known and most mature is that of the Forest Stewardship Council (FSC) and this scheme is discussed briefly as an example. FSC was founded in 1993 by representatives from environmental and conservation groups, the timber industry, the forestry profession, indigenous peoples' organizations, community forestry groups and forest product certification organizations from 25 countries. The Head Office is now in Bonn in Germany and it is funded by charitable foundations, government donors, membership subscriptions and accreditation fees. The standards of forest management set by FSC and against which forests are certified are outlined in their principles and criteria and the nine overarching principles are given in Box 6.1. The Forest Stewardship Council (FSC) offers third-party-assessed performance-based forest certification as a tool for promoting the sustainable management of forests. FSC also requires chain of custody to obtain a product label. Over the past 10 years, 42 million hectares in more than 60 countries (including about 5 million hectares of plantations) have been certified by FSC standards and several thousand products are produced using FSC certified wood and carrying the FSC label.

Certification has a long way to go before it makes a big impact on sustainable management of native forests. Even though the concept of certification was initially proposed to promote sustainable development of tropical forests, plantation managers have been eager clients. As yet only a very small proportion of forests are certified under any scheme and the success of certification in the marketplace is still an open question.

Managing Forests for Climate Amelioration

A current issue of great concern is the probability that increases in the concentration of carbon dioxide in the atmosphere from the burning of fossil fuels and the clearing of

---

Box 6.1. FSC principles

Principle 1: Compliance with laws and FSC Principles. Forest management shall respect all applicable laws of the country in which they occur, and international treaties and agreements to which the country is a signatory, and comply with all FSC Principles and Criteria.

Principle 2: Tenure and use rights and responsibilities. Long-term tenure and use rights to the land and forest resources shall be clearly defined, documented and legally established.

Principle 3: Indigenous people's rights. The legal and customary rights of indigenous peoples to own, use and manage their lands, territories, and resources shall be recognized and respected.

Principle 4: Community relations and worker's rights. Forest management operations shall maintain or enhance the long-term social and economic well-being of forest workers and local communities.

Principle 5: Benefits from the forest. Forest management operations shall encourage the efficient use of the forest's multiple products and services to ensure economic viability and a wide range of environmental and social benefits.

Principle 6: Environmental impact. Forest management shall conserve biological diversity and its associated values, water resources, soils, and unique and fragile ecosystems and landscapes, and, by so doing, maintain the ecological functions and the integrity of the forest.

Principle 7: Management plan. A management plan - appropriate to the scale and intensity of the operation - shall be written, implemented, and kept up to date. The long-term objectives of management, and the means of achieving them, shall be clearly stated.

Principle 8: Monitoring and assessment. Monitoring shall be conducted - appropriate to the scale and intensity of forest management – to assess the condition of the forests, yields of forest products, chain of custody, management activities and their social and environmental impacts.

Principle 9: Maintenance of high conservation value forests. Management activities in high conservation value forests shall maintain or enhance the attributes which define such forests. Decisions regarding high conservation value forests shall always be considered in the context of a precautionary approach.

Principle 10: Plantations. Plantations shall be planned and managed in accordance with Principles and Criteria 1 -9, and Principle 10 and its Criteria. While plantations can provide an array of social and economic benefits, and can contribute to satisfying the world's needs for forest products, they should complement the management of, reduce pressures on, and promote the restoration and conservation of natural forests.

---

vegetation, together with other gases, are warming the world at a rapid rate. The questions to be addressed are: is global warming occurring; are increased carbon dioxide concentrations to blame; and what does it matter anyway?

Is the Global Climate Changing

The answer to this question is, of course, yes: global climate is changing and is continually changing. Over geological time there have been cycles of warming and cooling and there are global and regional patterns within cycles within larger cycles. The Tables in Chapter 1 summarized some of the main climatic cycles of the past. The evolution of life on earth has been heavily influenced by climate change. There was a cold and dry period in Africa about 2.5 million years ago which assisted the development of grassland at the expense of forest, heralding a major advance in human evolution. The marked changes in climate since then, expressed as the ice ages, caused separation and rejoining of hominid populations, conditions ideal for human evolution and diversification. A more recent example is Greenland which supported a thriving agricultural community about 1000 years ago. Agriculture was abandoned in the 15th century because of falling temperatures and now some cultivation is resuming. Is it possible that the current warming trend is just part of a cycle like this and that falling temperatures can be expected soon?

There is good evidence, both anecdotal and scientifically robust, that the earth is in a rapid warming phase. There are a variety of predictions on how much warming will occur and what will be the effect of this on humans and other living things. Predicting climate change is a complex issue. There is no doubt that carbon dioxide concentrations are increasing but the impact of this on global climate is not so straight forward. Predictions can go from one extreme to the other. Some scientists in the 1970s predicted that the increased carbon dioxide concentrations would initially increase temperatures, which would increase the evaporation of water, which would increase cloud cover, which would increase snow on the high mountains, which would reflect more sunlight, which would lower the temperatures, which would precipitate an ice age. Other scientists have taken the other extreme and argue that increasing carbon dioxide concentrations will increase temperature, which will evaporate water, which will not form clouds, and therefore the temperature will rise further, and will evaporate even more water, causing even greater temperatures, and so on until finally all the oceans evaporate and the earth becomes a dead planet like Venus (Philander, 2003).

Figure 6.4 shows the globally averaged land and sea surface temperatures from 1860 to 2000 relative to the average for the period 1961 to 1990. This suggests that temperature rose from 1910 to 1945 and then steadied and perhaps even fell at times from 1940 to 1976 after which temperature rose again. The 1990s had the most rapid increase in global temperature of any decade in the last 1000 years (Houghton *et al.*, 2001). Uncertainty and differences of opinion among scientists is normal and healthy. This uncertainty, however, has assisted enterprises with a vested interest in not having to reduce their carbon emissions to cast doubt on current predictions of climate change and the credibility of the underpinning science. However, the balance of evidence strongly suggests that global warming is occurring, that it has serious consequences for the human race and that something needs to be done about it. Also, the balance of evidence strongly suggests that the current warming trend is not part of natural variation or a natural cycle but is caused by increasing concentrations of certain gases, called greenhouse gases, in the atmosphere. Global mean temperature has increased by 0.4 to 0.8°C during the 20th century (Houghton *et al.*, 2001). Areas of snow and ice are significantly decreasing, the sea level has risen by between 0.1 to 0.2 metres over the last few decades and heavy rainfall events have increased in middle latitudes (North, 2003). It is estimated that in the 21st century the mean global temperature will increase by 1.5 to 4.5°C, precipitation will increase by 3 to 5% and sea level will rise by about 45 cm. Models predict that temperature increases will be greatest at higher latitudes, particularly the northern latitudes, and that changes in rainfall patterns will be most pronounced at low latitudes. The evidence that the world is warming is

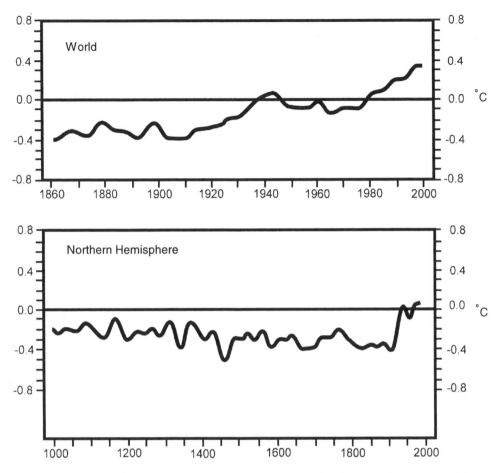

Fig. 6.4. Departures in temperatures (°C) from the average for the period 1961 to 1990 for the northern hemisphere from year 1000-2000 and of the world from 1860 to 2000 (adapted from IPCC, 2001).

stronger today than it was five years ago. Having said this, there is no guarantee that global warming will continue. Some scientists (a minority) do not acknowledge a link between global warming and greenhouse gas emissions.

The next question is why should it matter if there is global warming? Surely people in colder parts of the world would welcome warmer weather and appreciate not having to pay so much for heating their homes. Also the predicted increases in plant productivity should help rather than hinder the supply of food, fuel and fibre. However, the problem is that the changes are occurring more rapidly than any climate change over human history and the effects are not equally distributed around the earth. Some parts of the world will become colder, some wetter and others drier. Natural climatic disasters (hurricanes, floods, droughts) will change in frequency, severity and location. Wetlands may dry out, forest may disappear, fire frequency may increase, pests and diseases may increase, deserts may become hotter and drier, coral reefs may die, and ecosystems may change in structure and degrade. Melting of polar and glacial ice will raise sea levels and low lying communities will need to relocate. Species will shift in their geographical range, and some species may not be able to adapt quickly enough to changed conditions. UNEP World Conservation Monitoring Centre details the expected impacts of climate change on biodiversity on their web site.

What is the Greenhouse Effect and How is this Related to Global Warming

A greenhouse creates a warmer climate inside than outside because the incoming mainly visible radiation is reflected back as longer wave-length radiation which is not so readily transmitted through the greenhouse walls and roof. Heat therefore gets trapped inside the greenhouse. The earth and its atmosphere operate like a giant greenhouse with the atmosphere acting as the walls and roof. Solar radiation reaches the earth's surface mainly as visible radiation but is reflected back to space as longer wave-length radiation. This is impeded by atmospheric gases and trapped radiation is directed back to the earth's surface causing warming. The major greenhouse gas by far is water vapour. Without water vapour the temperature at the earth's surface would be an average of 33°C cooler. In recent times, however, certain gases generated by human activity have been increasing in concentration in the atmosphere, thereby increasing the efficiency with which the atmosphere reflects heat back to the earth's surface rather than into space. These greenhouse gases include methane, nitrous oxide, hydrofluorocarbons, perfluorocarbons and sulphur hexafluoride. However, the most significant is carbon dioxide which by itself is predicted to be responsible for half of the global warming that has occurred in recent times and which is yet to occur.

Carbon dioxide concentrations have fluctuated between 100 and 400 ppm for the last 20 million years (Pearson and Palmer, 2000) and in the last 2000 years until the time of the industrial revolution have remained between 270 and 290 ppm (Barnola *et al.,* 1995). The concentration of carbon dioxide in the atmosphere has been increasing exponentially since the industrial revolution (Figure 6.5) from about 275 ppm in 1750 to about 360 ppm in year 2000 and is predicted to increase to reach 550 ppm in 2050 and more than 700 ppm in year 2100 unless something is done about it (Schlesinger, 2003). Figure 6.5 gives the range of predictions based on several scenarios (see IPCC, 2001). All scenarios predict an increase in carbon dioxide.

Places where carbon is stored are called sinks and movements of carbon between sinks are called fluxes. Carbon sinks are the atmosphere, terrestrial and marine biomass (living and dead), carbon dioxide dissolved in the oceans, fossil fuels and carbonate rocks. The amount of carbon stored in terrestrial biomes is cyclic, being greater during warm periods and lesser in cooler periods. The amount of carbon (in carbon dioxide) in the atmosphere depends on the net flux of carbon moving to the atmosphere from other sinks and from the atmosphere to other sinks. The flux from the atmosphere to the land is mainly from photosynthesis (equation 4.1 in Chapter 4) and that from the land to the atmosphere is respiration, decomposition and combustion of carbohydrates from living or dead biomass (equation 4.2), combustion of fossil fuels such as oil (hydrocarbons) and coal (carbon), release from biogenic volatile organic compounds and from the release of carbon from carbonate rocks in the manufacture of cement. This all sounds relatively simple but the challenge of accounting for all of the carbon fluxes and storage at the national and global level is complex and challenging (Field and Raupach, 2004).

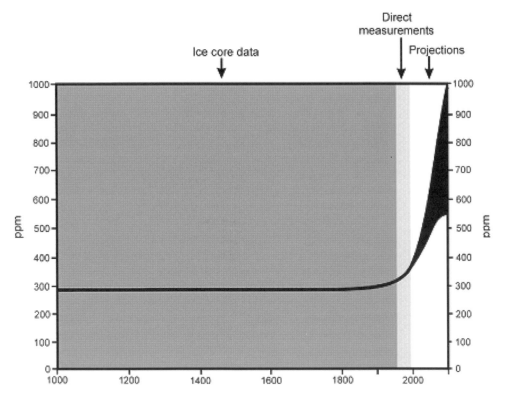

Fig.6.5. Atmospheric carbon dioxide concentrations from year 1000 to 2000 and the projected range from 2000 to 2100 (adapted from IPCC 2001).

Table 6.1 shows that most carbon is in carbonate rocks, that the oceans contain 10 times more carbon than terrestrial biomes, that carbon in fossil fuel reserves is about the same as the total of terrestrial biomes, that soil in terrestrial biomes contains five times more carbon than plants, that forests contain more carbon than other terrestrial biomes and that tropical forests contain the most carbon of all forests. The amount of carbon in the atmosphere is increasing at over 3 Gt per year. This is predominantly caused by burning fossil fuels. Deforestation and forest degradation contribute significantly to the flux from terrestrial biomes to the atmosphere. However, this is probably compensated for, or more than compensated for by other fluxes from the atmosphere to terrestrial biomes. The net flux from all forms of land use is from atmosphere to land, perhaps about 1.2 Gt of carbon per year (Table 6.1) but most likely greater than zero. Mahli *et al.* (1999) suggest that the increased growth of existing forests from increased photosynthesis, combined with the reforestation and regrowth of degraded forests, offsets global emissions from deforestation.

There are several possible reasons for the increased carbon in terrestrial biomes in recent years: abandonment of agricultural land in the developed world, forests growing faster than they are being cut in the developed world, better fire suppression, recovery of marginal forests and the 'fertigation' effect (Sabine *et al.,* 2004). Permanent yield plots throughout the world have shown a general increase in forest productivity in recent decades that cannot be explained in terms of better management alone. It is probably caused by a combination of increased photosynthesis from higher concentrations of carbon dioxide in the atmosphere and elevated temperatures from global warming. This has been called fertigation. It is very well demonstrated in Europe where, fortuitously, increased nitrogen and carbon dioxide, both from burning fossil fuels, have combined to significantly increase the productivity of European forests.

Table 6.1. Global carbon sinks and carbon fluxes to and from the atmosphere to terrestrial and oceanic sinks during the 1990s (adapted from Dixon *et.al.*, 1994; Jobbagy and Jackson, 2000; Prentice, 2001; Saugier *et al.*, 2001; Houghton, 2003; Field and Raupach, 2004; Sabine *et.al.*, 2004).

Sinks (Gt of carbon)		
Carbonate rocks		65,000,000
Ocean		39,000
Fossil fuel reserves		4,000
Terrestrial biomes (vegetation plus soil)		
Tropical forest plants	340	
Tropical forest soils	692	
Temperate forest plants	139	
Temperate forest soils	262	
Boreal forest plants	57	
Boreal forest soils	150	
Other biomes plants	116	
Other biomes soils	2,090	
Total terrestrial biomes plants	652	
Total terrestrial biomes soils	3,194	
Total terrestrial biomes plants plus soil		3,846
Atmosphere		750
Fluxes to (+) and from (-) the atmosphere to other sinks (Gt of carbon per year)		
Ocean to/from atmosphere		-1.9
Land to/from atmosphere		
Combustion of fossil fuels, cement production		+6.3
Deforestation, forest degradation and destructive land use		+2.2
Other land use		-3.4
Net carbon gain by the atmosphere (Gt of carbon per year)		3.2

    Even so, the net flux of carbon is from the land to the atmosphere and carbon dioxide concentrations in the atmosphere continue to increase (Table 6.1). There are three ways of addressing this: reduce the combustion of fossil fuels; reduce deforestation and forest degradation; and further increase the amount of carbon sequestered in terrestrial biomes, particularly in forest ecosystems and in agricultural soils. Brown *et al.* (1996) suggest that forests could be managed over the next 50 years to sequester 60 to 87 Gt of carbon and Cole *et al.* (1996) suggest that agriculture could be managed to sequester 20 to 40 Gt of carbon in agricultural soils.

What Role can Forests Play in Reducing or Reversing Global Warming

Forest ecosystems (which include the soil) store greater than 40% of terrestrial carbon (Table 6.1). However, many forests are close to carbon neutral (as for example a mature climax rainforest) where carbon gain from photosynthesis is nearly matched by carbon loss from respiration and decomposition. The 'ideal' tree from the perspective of sequestering carbon would be one that photosynthesizes profusely, respires sparingly and grows forever. No forests of course are anything like this. Trees slow down in growth as they get older and they die and decay, or else are harvested. If a forest is cleared and burnt then this will produce an immediate flush of carbon dioxide to the atmosphere. If a forest is cleared and the residues are left on the forest floor to decay, this will also add carbon dioxide to the atmosphere but at a slower and more prolonged rate (but of course as the forest regenerates it will sequester carbon from the atmosphere). If the trees are logged and processed into some long-lived product, the carbon in this product is protected against decay for many years. The process of making this product, however, involves the use of energy in felling the tree, transporting it to the processing plant and processing the product. Most often this energy is either from fossil fuel or from burning wood waste, both of which contribute carbon dioxide to the atmosphere. However, the process from tree to product usually will consume less energy and therefore contribute less carbon dioxide to the atmosphere than the same product fabricated from another material (see Chapter 4). All energy inputs into establishment, maintenance, harvesting and processing need to be accounted for when managing a forest for carbon sequestration. Simply speaking, any management that increases the productivity of existing forests, that repairs degraded forests and that increases forest area will increase carbon sequestration. Forests that maintain fast growth rates for as long as possible before harvest are the best prospects for carbon sequestration. Plantations meet these requirements if they are part of an estate where steady state biomass is constant or increasing over time.

The United Nations Framework Convention on Climate Change, the Kyoto Protocol and Carbon Credits

In 1990, the general assembly of the United Nations, responding to predictions of global warming, passed resolution 45/212 which launched an Intergovernmental Negotiating Committee (INC) which adopted the United Nations Framework Convention on Climate Change in 1991 and which was open to signature at the Rio de Janeiro Conference on Environment and Development in 1992. It entered into force in 1994 and currently has been joined by 188 states and the European Community. Annual Conferences of the Parties (COP) discuss how best to confront global climate change and have devised rules for implementation of the convention. However the COP soon realized that the convention alone could not be effective in arresting climate change and in 1997, the COP 3 meeting at Kyoto established the preliminary rules of the Kyoto Protocol. COP 4 at Buenos Aires in 1998, COP 6 at the Hague in 2000, COP 6 (resumed) in Bonn in 2001 and COP 7 at Marrakesh in 2001 refined the Protocol, concluding an intense international negotiating cycle which resulted in an agreement on the rules for implementation (UNFCCC, 2003). A list of the world's countries together with their relationship to the convention and to the protocol is given in Table 6.2.

The convention divides the countries into three main groups based on the degree of industrialization and development. These groups have different commitments under the convention and the protocol. Annex 1 countries are the OECD countries in 1992 including countries with economies in transition (EIT). Annex 1 parties were required to reduce their greenhouse emissions to 1990 levels by the year 2000, with some leniency being given to

Table 6.2. Countries and their relationship to the Framework Convention on Climate Change and the Kyoto Protocol (adapted from UNFCCC 2003, but also noting that Russia has since ratified the Kyoto Protocol). A, ratified UNFCC; B, ratified Kyoto Protocol; C, Annex 1 countries; D, Annex 1 countries with economies in transition.

COUNTRY	A	B	C	D	COUNTRY	A	B	C	D
Afghanistan	X				Fiji	X	X		
Albania	X				Finland	X	X	X	
Algeria	X				France	X	X	X	
Andorra					Gabon	X			
Angola	X				Gambia	X	X		
Antigua and Barbuda	X	X			Georgia	X	X		
Argentina	X	X			Germany	X	X	X	
Armenia	X				Ghana	X			
Australia	X		X		Greece	X	X	X	
Austria	X	X	X		Grenada	X	X		
Azerbaijan	X	X			Guatemala	X	X		
Bahamas	X	X			Guinea	X	X		
Bahrain	X				Guinea-Bissau	X			
Bangladesh	X	X			Guyana	X			
Barbados	X	X			Haiti	X			
Belarus	X				Holy See				
Belgium	X	X	X		Honduras	X	X		
Belize	X				Hungary	X	X	X	X
Benin	X	X			Iceland	X	X	X	
Bhutan	X	X			India	X	X		
Bolivia	X	X			Indonesia	X			
Bosnia and Herzegovina	X				Iran	X			
Botswana	X				Iraq				
Brazil	X	X			Ireland	X	X	X	
Brunei Darussalam					Israel	X			
Bulgaria	X	X	X	X	Italy	X	X	X	
Burkina Faso	X				Jamaica	X	X		
Burundi	X	X			Japan	X	X	X	
Cambodia	X	X			Jordan	X	X		
Cameroon	X	X			Kazakhstan	X			
Canada	X	X	X		Kenya	X			
Cape Verde	X				Kiribati	X	X		
Central African Republic	X				Kuwait	X			
Chad	X				Kyrgyzstan	X			
Chile	X	X			Lao People's Democratic Republic	X	X		
China	X	X			Latvia	X	X	X	X
Colombia	X	X			Lebanon	X			X
Comoros	X				Lesotho	X	X		
Congo	X				Liberia	X	X		
Cook Islands	X	X			Libyan Arab Jamahiriya	X			
Costa Rica	X	X			Liechtenstein	X		X	
Cote d'Ivoire	X				Lithuania	X	X	X	X
Croatia	X		X	X	Luxembourg	X	X	X	
Cuba	X	X			Madagascar	X			
Cyprus	X	X			Malawi	X	X		
Czech Republic	X	X	X	X	Malaysia	X	X		
Democratic Peoples Rep. of Korea	X				Maldives	X	X		
Democratic Republic of the Congo	X				Mali	X	X		
Denmark	X	X	X		Malta	X	X		
Djibouti	X	X			Marshall Islands	X			
Dominica	X				Mauritania	X			
Dominican Republic	X	X			Mauritius	X	X		
Ecuador	X	X			Mexico	X	X		
Egypt	X				Micronesia (Federated States of)	X	X		
El Salvador	X	X			Monaco	X		X	
Equatorial Guinea	X	X			Mongolia	X	X		
Eritrea	X	X			Morocco	X	X		
Estonia	X	X	X	X	Mozambique	X			
Ethiopia	X				Myanmar	X			
European Union	X	X	X		Namibia	X			

Country					Country				
Nauru	X	X			Slovakia	X	X	X	X
Nepal	X				Slovenia	X	X	X	X
Netherlands	X	X	X		Solomon Islands	X	X		
New Zealand	X	X	X		Somalia				
Nicaragua	X	X			South Africa	X	X		
Niger	X				Spain	X	X	X	
Nigeria	X				Sri Lanka	X	X		
Niue	X	X			Sudan	X			
Norway	X	X	X		Suriname	X			
Oman	X				Swaziland	X			
Pakistan	X				Sweden	X	X	X	
Palau	X	X			Switzerland	X	X	X	
Panama	X	X			Syrian Arab Republic	X			
Papua New Guinea	X	X			Tajikistan	X			
Paraguay	X	X			Thailand	X	X		
Peru	X	X			The Former Yugoslavia Macedonia	X			
Philippines	X				Togo	X			
Poland	X	X	X	X	Tonga	X			
Portugal	X	X	X		Trinidad and Tobago	X	X		
Qatar	X				Tunisia	X	X		
Republic of Korea	X	X			Turkey	X		X	
Republic of Moldova	X				Turkmenistan	X	X		
Romania	X	X	X	X	Tuvalu	X	X		
Russian Federation	X	X	X	X	Uganda	X	X		
Rwanda	X				Ukraine	X		X	X
Saint Kitts and Nevis	X				United Arab Emirates	X			
Saint Lucia	X				United Kingdom	X	X	X	
St Vincent and Grenadines	X				United Rep. of Tanzania	X	X		
Samoa	X	X			USA	X		X	
San Marino	X				Uruguay	X	X		
Sao Tome and Principe	X				Uzbekistan	X	X		
Saudi Arabia	X				Vanuatu	X	X		
Senegal	X	X			Venezuela	X			
Serbia and Monte Negro	X				Vietnam	X	X		
Seychelles	X	X			Yemen	X			
Sierra Leone	X				Zambia	X			
Singapore	X				Zimbabwe	X			

EIT parties. Annex 2 parties are Annex 1 parties less EIT parties and they are required to provide financial resources to assist developing countries to reduce emissions. The overall objective of the Framework Convention on Climate Change is to stabilize atmospheric concentrations of greenhouse gases. Parties to the convention are expected, among other things, to prepare and deliver national programmes that sustainably manage carbon sinks, develop environmentally friendly low or no carbon emission technologies, and promote research into climate change and public awareness about climate change. Clearly a reduction in the consumption of fossil fuels is a major plank in the convention strategy, but so also is the sustainable management of forest ecosystems to increase the amount of carbon sequestered in forests.

The convention continues to be the main avenue for governments to collectively fight climate change. However, it lacks teeth when it comes to controlling greenhouse gas emissions and the Kyoto Protocol aims to achieve this by applying strict legally enforceable emission controls, at least as far as international law can be enforced. Accordingly the response by countries to ratify the protocol has been less enthusiastic (see Table 6.2). The protocol cannot enter into force until at least 55 parties to the convention ratify it and this needs to include enough Annex 1 parties to cover at least 55% of the fossil fuel carbon dioxide emissions of all Annex 1 parties in 1990. The USA has said that it will not ratify the protocol because it is not in its national interest to do so. Another reason the USA has not ratified is that it considers the strong emerging economies like China, India and Brazil should also be required to accept mandatory emission controls. Because the USA had 36% of the Annex 1 parties' carbon dioxide emissions in 1990, some other big users needed to ratify the protocol in order for it to enter into force. The European Union (24% of 1990 emissions) has ratified and the Russian Federation (17% of 1990 emissions) has recently ratified (October 2004). The Kyoto protocol became enforceable in those countries that ratified it in February 2005. No single Annex 1 party can block the entry into force of the protocol. Annex 2 countries that have not elected to ratify the protocol (USA and Australia) are still working actively to reduce emissions. The USA, for example, is leading the world in developing technologies to develop cost-effective energy systems to replace fossil fuels.

The Kyoto Protocol requires that Annex 1 parties reduce their greenhouse emissions by at least 5% from the 1990 levels by 2008-2012. EIT parties will be given some leniency. Annex 1 parties to the Protocol are required to adopt policies that have a mitigating effect on climate change. They include: enhancing energy efficiency; promoting renewable energy; promoting sustainable agriculture and forestry; and protecting and enhancing carbon sinks. For reporting purposes, parties to the protocol must have robust and credible accounting systems for greenhouse gas emissions. Nations are required to compile a registry of emission reduction units, which can be traded as carbon credits (or Certified Tradeable Offsets) in the international market. Individual nations are responsible for how they account for the credits within their own borders. Carbon trading increases flexibility. For example, a nation that is having difficulty in reducing emissions in its own borders (or an industrial enterprise within that country) can invest in greenhouse gas reduction activities in another country to assist it to meet its obligations under the protocol. Forestry will be the major player in trading of carbon credits. Sustainable forest management, and particularly plantation management, should be a big winner out of the protocol. This should improve the financial performance of forest companies, providing of course that the cash goes to the company and is not dispersed across the national economy. Forest managers are becoming increasingly concerned that the latter may be the case. Trading in carbon credits is still in its infancy.

Forest management or mismanagement can contribute to both the problem (through deforestation and forest degradation) and the solution (by increasing the amount of carbon sequestered in forests). Forestry alone, however, is not the only or even the main solution to global warming. The primary focus should be on reducing carbon emissions from the combustion of fossil fuels.

References and Further Reading

Aplet, G.H., Johnson, N., Olson, J.T. and Sample, E.O. (eds) (1993) *Defining Sustainable Forestry.* Island Press, Washington, DC.

Barnola, J.M., Anklin, M., Porcheron, J., Raynaud, D., Schwander, J. and Stauffer, B. (1995) CO_2 evolution during the last millennium as recorded by Antarctic and Greenland ice. *Tellus* 47B, 264-272.

Baumert, K.A. (2002) *Building on the Kyoto Protocol: Options for Protecting the Climate.* World Resources Institute, Washington, DC.

Benecke, U. (1996) Ecological silviculture: the application of age old methods. *N.Z. Forestry* 41(2), 27-33.

Boyle, T.J.B. (2001) Interventions to enhance the conservation of biodiversity. In: Evans, J. (ed.) *The Forests Handbook*, Volume 2. Blackwell Science, Oxford, pp. 82-101.

Brown, S., Sathaye, J., Cannel, M. and Kauppi, P. (1996) Management of forests for mitigation of greenhouse gas emissions. In: Watson, R.T., Zinoyera, M.C. and Moss, R.H. (eds) *Climate Change 1995, Impacts, Adaptations and Mitigation of Climate Change: Scientific-Technical Analyses.* Contribution of Working Group II to the Second Assessment Report of the Intergovernmental Panel on Climate Change. Cambridge University Press, Cambridge, pp. 773-797.

Burgman, M.A. and Lindenmayer, D.B. (1998). *Conservation Biology for the Australian Environment.* Surrey Beatty and Sons, Chipping Norton, N.S.W., Australia.

Cerri, C., Minami, K., Mosier, A., Rosenberg, N. and Sauerbeck, D. (1996) Agricultural options for mitigation of greenhouse gas emissions. In: Watson, R.T., Zinoyera, M.C. and Moss, R.H. (eds) *Climate Change 1995, Impacts, Adaptations and Mitigation of Climate Change: Scientific-Technical Analyses.* Contribution of Working Group II to the Second Assessment Report of the Intergovernmental Panel on Climate Change, Cambridge University Press, Cambridge, pp. 745-771.

Diamond, J.M. (1975) The island dilemma: lessons of modern biogeographical studies for the design of natural preserves. *Biological Conservation* 7, 129-146.

Dixon, R.K., Brown, S., Houghton, R.A., Solomon, A.M. Trexler, M.C. and Wisniewski, J. (1994) Carbon pools and flux of global ecosystems. *Science* 263, 185-190.

FAO (2001) *Global Forest Resources Assessment 2000 - Main Report.* FAO Forestry Paper 140, Food and Agriculture Organization of the United Nations, Rome.

FAO (2003) *Report on International Conference on the Contribution of Criteria and Indicators for Sustainable Forest Management: The Way Forward.* 3-7 February 2003, Guatemala City, Guatemala, www.fao.org (accessed 18 September 2004).

FAO (2004) *Report on Expert Consultation on Criteria and Indicators for Sustainable Forest Management,* 2-4 March 2004, Cebu City, Philippines, www.fao.org (accessed 18 September 2004).

Ferguson, I.F. (1996) *Sustainable Forest Management.* Oxford University Press, Oxford.

Field, C.B. and Raupach, M.R. (eds) (2004) *The Global Carbon Cycle – Integrating Humans, Climate, and the Natural World.* Island Press, Washington, DC.

Franklin, J.F., Berg, D.R., Thornburgh, D.A. and Tappeiner, J.C. (1997) Alternative silvicultural approaches to timber harvesting: variable retention harvesting systems. In: Kohm, K.A. and Franklin, J.A. (eds) *Creating a Forestry for the 21st Century.* Island Press, Washington, DC, pp. 111-139.

Griffin, J.M. (ed.) (2003) *Global Climate Change – The Science, Economics and Politics.* Edward Elgar, Cheltenham.

Helles, F., Holten-Anderson P. and Wichmann, L. (eds) (1999) *Multiple Use of Forests and Other Natural Resources.* Kluwer Academic Publishers, Dordrecht.

Houghton, J., Ding, Y., Griggs, D.J., Noguer, M., Linden, P.J. van der, Dai, X., Maskell, K. And Johnson, C.A. (eds) (2001) *Climate Change 2001: The Scientific Basis.* Contribution of Working Group I to the Third Assessment Report of the Intergovernmental Panel on Climate Change, Cambridge University Press, Cambridge.

Houghton, R. (2003) Why are estimates of the terrestrial carbon balance so different? *Global Change Biology* 9, 500-509.

Hunter, M.L. Jr. (2002) *Fundamentals of Conservation Biology,* 2nd edition. Blackwell Science, Oxford.

Hunter, M.J. Jr. (ed.) (1999) *Maintaining Biodiversity in Forest Ecosystems.* Cambridge University Press, Cambridge.

Ierland, E.C. van, Gupta, J. and Kok, M.T.J. (eds) (2003) *Issues in International Climate Policy – Theory and Policy.* Edward Elgar, Cheltenham.

IPCC (2001) *Climate Change 2001: The Synthesis Report.* Cambridge University Press, Cambridge.

Jobaggy, E. and Jackson, R. (2000) The vertical distribution of soil organic carbon and its relation to climate and vegetation. *Ecological Applications* 10, 423-436.

Karnosky, D., Ceulemans, R., Scarascia-Mugnozza, G. and Innes, J.L. (2001) *The Impact of Carbon Dioxide and Other Greenhouse Gases on Forest Ecosystems.* CAB International, Wallingford.

Kauffmann, M.R., Graham, R.T., Boyce, D.A. Jr., Moir, W.H., Perry, L., Reynolds, R.T., Bassett, R.L., Mehlhop, P., Edminster, C.B., Block, W.M. and Corn, P.S. (1994) *An Ecological Basis for Ecosystem Management.* USDA, US Forest Service, Rocky Mountain Forest and Range Experiment Station, General Technical Report RM-246, Fort Collins, Colorado.

Kimmins, H. (1992) *Balancing Act: Environmental Issues in Forestry.* University of British Columbia Press, Vancouver.

Mencuccini, M., Grace, J., Moncrieff, J. and McNaughton, K.G. (eds) (2004*) Forests at the Land-Atmosphere Interface.* CAB International, Wallingford.

Nambiar, E.K.S. (1996) Sustaining productivity of forests as a continuing challenge to soil science. *Soil Science Society of America Journal* 60, 1629-1642.

Mahli, Y., Baldocchi, D.D. and Jarvis, P.G. (1999) The carbon balance of tropical, temperate and boreal forests. *Plant Cell and Environment* 22, 715-740.

Maser, C. (1994) *Sustainable Forestry - Philosophy, Science and Economics.* St Lucie Press, Delray Beach, Florida.

North, G.R. (2003) Climate change over the next century. In: Griffin, J.M. (ed.) *Global Climate Change – The Science, Economics and Politics.* Edward Elgar, Cheltenham, pp. 45-66.

Norton, D.A. (1999) Forest Reserves. In: Hunter, M.J. Jr. (ed.) *Maintaining Biodiversity in Forest Ecosystems.* Cambridge University Press, Cambridge, pp. 525-555.

O'Riordan, T and Stoll-Kleeman, S. (2002) *Biodiversity, Sustainability and Human Communities.* Cambridge University Press, Cambridge.

Pearson, P.N. and Palmer, M.R. (2000) Atmospheric carbon dioxide concentrations over the past 60 million years. *Nature* 406, 695-699.

Powers, R.F. (2001) Assessing potential sustainable wood yield. In: Evans, J. (ed.) *The Forests Handbook,* Volume 2. Blackwell Science, Oxford, pp. 105-128.

Philander, S.G. (2003) Why global warming is a controversial issue. In: Rodó, X. and Comín, F.A. (eds) *Global Climate – Current Research and Uncertainties in the Climate System.* Springer-Verlag, Berlin, pp. 25-33.

Prentice, C., Farquhar, G.D., Fasham, M.J.R. Goulden, M.L., Heimann, M., Jaramillo, V.J., Kheshgi, H.S., Le Quéré, C., Scholes, R.J. and Wallace, D.W.R. (2001) The carbon cycle and atmospheric carbon dioxide. In: Houghton, J., Ding, Y., Griggs, D.J., Noguer, M., Linden, P.J. van der, Dai, X., Maskell, K. and Johnson, C.A. (eds*) Climate Change 2001: The Scientific Basis.* Contribution of Working Group I to the Third Assessment Report of the Intergovernmental Panel on Climate Change. Cambridge University Press, Cambridge, pp. 183-287.

Raison, J.R., Brown, A.G. and Flinn, D.W. (2001) *Criteria and Indicators for Sustainable Forest Management.* IUFRO Research Series 7, CAB International, Wallingford.

Rametsteiner, E. and Widgewardana, D. (2003) Key Issues in the future development of international initiatives on forest-related criteria and indicators for sustainable development. In: *Report from International Conference on the Contribution of Criteria and Indicators for Sustainable Forest Management: The Way Forward.,* Volume 2, 3-7 February 2003, Guatemala City, Guatemala, www.fao.org/DOCREP/005/J0077E/J0077E00.HTM (accessed 18 September 2004).

Rodó, X. and Comín, F.A. (eds) (2003) *Global Climate – Current Research and Uncertainties in the Climate System.* Springer-Verlag, Berlin.

Sabine, C.L., Heimann, M., Artaxo, P., Bakker, D.C.E., Chen, C-T.A., Field, C.B., Gruber, N., Le Quéré, C., Prinn, R.G., Richey, J.E., Lankao, P.R., Sathaye, J.A. and Valentini, R. (2004) Current status and past trends of the global carbon cycle. In: Field, C.B. and Raupach, M.R. (eds) *The Global Carbon Cycle – Integrating Humans, Climate, and the Natural World.* Island Press, Washington, pp. 17-44.

Sands, R. (2003) Professional forestry education in Australasia. *New Zealand Journal of Forestry* 48, 20-26.

Saugier, B., Roy, J. and Mooney, H.A. (2001) Sinks for anthropogenic carbon. *Physics Today* (August) 30-36.

Sawa, T. (ed.) (2003) *International Frameworks and Technological Strategies to Prevent Climate Change.* Springer-Verlag, Tokyo.

Schlesinger, W.H. (2003) The carbon cycle: human perturbations and potential management options. In: Griffin, J.M. (ed) *Global Climate Change – The Science, Economics and Politics*. Edward Elgar, Cheltenham, pp. 25-44.

Sheppard, S.R.J. and Harshaw, H. (eds) (2001) *Forests and Landscapes: Linking Ecology, Sustainability and Aesthetics*. CAB International, Wallingford.

Simberloff, D.S. and Abele, L.G. (1982) Island biogeography theory and conservation: effects of fragmentation. *American Naturalist* 120, 41-50.

Smith, D.M., Larson, B.C., Kelty, M.J. and Ashton, P.M.S. (1997) *The Practice of Silviculture: Applied Forest Ecology*. Wiley, New York.

Smith, W. and Maser, C. (2001) *Forest Certification in Sustainable Development – Healing the Landscape*. Lewis Publishers, Boca Raton.

Smith, J.B., Klein, R.J.T. and Huq, S. (eds) (2003) *Climate Change – Adaptive Capacity and Development*. Imperial College Press, London.

UNFCCC (2003*) Caring for Climate: A Guide to the Climate Change Convention and the Kyoto Protocol.* Climate Change Secretariat (UNFCCC), Bonn, Germany.

Viana, V.M., Ervin, J., Donovan, R.Z., Elliott, C. and Gholz, H. (1996) *Certification of Forest Products*. Island Press, Washington, DC.

Vogt, K.A., Larson, B.C., Gordon, J.C., Vogt, D.J. and Fanzeres, A. (1999) *Forest Certification – Roots, Issues, Challenges, and Benefits*. CRC Press, Boca Raton.

Wijewardana, D. (2005) Sustainable forest management. In: *Forestry Handbook*. NZ Institute of Forestry Handbook, Wellington, (in press).

On-line resources
www.unep-wcmc.org
www.ipcc.ch
www.unfccc.int

Chapter 7

Forest Plantations

The species grown in plantations have been discussed in Chapter 2, the role of plantations in providing fuel and wood products in Chapter 4, and the role of plantations in arresting tropical deforestation in Chapter 5.

What is a Forest Plantation

There is no sharp boundary that separates a forest plantation from other forest types. The extremes are easy to differentiate. An area that is planted with a monoculture of an exotic species and managed as even-aged blocks is clearly a plantation. A forest that has minimal human impact and where regeneration is 'natural' clearly is not a forest plantation. What about a forest plantation of an exotic species that is so well adapted to its new environment that it naturally regenerates and in subsequent rotations the natural regeneration is used to form the next crop rather than replanting? This is probably a forest plantation, for a while anyway. What about a forest that contains native species but where natural regeneration is replaced to a greater or lesser degree by seeding or planting at the hands of humans? This is trickier. There has been a recent tendency to confine the term 'forest plantation' to exotic even-aged largely monocultural forests and to use the term 'planted forests' more widely to encompass forests having a significant component of human intervention in regeneration. What about the situation where an exotic species is planted as a even-aged stand solely for protection against soil erosion compared to a native forest of just one or a few species that is intensively managed for timber production and where regeneration is greatly assisted by planting? Which one is the forest plantation? The first should qualify but the second is a matter of opinion on where the line is drawn between a planted forest and a semi-natural forest. What about enrichment planting in cut lines or gaps in tropical rainforest? Many would consider this to be a plantation. What about a plantation of either exotic or native trees that is established for conservation purposes and is not used for timber production and where natural regeneration is encouraged, no further planting ever being carried out? There are differences of opinion about whether or not this is a plantation. A plantation of trees where the main use is for agricultural production, such as palm oil and coconuts, is not a forest plantation (even though coconut trees can be used as a wood source). However, plantations where an important activity is producing resins, rubber, cork, Christmas trees and pine seeds probably are considered to be forest plantations (rubber trees also being used to produce wood). This may all seem to be somewhat arbitrary and obscure but it is important that there is some form of international parity for reporting purposes. In the developed world, particularly in Europe, it is sometimes difficult to define the boundary between plantations and 'semi-natural' forests and Europe probably reports areas as natural forest that other countries would report as plantations. Carle and Holmgren (2003) further discuss the definition dilemma.

FAO (2001a) defines a forest plantation as: *'those forest stands established by planting or/and seeding in the process of afforestation or reforestation. They are either introduced or indigenous species which meet a minimum area requirement of 0.5ha; tree crown cover of at least 10 percent of the land cover; and total height of adult trees above 5 m.'* Reforestation is where an area cleared of forest is regenerated (naturally or by seeding or planting) and afforestation is where forest is established on non-forested land. If natural forest is cleared and then reforested to plantations, FAO does not classify this as deforestation but it does contribute to overall loss of the area of natural forest (Table 7.1).

How Much Forest Plantation is There and Where is it

Table 7.1 estimates global changes in forest area from 1990-2000. This shows that 1.9 million hectares per year of plantations were established in tropical areas and 1.2 million hectares in non-tropical areas. The areas of natural forest converted to plantations were 53% in tropical areas and 42% in non-tropical areas. If this represents deforestation of natural forest in order to establish plantations, there is cause for concern. If it represents reclassifying natural forests to plantations through greater human intervention in semi-natural forests, or if it represents plantation establishment on severely degraded natural forests, there is less cause for concern.

Table 7.1. Annual change in forest area, 1990-2000 (million ha/year) (from FAO, 2001b).

Domain	Natural forest					Forest plantations			Total forest
	Loss			Gain	Net change	Gain		Net change	Net change
	Deforestation	Conversion to forest plantations	Total loss	Natural expansion of forest		Conversion from natural forest	Afforestation		
Tropical areas	-14.2	-1.0	-15.2	+1.0	-14.2	+1.0	+0.9	+1.9	-12.3
Non-tropical areas	-0.4	-0.5	-0.9	+2.6	+1.7	+0.5	+0.7	+1.2	+2.9
World	-14.6	-1.5	-16.1	+3.6	-12.5	+1.5	+1.6	+3.1	-9.4

The total areas of plantations in year 2000 and the annual rates of planting are given in Table 7.2. The plantation areas and their proportion to total forest area of the ten largest plantation development countries are provided in Table 7.3. Between-country comparisons should be considered with the problems of definition in mind.

Asia has by far the most plantations of any region (Table 7.2). The large areas of plantations in China and India (Table 7.3) are of relatively low productivity owing to low site quality, poor stocking or inadequate maintenance (Evans and Turnbull, 2004). The global plantation estate is 4.8 % of total forest area and is expanding at about 4.5 million ha per year. Forty-eight per cent of world plantations have been designated for industrial purposes. The remainder has been established for conservation of soil and water, the provision of fuelwood, the shelter of livestock, the provision of habitat for wildlife, for aesthetic purposes or for otherwise unspecified purposes. The proportion of plantation that

Table 7.2. Plantation areas (000 ha) by region and species group in year 2000, and plantation establishment rates (000 ha/y) (adapted from FAO, 2001a).

Region	Total area	Annual rate	*Acacia*	Eucalypt	*Hevea*	*Tectona*	Other non-conifer	*Pinus*	Other conifer	Un-specified
Africa	8,036	194	345	1,799	573	207	902	1,648	578	1,985
Asia	115,847	3,500	7,964	10,994	9,058	5,409	31,556	15,532	19,968	15,365
Europe	32,015	5	-	-	-	-	15	-	-	32,000
Nth. and Cent. America	17,533	234	-	198	52	76	383	15,440	88	1,297
Oceania	3,210	50	8	33	20	7	101	73	10	2,948
South America	10,455	509	-	4,836	183	18	599	4,699	98	23
World	187,086	4,493	8,317	17,860	9,885	5,716	33,566	37,391	20,473	53,618

is privately owned is probably somewhere between 40% and 50%. There are little reliable data on the quality of plantations and anecdotal evidence suggests there may be large areas of plantations in poor condition.

What Role do Forest Plantations Play in Sustainable Forest Management

This question was introduced in Chapter 6. There are differences of opinion about the environmental credibility of plantations with various arguments for and against. To some extent a person's attitude depends on whether she or he views a plantation as a branch of agriculture (tree farming) or of forest management. FAO classifies plantations as forest because even the most intense example of plantation forestry, an exotic even-aged monoculture managed for maximum wood production, is more complex, biodiverse and long-lived than an agricultural system. Also there is greater public expectation that forest plantations should provide most of the environmental values expected from more 'natural' forests.

Arguments Used Against Plantations

Biodiversity and Environmental Services

If native forest is felled to establish a plantation, biodiversity may be sacrificed and the environmental services offered by the forest downgraded. Unfortunately, native forest is still being cleared to establish plantations (Table 7.1) and there is no real excuse for this. There is plenty of degraded land and unused cleared land available for the purpose. This requires a strong response from the regulators of land use. Biodiversity may be reduced in plantation establishment, but not always. Biodiversity may be increased when plantations replace degraded forest or when they are established on cleared land.

Poor plantation management can cause soil erosion, pesticide contamination, nutrient accession to ground water, soil compaction, soil acidification, and increased sedimentation of soil in streams. This can be minimized or eliminated by careful management based on knowledge and continuous monitoring. The techniques for avoiding damage are well-known. Indeed one of the reasons why plantations managers are keen to have their forests

Table 7.3. Plantation area, % plantation area of total forest area and main plantation species in countries having more than 1 million hectares of plantation (data from various sources including FAO, 2001b; and Turnbull, 2002).

Country	Plantation area (000 ha)	% plantation area of total forest	Species	Exotic (E) or Native (N)	Tropical/subtropical (Tr), or Temperate (Te)
China	45,083	27.6	*Cunninghamia lanceolata*	N	Tr and Te
			Pinus massoniana	N	
			Pinus koraiensis	N	
			Pinus tabulaeformis	N	
			Eucalyptus spp.	E	
			Larix spp.	E	
			Pinus elliotiii	E	
			Pinis taeda	E	
			Populus spp.	E	
India	32,478	50.7	*Tectona grandis*	N	Tr
			Eucalyptus spp.	E	
			Acacia spp.	N and E	
			Pinus spp.	N	
			Dalbergia sissoo	N	
Russian Federation	17,340	2.0	*Picea* spp.	N	Te
			Quercus spp.	N	
USA	16,238	7.2	*Pinus taeda*	N	Tr and Te
			Pinus elliotii	N	
Japan	10,682	44.4	*Cryptomeria japonica*	N	Te
			Chamaecyparis obtusa	N	
			Pinus densiflora	N	
			Pinus thunbergii	N	
			Larix kaemferi	N	
Indonesia	9,871	9.4	*Acacia mangium*	N	Tr
			Acacia auriculiformis	E	
			Hevea brasiliensis	E	
			Tectona grandis	N	
			Eucalyptus spp.	E	
Brazil	4,982	0.9	*Eucalyptus* spp.	E	Tr
			Gmelina arborea	E	
			Pinus caribaea	E	
Thailand	4,920	33.3	*Tectona grandis*	N	Tr
			Eucalyptus spp.	E	
			Acacia spp	E	
Ukraine	4,425	8.7	*Pinus* spp.	N	Te
			Picea abies	N	
			Quercus spp.	N	
Iran	2,284	3.4	*Pinus* spp.	N and E	Te
Chile	2,017	13.0	*Pinus radiata*	E	Te
			Eucalyptus spp	E	
			Populus spp.	E	
			Salix spp.	E	
United Kingdom	1,928	69.0	*Pinus sylvestris*	N	Te
			Picea sitchensis	E	
			Pseudotsuga menziessii	E	
			Picea abies	E	
Spain	1,904	13.2	*Eucalyptus* spp.	E	Te
			Pinus radiata	E	
Turkey	1,854	18.1	*Pinus* spp.	N	Te
			Eucalyptus spp.	E	
			Populus spp.	N and E	
New Zealand	1,814	22.8	*Pinus radiata*	E	Te
			Pseudotsuga menziesii	E	
Malaysia	1,750	9.1	*Hevea brasiliensis*	E	Tr
			Acacia mangium	E	
			Eucalyptus spp.	E	
Vietnam	1,711	17.4	*Eucalyptus* spp.	E	Tr
			Acacia mangium	E	
			Pinus spp.	N and E	
Australia	1,600	1.0	*Pinus radiata*	E	Tr and Te
			Pinus taeda	E	
			Pinus elliottii x caribaea	E	
			Araucaria cunninghamii	N	
			Eucalyptus spp.	N	
South Africa	1,554	17.4	*Eucalyptus* spp.	E	Te
			Pinus spp.	E	
			Acacia spp.	N and E	
Total	164 435				
World total	187 086	4.8			

certified is that the plantation manager can demonstrate robust and environmentally responsible management. This is not to say, of course, that all plantations are responsibly managed. Many are not and some are in poor condition.

If a highly productive plantation replaces less productive forest or open grassland, interception and transpiration of water are likely to increase while catchment yield of water will decrease. This needs to be considered when establishing plantations in drier areas. This of course equally holds for management that increases the productivity of native forests.

Not all plantations are established to provide wood. More than 50% of plantations have been established for non-industrial reasons. Plantations have been extensively used to stabilize sand dunes (*Pinus pinaster* in France; *Acacia* in Africa; *Haloxylon* in Iran; *Tamarix* in Israel; and *Casuarina* in China, Senegal, India and Vietnam). Plantations have been established in upper catchments to arrest erosion and improve water quality and as windbreaks to protect livestock and crops. Plantations are also established for aesthetic and recreational purposes (see Turnbull, 2002) and are increasingly seen as valuable carbon sinks.

The Impact of Plantations on Local Communities

Plantation establishment, particularly in the developing world, can displace farmers and local communities, aggravate their impoverishment and cause them to relocate into areas where they will clear native forests. If plantation establishment displaces peasant farmers from the land they need to feed themselves, this is blatantly unjust. There are many examples across the world where local communities have protested against pulp companies establishing plantations on their patch. The situation is further complicated if the local peasant farmers have no legal title over their land (see Chapter 5). Whether or not the employment opportunities offered by a plantation programme will compensate for the dislocation will depend on the particular circumstances. The objective must be to provide sustainable and not transient employment opportunities. Agencies establishing plantations are usually private or public enterprises that have the tendency to remain aloof and unaware of the impacts of their actions on local communities. The ideal situation of course is to have the local communities involved in the planning of the plantation enterprise, better still as partners.

Obviously plantation establishment will cause deforestation of natural forest if natural forest is cleared for the purpose. However, plantation establishment may also be indirectly responsible for additional deforestation. Any activity that improves road access into the native forest, and this includes plantation establishment, may increase the movement of shifting agriculturalists, farmers and land speculators into the area. However, road access into an area can also assist in development that can generate employment and alleviate poverty. The relationship between development and deforestation is a complex one and has been discussed in Chapter 5. Much of the controversy around plantations has a social, cultural and ethical dimension that cannot be resolved by science and technology.

Nutrient Depletion

Another argument is that intensive short rotation plantation forestry will deplete the soil of nutrients the same way as does shifting agriculture, by removing nutrients repeatedly from the site and not allowing sufficient time for the soil to recover. Cossalter and Pye-Smith

(2002) estimate that there are about 10 million ha (about 5% of world plantation area and 0.25% of total forest area) and increasing at 10% per year, that fall into the category of fast grown plantations. They classified fast grown plantations as those with a mean annual increment (MAI) of >15 m^3/ha/y and a rotation length of <15 years. Extra short rotations (<10 years) of highly productive species, such as the eucalypt coppice plantations in South America, are approaching the intensity of agricultural systems and special care needs to be taken to leave as much organic matter on the ground after harvest as possible. Also, as in agriculture, the use of fertilizers may be required to compensate for nutrients taken off the site in harvest. There are several historical reports of declining productivity in plantations attributable to degradation in soil properties (*Picea* and *Populus* plantations in Europe, *Pinus radiata* plantations in Australia, *Cunninghamia* plantations in China and *Eucalyptus* plantations in India). Often the argument has become confused with the issue of exotics and monocultures and this will be discussed as a separate issue. If management promotes nutrient depletion and soil erosion, inevitably the soils will become degraded. Intensive management for high productivity has the capacity to degrade soils but it does not have to and should not be allowed to.

There is a considerable literature on the effect of intensive plantation management on soil properties and in most instances increasing the tree productivity enriches the soil rather than the opposite, providing litter and harvesting debris are not burnt or removed from the site and nutrients are added to the site in such a way that they are incorporated into biomass and not leached or eroded. A good example comes from *Pinus radiata* plantations established on sandy soils in South Australia where marked declines in the productivity of the second rotation were noted in the 1960s (Keeves, 1966). The cause of this was found mainly to be the loss of nutrients associated with fierce broadcast burns of residual biomass lying on the ground following harvesting, prior to preparing and planting the next rotation. Current nutritional management conserves soil organic matter, with tree productivity and soil nutritional status now increasing in subsequent plantations (Nambiar, 1996). Certainly plantation forestry is more environmentally benign than cultivated agriculture and probably also more than grazing livestock. Examples of comparative nutrient removals between crop types and the costs of managing pests and diseases are given in Tables 7.4 and 7.5, respectively.

Table 7.4. Quantities of nutrients removed (kg/ha/y) by harvested plant crops in New Zealand (from Will and Ballard, 1976).

Crop	Nitrogen	Phosphorus	Potassium	Calcium
Pinus radiata (averaged over one rotation)	4.4	0.06	5.4	3.6
Grain crops	60-70	10-15	60-70	6-21
Potatoes	80	5	100	3
Lucerne	155	16	110	120

Fast growing eucalypts inevitably will consume a lot of water and they are very efficient competitors with other plants for available soil water. Eucalypts can deplete water storages and overtake other species in the process. When this occurs, it is a case of inappropriate management, putting the wrong tree in the wrong place for the wrong reason. The natural distribution of *Eucalyptus* is almost entirely confined to the driest continent on earth. This has led to the mistaken expectation by some that Australia can provide a highly productive tree for the semi-arid zone. This is physiologically impossible. Australia does have many trees adapted to semi-arid areas but they do not grow quickly unless they have access to groundwater or are irrigated. The fast growing species naturally occur in wetter areas.

Table 7.5. Annual costs of pests and diseases expressed as a percentage of the annual returns for different crops in New Zealand (from Sweet, 1989).

Crop	Pest/disease cost (%)
Wheat	12.2
Oilseed rape	10.0
Barley	7.2
Rye-grass	5.6
Goats	4.5
Sheep (breeding ewes)	4.4
White clover (seed)	4.1
Dairy	3.6
Deer	3.1
Plantation forestry	3.1
Sheep (2-year-flock)	2.4
Beef	2.4

Even-aged Monocultures of Exotics

This argument sees plantations as disease-prone unattractive landscapes that are periodically interspersed with even more ugly clear cuts. The argument against even-aged monocultures of exotics is a mixed argument and the different components – even-aged, monocultures and exotics – need to separated for analysis.

Many natural forests are monocultures. Likewise, large areas of natural forests can be even-aged or close to it (Chapter 6). Clear-cuts may be ugly but so then are clearings caused by disturbances such as wildfire and wind. People find plantations unattractive because of regimentation and uniformity: straight lines of rows, sharp edges, and all trees about the same size. The ugliness of clear-cuts can be hidden by having tree and other plant screens that hide the clear-cut. The unattractiveness of plantations can be alleviated by shaping plantation edges so they are not uniformly straight and do not cause sudden discontinuities in the landscape. Many people find older plantations with widely-spaced large trees to be attractive and are often surprised to learn that they are actually plantations.

A common argument about monocultures is that they degrade the soil through processes of acidification and podzolization. This argument is confused with the one above about nutrient depletion from highly productive plantations. Providing organic residues are not removed from the site at too great a frequency and fertilizers are judiciously administered if required, there is no good reason to believe that monocultures will degrade the soil. Certainly there are examples where monocultures of a particular species planted on infertile soils with a history of litter removal have resulted in deterioration of soil properties. These are more than compensated for, however, by examples of well managed plantations enriching the soil.

Monocultures are more susceptible to disease and insect attack than multi-species forests. This is the case for both natural monocultures as well as planted forests. If disease or insect attack seriously affects a monocultural species, this may be a disaster (from the human perspective). If disease or insect attack affects just one species of many in a multi-species forest, the overall damage is proportionately less or maybe not even obvious. Consequently there is a risk in having just the one species in a plantation estate. This risk can be reduced by having several species spread across the plantation estate. Plantation managers are often prepared to take (or ignore) the risk of using just the one species across their estate if this species shows superior growth and performance. They argue that economies of scale, proven technologies, reliable seed supply, more focused research, and high investment in pest management can balance out the extra risk associated with having to trial, manage and market a number of species. Some shade trees (see Chapter 2) from multi-species forests are readily attacked by insects or fungi when grown as a monoculture. *Toona australis* (Australian red cedar) is just one example of many. Non-pioneer shade trees are usually unsuitable as plantation species. Species that are prone to disease and/or insect attack when planted as monocultures do not become long-standing successful plantation species. An example is *Leucaena leucocephala* which was widely planted in the tropics but now is rarely planted because of defoliation by the psyllid *Heteropsylla cubana*. Successful plantation species have proved that they are not prone to pests and diseases when grown as a monoculture. There are, of course, examples of diseases and insect attack in established plantations but there are many examples of large scale plantation programmes that, up until the present time, have been remarkably free of disease. Indeed Gadgil and Bain (1999) argue that the health of intensively managed plantations of proven successful plantation species is better than the health of trees in natural forests. A catastrophic outbreak, however, is always a threat and a monoculture is more susceptible to this than a multi-species uneven-aged forest. Trees that are water and/or nutrient stressed, over stocked or neglected are more prone to insect and fungal attack. Good management and hygiene can reduce the risk significantly.

Native species are to be preferred if they can meet requirements. Native species are likely to have greater conservation value and carry more local biodiversity than exotics. They are also more likely to be at equilibrium with associated fungi and insects. Exotic species have often been considered to be at greater risk from insect attack and fungal disease than native forests. *Eucalyptus* spp. grown outside of their native Australia, often carry more leaf area and grow faster than the species does in Australia. The increased leaf area is because eucalypts grown as an exotic do not carry the plethora of leaf eating insects that co-exist with eucalypts in Australia. Plantation owners of exotic eucalypt plantation species are rightly concerned that Australian leaf eating insects should not also become successful exotic species. Even so there is little actual evidence to suggest that successful exotic species have been more prone to insect attack and fungal diseases than native species. This is a somewhat circuitous argument because a successful plantation species must by definition have proved that it is relatively immune to pests and diseases. Nair (2001) compared a range of native and exotic tropical plantation species and concluded that monocultures caused an increase in pest problems but that there was no evidence to suggest that the exotics as a group were any more or less prone to pests than the natives.

Some argue that exotics should be avoided because they have the potential to become invasive species that out-compete native vegetation. There is the risk of weediness and this has caused considerable conflict in various parts of the world where an introduced species has invaded the surrounding territory, for example *Acacia melanoxylon* in South Africa, *Cinnamomum camphora* in Australia, *Melaleuca quinquenervia* in Florida and *Pinus contorta* in New Zealand. The whole concept of introducing an exotic species into a native environment is a complex mixture of protecting native biodiversity and of nationalism. Most people identify strongly with their native flora and fauna and this is a value that is independent of scientific analysis.

In summary, there are legitimate arguments against poorly managed plantations but most can be countered with appropriate management.

The Arguments in Favour of Plantations

The most convincing argument in favour of plantations is that, because they are highly productive, they will reduce the amount of wood taken from native forests. At the macro level this must be true. Plantations in year 2000 occupied less than 5% of forest area but supplied 35% of global roundwood. This is predicted to increase to nearly 50% by 2020. Many plantations managed for timber production achieve yields of between 10 and 40+ m^3/ha/y whereas the average for native forest is somewhere around 2 m^3/ha/y. This does not necessarily mean that deforestation will be arrested by establishing plantations. The major cause of deforestation is clearing for agriculture and grazing (Chapter 5). However, unsustainable logging degrades native forests. Plantations should permit lower intensity sustainable management regimes in native forests.

Even-aged monocultures are easier to manage and therefore are more cost-effective than multi-species uneven-aged forests. They are better suited to more intensive practices such as cultivation, fertilization and weed control. Forest operations can be carried out at regular and predetermined intervals and many forest operations can be mechanized. Regulation and control of yield is much easier in a plantation. There are usually just one or two major species in a plantation programme, providing uniformity of product, familiarity with the product and economies of scale in processing Plantations can be established on degraded land that is unsuitable for agriculture. Economies of scale can support expensive tree breeding programmes and the development of clonal propagation of superior genetic stock. For example, superior clones of *Eucalyptus grandis* x *Eucalyptus urophylla* hybrids in the Aracruz plantations in Brazil consistently yield 45 m^3/ha/y over a seven year rotation (Camphinos, 1999). Plantations can be a profitable business. For example, *Pinus radiata* plantations in New Zealand provide a utility bulk commodity that can return an average of about 7% on investment. Natural forests cannot achieve this unless they are ruthlessly high graded for high value timber or unless the real value of the forest is not acknowledged as a cost. Plantations provide a construction material that is environmentally superior to its competitors (cement, steel, aluminium) and a source of fuel that is carbon neutral (Chapter 4). Plantations have been used for stabilizing sand dunes, arresting soil erosion, improving water quality, providing shelter for crops and livestock, encouraging wildlife, and for recreation and amenity. Plantations are also ideal for sequestering carbon and can play a very important part in mitigating adverse effects of global climate change. Environmental and social benefits from plantations can be substantial.

The main concern from environmental groups about plantations has been clearing native forest for the purpose and this is a legitimate concern. Currently about half of plantations are established on deforested land (Table 7.1). The challenge is to stop clearing native forest for whatever purpose, whether this is for agriculture, grazing or plantation establishment (see Chapter 5). Some argue that a small amount of clearing of native forest for plantations can be justified on the grounds that it saves a much larger area of native forest from over-exploitation. Often this is used as an excuse to save money. Ideally plantations should be established on already cleared land and on degraded lands, in which case it will be an environmental gain. There is of course a grey area about defining degraded land, some arguing that degraded land should be restored to a native forest similar to its original structure. This may be technically difficult and financially overwhelming. If the objective is to maximize the chance of restoration and of capturing biodiversity, plantation establishment may well be the better option (a greater chance of success). Restoration of degraded lands comes at a cost and one of the reasons plantation managers sometimes avoid degraded lands is that of the additional cost. Planted forests can make a major contribution to ecological restoration and landscape protection (Maginnis and Jackson, 2003).

The balance of the argument is firmly in favour of establishing and managing forest plantations, providing: it is done in an environmentally responsible manner; that native forest is not cleared for the purpose; and that the human dimensions are satisfactorily accommodated. Quite probably, sustainable management of native forests could not be achieved without plantations. Perhaps an extreme possibility for the future is that all industrial wood will be supplied from plantations and all native forests can be placed in protected areas or managed for other values. New Zealand is close to this situation already. It is open to debate, however, whether this is the best approach. There is the counter argument that sustainably managing native forests protects them from deforestation (for agriculture and grazing) as well as from neglect.

Silviculture and Management

The nature and purpose of plantations can be quite variable and a comprehensive account of their silviculture and management will not be attempted here. Planted forests are the subject of several books including Bowen and Nambiar, 1984; Maclaren, 1993, 1996; Nambiar and Brown, 1997; Boyle *et al.*, 1999; Evans and Turnbull, 2004. In this chapter, the silviculture and management of plantations will be confined to intensively managed, even-aged monocultures established to provide industrial wood products. Industrial plantations occupy less than 50% of the global plantation estate and intensively managed plantations probably less than one-half of this. There is a natural sequence of events that needs to be considered: species and site selection; site preparation; conservation of soil, water and biodiversity; planting (seed collection; raising seedlings or cuttings in nurseries, planting in the field); stocking; nutritional management; weed control; management of pests and diseases; residue management; thinning; pruning; rotation length; the economics of plantations; yield regulation; whole estate planning; and harvest planning. Post-harvest technology (transport, processing, manufacturing and marketing) will not be considered in this chapter but the whole process, seedling to market, should be considered as a continuum. Species selection should commence by considering what product is needed for what market. Indeed some lecturers in plantation silviculture teach the subject in reverse order, starting with defining the market and then working backwards to devise the most appropriate plantation programme to meet the market.

Species Selection

The ideal species for a plantation is one that is physiologically suited to establishing on exposed sites and growing in competition with neighbours of the same species and age. For industrial plantations a fast growth rate is also necessary. Pioneer species are the first to establish after a disturbance event such as glaciation, wind and wildfire and these species are often the best suited to plantations. This significantly constrains the choice of species and sometimes means that a native species may not be suitable (see Chapter 2). The main plantation genera are *Pinus* (20% of world total) and *Eucalyptus* (10% of world total) although *Eucalyptus* is the most important genus in the tropics (50% of total). There are many examples of where native species meet the physiological requirements and are grown successfully in plantations (Table 7.3). For example, the USA, Japan and the Russian Federation have plantations almost entirely of native species. Other countries cannot find native species that meet their requirements and have chosen to plant exotics. For example, New Zealand, Chile and Brazil have plantations of almost entirely exotic species. The dominant exotic genera are *Eucalyptus* (from Australia) and *Pinus* (mainly from North and Central America). Important exotic pine plantation species are *Pinus taeda, Pinus elliottii, Pinus radiata,* and *Pinus caribaea.* The most important eucalypt plantation species grown in temperate areas are *Eucalyptus globulus* and *Eucalyptus nitens.* Significant species in tropical and subtropical areas are *Eucalyptus grandis* and its hybrids, *Eucalyptus urophylla, Eucalyptus saligna, Eucalyptus tereticornis, Eucalyptus deglupta* and *Eucalyptus camaldulensis.* Seventy-five per cent of plantations are in the temperate zone and 25% in the tropics and subtropics (Turnbull, 2002).

The most appropriate species for a region will depend on climate, topography and soil. Advantage should be taken of the natural genetic variation in a species that occurs across its natural range of distribution. Selection in the first instance should be based on comprehensive species and provenance trials. This may naturally progress to a tree breeding programme designed to improve traits such as growth rate, stem straightness and wood quality. (see later in this chapter). Fast growth is not the only factor that should be considered when choosing a species. It is also necessary to have appropriate wood quality for the purpose. Fast growing plantations of poor quality wood can be a bad investment. Wood from plantations is used for a range of purposes: energy, pulp, panels and sawnwood. Clearly a species that can meet all of these functions is more versatile and a better risk than one which is more limited in end use. *Pinus* most often is versatile enough to cover the whole range of end uses except the top end of fine furniture timbers, appearance grades and appearance veneers. Fast growing *Eucalyptus* plantations were originally established as a pulp wood species and quickly gained the reputation that they were only good for firewood, charcoal and pulp. They have been difficult to saw because of growth stresses, but recent technology is resolving this. There is an increasing tendency for more *Eucalyptus* plantations to be established under saw log regimes. The reputation that eucalypts have for lower quality wood is based on the few fast growing plantation species that were established primarily for pulp. Genus *Eucalyptus* has many hundreds of species, some of which have exceptionally high quality timber suited to the most demanding of end uses, but which grow more slowly.

Experience has shown that a plantation species may be planted outside the climate of its natural range, but not too far outside. Clearly there is an ecological risk in straying too far. *Pinus radiata* is exceptional in this regard. The natural occurrence of *Pinus radiata* is restricted to fragments totalling less than 8000 ha in the fog belt of Monterey in California and the adjacent Mexican Pacific Islands of Guadalupe and Cedros where it is in poor health and of little commercial significance. It is virtually a relict species in its native environment. However, there are more than four million ha of *Pinus radiata* plantations distributed mainly throughout New Zealand, Australia, Chile and Spain. Even so, the large areas of occurrence as an exotic are confined to temperate maritime climates. If it is planted in a humid subtropical climate it very easily succumbs to disease.

Table 7.6. Indicative MAI and rotation length for *Eucalyptus* and *Pinus* plantations throughout the world (adapted from Camphinos, 1999 and other sources).

Species	Country	Rotation (y)	MAI (m³/ha/y)
Eucalyptus hybrids	Brazil (Aracruz)	7	40
Eucalyptus hybrids	Congo	8	25
Eucalyptus grandis	South Africa	8-10	20
Eucalyptus grandis (saw logs)	Uruguay (Rivera)	16	35
Eucalyptus globulus (pulp)	Uruguay	8	20
Eucalyptus globulus	Chile	10-12	20
Eucalyptus globulus	Portugal	12-15	10
Pinus radiata	New Zealand	25	23
Pinus radiata	Chile	23	22
Pinus patula	South Africa	25	18
Pinus caribaea	North Queensland	25-30	18
Pinus patula	South Africa	25	18
Pinus radiata	Australia	30	17
Pinus caribaea	Latin America	15	15
Pinus caribaea x *elliottii*	South Queensland	25-30	15
Pinus taeda, P. elliottii	USA	23	13
Pinus elliottii	South Africa	30	12

It is unreasonable to expect high productivity plantation species to thrive on very infertile soils without the addition of fertilizers and/or in dry areas. Generally speaking, for a given amount of radiation, productivity is directly proportional to water and nutrient availability. Matching species to site (climate, soils, topography) is fundamentally important in plantation management. There may be considerable genetic variability between different provenances of a species from different regions across its genetic range. It is logical to test and exploit this variability. Plantations established for wood production need to be sufficiently productive to get a return on investment after carrying the cost of establishment and maintenance through to rotation age. Usually, if unsubsidized, this requires an MAI of more than 10 m³/ha/y over a rotation length of less than 40 years. Table 7.6 summarizes MAI and rotation length for *Eucalyptus* and *Pinus* plantations around the world.

Planting

The definition of a plantation and the fuzzy boundary between a plantation and a semi-natural forest is centred on regeneration. Planted forests have human-assisted seeding or planting for some or all of its regeneration. The most extreme and the most usual example is an even-aged plantation where a cleared area is planted with one species, usually planted in lines, and to a predetermined planting density (number of tree per hectare expressed as distance between trees in a row and distance between rows). Some species (for example some tropical eucalypts) freely coppice from their cut stumps and after the first rotation they are managed as coppice plantations for the next two or three rotations before replanting. Coppice plantations can be very productive because they do not have to re-establish a new root system between rotations. Sometimes, rather than replanting, the natural regeneration from a plantation is managed as the next rotation. The use of direct seeding to establish a plantation in the field is rare. Mostly propagules are planted in the field, either mechanically or by hand. These need to be raised in nurseries either as seedlings from seed or vegetatively propagated as cuttings.

Propagation from Seed

Seedlings from seed are sexually reproduced and the individual seedling has its own unique genetic identity but shares genetic information of its parents. Large plantation programmes may justify having their own seed orchards to provide genetically superior seed. Alternatively improved seed from orchards owned by cooperatives or by private enterprises can be sold on the open market. Orchard seed can be derived from cross-pollination of selected superior parents where both parents are known (full-sib), or from open-pollination where just the female parent is known (half sib). Seedlings are produced in nurseries. Because many thousands to millions of seedlings are produced each year in large plantation enterprises, nursery operations can be quite capital intensive and highly mechanized compared to other plantation operations. Smaller, less capital intensive and perhaps non-permanent nurseries are more appropriate for smaller enterprises and particularly for community-based forestry programmes.

Seedlings raised and planted in climates that have a pronounced winter season may be planted in large beds and lifted as 'bare-rooted' seedlings from the nursery bed in winter while they are physiologically sluggish after which they can be planted in the field or cold-stored until planting. It is very important to reduce any water loss from the seedlings between lifting and planting. Care needs to be exercised in planting open-rooted stock to ensure that there is adequate contact between the root and the soil: over compaction of the soil will cause root damage and under compaction will provide poor root to soil contact. This can induce water stress in the seedling, even if the soil has adequate water supply and the vapour pressure deficit is low (Sands, 1984). (Vapour pressure deficit is a measure of the dryness of the atmosphere.) Root systems of open-rooted seedlings planted in the field are inefficient in the uptake of water and nutrients because of poor root to soil contact. Success in establishment will depend on the rate of new root growth after transplanting and this is directly related to the number of root apices. The number of root apices determines the 'root regeneration potential' and can be increased by root pruning, the periodic severing of roots at a depth of 10 to 20 cm to encourage a multi-branched fibrous root system. Seedlings in the nursery beds should be healthy, vigorous, and uniform in size with an optimal ratio between tops and roots. This means that the seedlings should be raised under optimal water and nutrient supply but should not be lifted when they are 'soft' and unable to withstand the rigours of transplanting into a more hostile environment. Seedlings need to be 'conditioned' by a programmed withdrawal of water and nutrients prior to lifting. If seedlings are grown under shade cloth, the shade also needs to be progressively removed.

Seedlings, however, should be well-watered just prior to lifting. Root-pruning also assists in conditioning seedlings and providing the optimal root/shoot ratio.

It is increasingly difficult to raise open-rooted seedlings as the weather becomes warmer and the distinction between winter and the other seasons becomes less pronounced. Open-rooted seedlings of some species can be raised in subtropical areas under certain conditions. This is more demanding and the risks associated with transplanting are greater. Transplanting risks can be significantly reduced, in both temperate and tropical zones, by raising seedlings in containers but this is more expensive. Economies of scale, risk reduction, and flexibility in management, however, may make containerized seedlings the better option under certain circumstances. For some species in warmer areas there may be no option. Containerized seedlings also need to be conditioned by the progressive decrease in water and nutrients and increase in sunlight. Because the roots are confined within containers, it is important to ensure that top growth does not become excessive. For both open-rooted and containerized seedlings it is important not to distort the root systems when transplanting as these distortions will remain for the life of the tree.

A well-managed plantation programme should expect at least 80% survival of seedlings planted in the field. Nursery soil can be inoculated with mycorrhizas if it does not contain the most appropriate mycorrhizal environment for the species. This may be the case for an exotic species introduced into a new environment.

Propagation from Cuttings

Cuttings from portions of stem, root or sprout can root in soil. Rooted cuttings have been used for propagating plants for thousands of years. Cuttings for plantations can be set directly in the field, but this is rare. Usually cuttings are set in a nursery (in which case they are sometimes called stecklings) and subsequently transplanted as rooted cuttings into the field. Because rooted cuttings are vegetatively produced, they are the basis of clonal forestry where the same superior genotype can be reproduced more or less indefinitely. Rooted cuttings are usually more expensive to produce than seedlings, but these extra costs can be compensated for by greater uniformity and the ability to match highly productive clones to site. Clonal forestry has become particularly prominent in the tropics, particularly with eucalypts. Cuttings from eucalypts root most easily on young material in warmer climates. For example, it is extremely difficult to root cuttings of the cool montane *Eucalyptus nitens*, difficult but manageable for the temperate *Eucalyptus globulus* and easy for the tropical *Eucalyptus grandis*. There are extensive clonal plantations of fast grown eucalypt pulp plantations, well-known examples being the highly productive clonal plantations of the *Eucalyptus grandis* x *Eucalyptus urophylla* hybrid in Brazil. Globally, propagation by seedlings is still the main form of reproduction in plantations but the contribution of rooted cuttings is increasing. Tissue culture and somatic embryogenesis are other forms of vegetative production. Their contribution at present to plantation establishment is small for tissue culture and negligible for somatic embryogenesis (Libby, 2004).

Open-rooted seedlings, containerized seedlings and cuttings should be planted when the soil has plenty of water and the vapour pressure deficit is low. For temperate regions this is late winter and spring. For tropical regions these conditions are more variable and less certain. Consequently transplanting is more risky.

Early Silviculture

Residue Management

Arguably the most important silvicultural treatment at the time of site preparation from both an environmental and productivity perspective is the maintenance of soil organic matter and the retention of litter and residues. Nutrients are relatively tightly cycled within biogeochemical pathways in forested ecosystems. Any management that breaks this cycle and removes significant amounts of nutrients off-site (as leachate to groundwater and streams, as eroded soil, or as biomass removed from the site) has the potential to reduce soil fertility and therefore plantation productivity. Logs are relatively low in nutrients and their removal in harvesting has a small impact on nutrient loss (except for very short rotations). Twigs, leaves and other litter on the forest floor have a higher nutrient content and their removal should be avoided. De-barking in the forest is recommended if feasible.

Sites can be cleared by hand, mechanically, by burning, by grazing or by using herbicides. The choice will depend on topography and scale of operation among other things. As with minimum tillage systems in agriculture, clearing with herbicides will minimize soil disturbance and prevent erosion. Usually, however, the soil is cultivated or disturbed to some extent. Site preparation on non-forested land, which is often the case for the first rotation, is usually less difficult than between rotations when there is a lot of logging debris on the ground. Site preparation between rotations is more challenging. Burning is an inexpensive and easy (although risky) way to clear the site of tonnes of debris. This was routinely done in the past in many places but has been shown to reduce soil fertility in some instances. Burning between rotations should be avoided, particularly on inherently infertile soils. Plantation managers ideally would like a clear site that they can cultivate to control weeds and improve soil condition, mound to improve drainage, or rip to break up compacted soils. It is easier to move machinery across a clean site than one that is covered in debris and compromises are often made for convenience. One such compromise is leaving the finer debris on the surface and raking the larger material into windrows or heaps which are then burnt. This is decidedly better than intense broadcast burning, but is still not ideal. This treatment also has the unfortunate effect of unevenly distributing residual nutrients and ash across the site. Treatments that smash up the logging debris into small pieces that will carry a light burn, or better still left on the ground to decay, are to be preferred. If residues are sufficiently broken down, machines can traverse this without great difficulty. Machines have been designed that can cultivate a defined small area around a planting spot and leave the rest of the ground residues in place.

Extreme harvesting practices that remove a wide range of material such as in biomass plantations and whole-tree harvesting have the capacity to reduce soil fertility. Attention has already been given to the possibility that extremely fast growing short rotations for pulpwood or fuel may also have the capacity to reduce soil fertility. There have been many studies on the impact of removing harvesting residues from forests. Not surprisingly the results depend on the intensity of management and the fertility of the soil. Some plantation systems are very robust and others less so. The addition of nutrients through mineral fertilizers and, in the case of nitrogen, by using nitrogen-fixing species, may not only improve tree productivity but also improve soil properties for subsequent plantations, providing the nutrient cycle does not unduly leak.

Weed Control

Weeds compete with plantation trees for water, nutrients and light. Removing weeds can greatly increase productivity. Weeding is most effective in increasing plantation productivity if it is carried out just prior to and in the first year or two after planting. Fertilizing soon after planting may be ineffective unless the weeds are controlled as well, as this will also fertilize the weeds and increase their competitiveness. Weed removal is often no longer necessary after the first year or two and indeed fertilization of weeds at this stage may have the beneficial effect of increasing the amount of nitrogen held in the site. Weeds are mainly controlled by cultivation or by herbicides. The appropriate timing and intensity of weed control depends on the particular circumstances (tree and weed species, climate and soil). Weeds may be herbaceous or woody, native or exotic. Another control option when water is not the major limiting resource is to replace the weeds with a legume. This will protect the soil from erosion, increase soil nitrogen, reduce fire risk and generally make the plantation look more attractive. Oversowing with legumes has been routinely practised when establishing radiata pine plantations in New Zealand and this has significantly increased tree growth.

Table 7.7. Aboveground biomass (Mg/ha) of 6-year-old *Pinus elliottii* and *Pinus taeda* in Florida (abridged from Dalla-Tea and Jokela, 1991).

Treatment	*Pinus elliottii*	*Pinus taeda*
Control	11.9	7.1
Weeded	26.2	35.2
Fertilized	26.0	38.6
Weeded and fertilized	34.2	57.6

Figure 7.1 represents growth of *Pinus radiata* over a hot dry summer in southern Australia. The growth of the pine was reduced over the hot summer period but not as much for those trees that had the weeds removed from around them because they were less water stressed (leaf water potentials closer to zero) and consequently their stomata were more open (stomatal conductance greater) and therefore they had a greater rate of net photosynthesis. Weed control can induce considerable productivity gains. Table 7.7 shows the effect of weed control and fertilization, singly and in combination.

Plants require certain essential nutrients if they are to survive and grow. Some of these nutrients (micronutrients) may only be needed in very small quantities. Table 7.8 shows the essential nutrients ranging from most concentrated in plant tissue to least concentrated.

Management of Nutrition

The long-term productivity of plantations depends on organic matter management and judicious application of nutrients where required. Globally, increases in productivity from applying nitrogen are common. Phosphorus can be a major limiting nutrient in many places and a variety of other deficiencies are known. Deficiencies can usually be corrected by adding mineral fertilizers and in the case of nitrogen by exploiting nitrogen fixation from legumes as well. Composted municipal waste, fire ash, and treated sewage sludge and effluent are also occasionally used. Nutritional management based on empirical fertilizer trials that have a combination of nutrients and rates is very useful. This alone, however, is a 'shot gun' approach that has the capacity to be inefficient and wasteful. It is important not to add unnecessary nutrients because this is costly, wasteful of resources and potentially damaging to the environment if the nutrients move off-site. Usually when nutrients are most available to plant roots they are also most vulnerable to leaching. Careful nutritional management needs to consider this. There has been considerable research into the nutritional physiology of plantation species and of soil management to optimize tree nutrition. Knowledge-based nutritional regimes are overtaking those based solely on empirical trials (Nambiar, 1996; Nambiar and Brown, 1997).

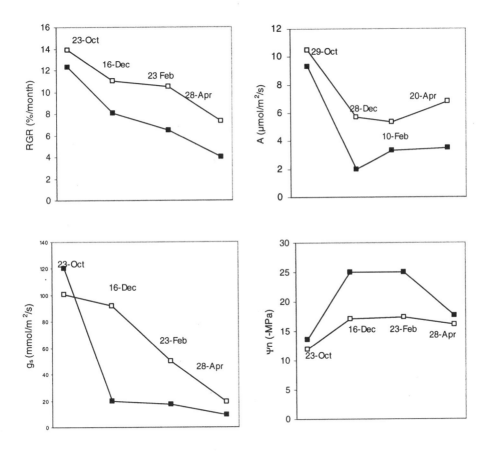

Fig 7.1. Relative growth rate in stem volume (RGR), net photosynthesis (A), stomatal conductance (g$_s$) and needle water potential (Ψ_n), all measured near midday during the period October 1990 to April 1991 in southern Australia, of 2-year-old *Pinus radiata* with no weed control (■) or with weeds removed in a 1.5 x 1.5m square around each tree (□) (from data of Hadryanto in Nambiar and Sands, 1993).

Table 7.8. The nutrients obtained from soil that are essential for plant growth together with their average concentration in dry plant matter (abridged from Taiz and Zieger, 1998). The essential status of nutrients in parentheses is still open to question. Nutrients in bold type are the most commonly reported deficiencies in forest plantations.

Nutrient	Concentration μmol/g dry weight
Macronutrients	
Nitrogen	1,000
Potassium	250
Calcium	125
Magnesium	80
Phosphorus	60
Sulphur	30
(Silicon)	30
Micronutrients	
Chlorine	3.0
Iron	2.0
Boron	2.0
Manganese	1.0
(Sodium)	0.40
Zinc	0.30
Copper	0.10
Nickel	0.002
Molybdenum	0.001

Fertilizers can greatly improve productivity on infertile soils (Table 7.7). Visual deficiency symptoms can often be used to diagnose the particular nutrient that is deficient. The most common visual symptoms are yellowing (chlorosis) or dying (necrosis) of foliage. These symptoms can be similar to symptoms of disease or of water stress and local knowledge and experience is required to distinguish between these. A plantation tree may be deficient in nutrients and respond to fertilization when no visual symptoms are evident. Other diagnostics such as sampling foliage for critical levels of nutrients and soil survey are used to determine whether or not a plantation will respond to fertilization. It does not follow that fertilizer should always be used, even if it promotes an increase in productivity. The decision whether or not to fertilize will be both an economic and an environmental one. Sometimes fertilization can produce a very large response that is both economically and environmentally justified. Indeed, in some instances plantation establishment will fail without fertilization and there are many instances where fertilization has demonstrated a permanent improvement in soil nutritional status. Fertilizing plantations is common in southern USA, Brazil, Chile, South Africa, Australia and New Zealand where the most common deficiencies have been nitrogen, phosphorus, potassium, magnesium, zinc, copper and boron. Routine application of fertilizers is not common in tropical countries but is increasing (Gonçalves *et al.,* 1997). Fertilizing is usually done at the beginning of a rotation but later-age fertilization, particularly with nitrogen and following thinning, can also be beneficial. Increasing productivity with fertilizers depends on other resources, particularly water, being present in sufficient quantity to support the increased growth rate. Soil water status also is important in determining nutrient availability to tree roots. Figure 7.2 shows the effect of irrigation and fertilization, together and in combination, for four different plantation species in four different countries. When water and/or nutrients are applied to a soil that is deficient in water and/or nutrients, any increased photosynthate (and therefore growth) usually is preferentially directed towards aboveground at the expense of below ground. Fertilizer responses are a combination of increased total biomass and an increased allocation to above ground (Table 7.9).

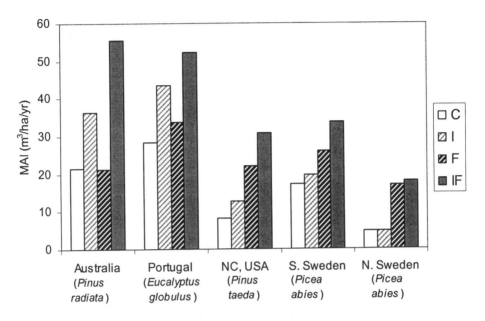

Fig. 7.2. The effect of no irrigation or fertilization (C), irrigation (I), fertilization (F) and irrigation and fertilization in combination (IF) on aboveground volume production of four plantation species from different parts of the world (from Sune Linder).

Plantation species have a wide range of growth rates (Table 7.6). These are caused by differences between species, differences between climate, and differences in soil and site. Figure 7.3 shows inherent differences between species when grown in the same climate on soil with adequate water and nutrients. Figure 7.2 shows the effect of water and nutrients but also the effect of climate and species. The productivity of a particular species is directly proportional to the amount of radiation intercepted by the leaves for photosynthesis. This is roughly proportional to latitude.

Table 7.9. Current net photosynthesis and fluxes of dry matter above and below ground (t/ha/y) of 20-year-old *Pinus sylvestris* in Sweden after 6 years of irrigation and nutrient addition (derived from Linder and Axelsson, 1982).

	Control	Fertilized and Irrigated
Above ground	4.7	15.3
Below ground	6.9	6.9
Total	11.6	22.2
Above/below	0.7	2.2

Stand Dynamics

Stand Density, Thinning and Pruning

Almost invariably many more trees regenerate in a native forest than reach maturity. Trees at initial high stand density (number of stems per hectare) compete with each other for water, nutrients and light, the stronger persisting while the weaker trees may die prematurely owing to lack of growing space. The plantation manager can choose to plant at densities low enough to preclude competition-induced mortality but often will plant at higher densities to promote early height growth, to inhibit the development of large side branches, to compensate for poor survival in transplanting, and to provide a large pool of seedlings from which to select the final crop. If the primary objective is to produce biomass then the optimum stand density at harvest is the highest that the site can support without causing mortality from competition. Any thinning, if required, would be to pre-empt this mortality. If a higher value product is required, such as a saw-log, then further thinning beyond this point is required to promote growth on fewer stems. This will be at the expense of total productivity per hectare. Initial densities may be as high as 18,000 stems per hectare in cold-temperate areas. Stand densities of around 2000 stems per hectare are common for pulpwood and fuelwood regimes (Theron and Bredenkamp, 2004). Initial stand densities for high quality saw-log regimes from high productivity plantations are more often of the order of 500-1500 stems/ha, depending on species and circumstances. Cuttings and genetically improved seedlings are more uniform in performance and can be planted at lower stand densities than routine unimproved seedlings. For saw-log regimes there is usually at least one and often a series of thinnings with the objective of concentrating the growth onto fewer straighter larger trees at rotation age. The first thinning may be a non-commercial thinning where the thinnings are left on the ground to decay and return nutrients and organic matter to the soil. Subsequent thinnings can be used for pulp and for small round timber and later thinnings for sawnwood. The number and timing of thinnings varies between species and product profile. Typically, however, a stand at rotation age will have about 200-400 stems/ha. There are many thinning regimes and the figures given here provide a rough guide only.

Light penetrates a thinned stand and illuminates the lower branches. This stimulates the photosynthesis of the lower live branches which can become quite thick. This will leave a large and unsightly knot in the sawn timber. The lower branches can be periodically pruned to confine the knots to a 'knotty core' with clear knot-free wood being produced outside of this core (Chapter 4). Clearly pruning and thinning should be considered together in planning. Whether or not to prune will depend on the nature of the product and whether it can support a premium in price to cover the cost of the pruning, which is expensive. When to prune, which stems to prune, how high to prune, and how many pruning lifts are required depend on species, site and market. Thinning and pruning regimes are subjects of great debate and passion among plantation managers. Thinning and pruning both reduce total tree volume per hectare. The objective of thinning and pruning is to produce larger higher value individuals.

Rotation Length

The rotation length is the age of the trees at final harvest. Rotation ages vary between 3 for some firewood crops to over 100 years for some slow growing species in cool climates. Usually the faster growing the species the shorter the optimal rotation. Fast growing eucalypts are usually grown at rotation lengths of less than 15 years. Temperate pines are usually grown at rotation lengths of 15 to 40 years. Typically, the MAI of a plantation increases with stand age relatively rapidly to a maximum value after which it slowly declines (Fig. 7.3). If volume production was the only criterion, the age at maximum MAI would be the logical rotation age. There are many other factors to be considered however. Figure 7.3 shows that *Pseudsotsuga menziesii* is potentially more productive than *Pinus radiata* providing you wait long enough. The reason why *Pinus radiata* and particularly some *Eucalyptus* are better high productivity plantation species is that they grow quickly early which means they can be grown over shorter rotations. This is important from an economic perspective. The costs of establishment and maintenance have to be carried with interest over the length of the rotation. High interest rates will favour shorter rotations. Also longer rotations carry greater risk of catastrophic damage from fire, disease and wind. However, the wood quality from short rotations is inferior and the logs are smaller which limits their utility. Wood quality improves with tree age: wood density and stiffness increase with distance from the centre of the tree (the pith) and the outer wood is of better quality than the younger inner wood (juvenile wood). The market determines the size and nature of product required. Pulpwood and fuelwood can be grown on a shorter rotation than saw-logs. Sawmills may have a minimum and maximum log size that they can handle. Rotation ages need to be flexible to meet real world conditions. Plantations rarely have evenly distributed age-classes and the market conditions that determine demand and supply may vary considerably. There is a growing movement favouring 'continuous canopy forestry' for plantations, which means that at no stage is there a clear-cut between rotations. This has aesthetic advantages and probably environmental advantages. However, there are probably economic disadvantages. The application of continuous cover forestry to plantations is in its infancy but is increasing in appeal. Its success depends on satisfactory growth of seedlings in semi-shade. It is better suited to managing natural regeneration and this revisits the argument of 'when is a plantation not a plantation'.

Plantation Protection and Maintenance

Plantations face many hazards, biotic and abiotic, anthropogenic and natural, such as insect attack, browsing by vertebrates, disease infections, destruction by fire, damage by wind or flood, salt affected soils, air pollution, toxic chemicals in the soil, climate change and in some places volcanism. Control measures are outside of the scope of this book, but this does draw attention to the fact that plantations are a long-term investment that carries considerable risk. Sometimes plantations can be insured against some of these hazards. Regrettably there are many instances globally where plantations have been established and essentially abandoned, presumably under the mistaken impression that once planted nature will take over and the plantation will look after itself. Plantations need to be actively managed and to be tended and cared for over their rotation length. All too often plantations

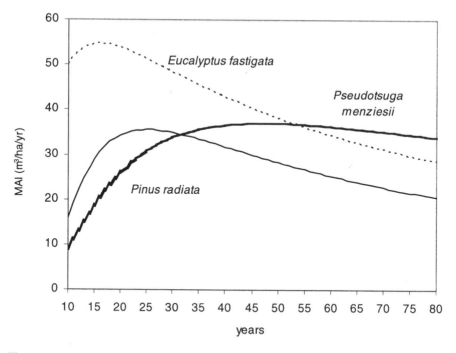

Fig. 7.3. MAI versus age for *Eucalyptus fastigata*, *Pinus radiata* and *Pseudotsuga menziesii* grown under similar climate on fertile well watered sites in New Zealand. Stands were planted at 1000-1200 stems per hectare and were unthinned (compiled from permanent sample plot data by Richard Woollons).

are thinned too late or not at all. Overstocked plantations riddled with mortality are an eyesore and a waste of money. Plantations in poor condition are prime candidates for insect attack and disease.

Tree Breeding

There are a very large number of tree species and they can be very different in morphology, growth habit and suitability to grow and survive in radically different climates and soil conditions. Differences within a species are in a narrower range but they can still be substantial. There may be considerable variation in the growth characteristics of a tree species across its range of natural distribution. Plantation managers often attempt to obtain 'local' seed when planting a species within its natural range. When the species is planted in an exotic environment, they will seek seed from origins (provenances) that have similar climate and soils. The fundamental difference between individuals within provenances of a given species is determined by their genetic identity, or genotype. Each genotype has a different potential to function and grow, but this potential interacts with the local environment. The environmental and genetic components together determine the appearance and performance (called the phenotype) of an individual. For example, genetically identical twin children will have the same genotype, but if one twin consumes only bread and water and the other has a rich and varied diet they will exhibit quite different phenotypes. The genetic variability within a species can be exploited to improve traits of interest providing these traits are heritable and genetically variable, therefore capable of providing a genetic gain between generations of artificial selection. This forms the basis of tree breeding which in its simplest form is selecting superior phenotypes from wild populations and testing their open or cross pollinated progenies for traits of interest. This will identify superior genotypes that can be used in breeding in subsequent generations for successive tree improvement. Genetically superior seed can be raised in seed orchards and/or genetically superior individuals can be clonally propagated as cuttings. Figure 7.4 is a schematic of a typical tree breeding programme.

Progeny evaluation trials should be established over the range of environments expected in the plantation estate. Traditionally, the traits of most interest to plantation managers have been growth rate, tree form, wood quality and pest and disease resistance. Other traits such as the ability to overcome environmental hazards may also be important. Tree breeding can be used to match the most appropriate genotype to water and nutrient limiting environments. The long time between generations makes tree breeding a slow process. However, the study of early to late age correlations, early flower induction, marker-aided selection, physiological testing and modelling has become increasingly important in improving the efficiency and reducing the times required in tree breeding (Boyle *et al.*, 1997; Burley, 2004).

Gains from tree breeding can be very impressive (Fig. 7.5) but gains from site and silviculture are invariably much larger (Carson *et al.*, 1999). Despite this, the performance rankings of improved genotypes, particularly for wood quality and disease resistance, do not change much over a range of site and silviculture. Accordingly, single national breeding programmes have been successful for a broad range of breeding objectives (Carson, 2002). One of the dangers in tree breeding is the potential to narrow the range of genetic variation to the extent that the subset of genotypes deployed to plantations may have increased vulnerability to pests and diseases. Also reducing genetic variability in the underpinning breeding population may limit future but as yet unknown options, and, at the extreme, may lead to problems associated with inbreeding depression. Tree breeding programmes recognize this danger and ensure that breeding populations are large and broadly-based, and that the genetic range of genotypes deployed to plantations is not unduly restricted. Tree breeding is mainly a relatively recent phenomenon (50 years old) and populations are still relatively wild compared to some agricultural crops whose genetic range may have been severely confined through intense breeding over many generations. Sexual reproduction between individuals from different species is usually not possible because the genetic differences are too great. However, the genetic composition is similar enough between some tree species to permit hybridization which may result in superior performance. Examples of successful hybrids in plantation forestry are *Eucalyptus grandis* x *Eucalyptus urophylla* in Brazil and *Pinus elliottii* variety *elliottii* x *Pinus caribaea* variety *hondurensis* in Queensland, Australia.

There is some research being carried out into producing transgenic trees through genetic modification to provide pest resistance, herbicide resistance and desired wood properties. The future of this technology in tree breeding is not known. It is being strongly resisted by some organizations (e.g. Forest Stewardship Council) and its adoption may be restricted

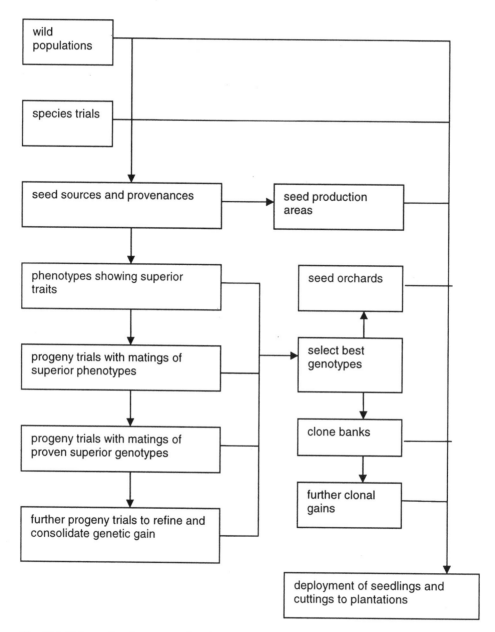

Fig. 7.4. Schematic of a conventional tree breeding programme.

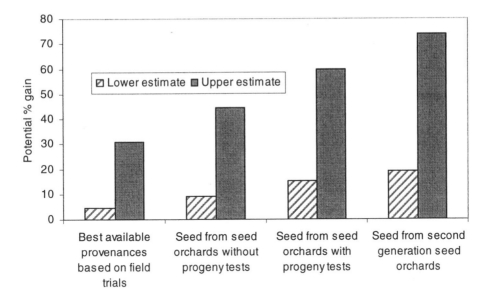

Fig. 7.5. The potential percentage gain in value production with increasing intensity of improvement from tree breeding in tropical plantations (redrawn from Foster *et al.,* 1995; and Kjaer, 2004).

by regulation. Whether or not transgenic trees are used in plantations, there are molecular techniques (e.g. marker-aided selection) that are likely to prove useful in refining classical tree breeding.

References and Further Reading

Bowen, G.D. and Nambiar, E.K.S. (eds) (1984) *Nutrition of Plantation Forests.* Academic Press, London.

Boyle, J.R., Winjum, J.K., Kavanagh, K. and Jensen E.C. (eds) (1999) *Planted Forests: Contributions to the Quest for Sustainable Societies.* Kluwer Academic Publishers, Dordrecht.

Boyle, T.J.B., Cossalter, C. and Griffin, A.R. (1997) Genetic resources for plantation forestry. In: Nambiar, E.K.S. and Brown, A.G. (eds) *Management of Soil, Nutrients and Water in Tropical Plantation Forests.* Australian Centre for International Agricultural Research, Canberra, pp. 25-63.

Burley, J. (2004) A historical overview of forest tree improvement. In: Burley, J., Evans, J. and Younquist, J.A. (eds) *Encyclopedia of Forest Sciences.* Elsevier Academic Press, Amsterdam, pp. 1532-1538.

Camphinos, E. Jr. (1999) Sustainable plantations of high-yield *Eucalyptus* trees for production of fiber: the Aracruz case. *New Forests* 17, 129-143.

Carle, J. and Holmgren, P. (2003) Definitions related to planted forests. In: *UNFF Intersessional Expert Meeting on the Role of Planted Forests in Sustainable Forest Management,* 24-30 March 2003, 11 pages, Ministry of Agriculture and Forestry, Wellington, New Zealand.

Carson, M.J. (2002) Intensive tree breeding to enhance fibre production. In: Simspon, J.D (ed) *Proceedings of 28th Meeting Canadian Tree Improvement Association,* Edmonton, Alberta.

Carson, S.D, Garcia, O. and Hayes, J.D. (1999) Realized gain and prediction of yield with genetically improved *Pinus radiata* in New Zealand. *Forest Science* 45(2), 186-200.

Carrere, R. and Lohmann, L. (1996) *Pulping the South: Industrial Tree Plantations and the World Paper Economy.* Zed Books, London.

Cossalter, C. and Pye-Smith, C. (2003) *Fast-Wood Forestry: Myths and Realities.* Center for International Forestry Research, Bogor, Indonesia.

Dalla-Tea, F. and Jokela, E.J. (1991) Needlefall, canopy light interception, and productivity of young intensively managed slash and loblolly pine stands. *Forest Science* 37, 1298-1313.

Evans, J. and Turnbull, J.W. (2004) *Plantation Forestry in the Tropics* (3rd Edition). Oxford University Press, Oxford.

FAO (2001a) *Global Forest Resources Assessment 2000 - Main Report.* FAO Forestry Paper 140, Food and Agriculture Organization of the United Nations, Rome.

FAO (2001b) *State of the World's Forests.* Food and Agriculture Organization of the United Nations, Rome.

Foster, G.S., Jones, N. and Kjaer, E.D. (1995) Economics of tree improvement in development projects in the tropics. In: Shen, S. and Converas-Hermosilla, A. (eds) *Environmental and Economic Issues in Forestry: Selected Case Studies in Asia.* World Bank Technical Paper No. 281. The World Bank, Washington, DC.

Gadgil, P.D. and Bain, J. (1999) Vulnerability of planted forests to biotic and abiotic disturbances. *New Forests* 17, 227-238.

Gonçalves, J.L.M., Barros, N.F., Nambiar, E.K.S. and Novais, R.F. (1997) Soil and stand management for short-rotation plantations. In: Nambiar, E.K.S. and Brown, A.G. (eds) *Management of Soil, Nutrients and Water in Tropical Plantation Forests.* Australian Centre for International Agricultural Research, Canberra, pp. 379-417.

Keeves, A. (1966) Some evidence of loss of productivity with successive rotations of *Pinus radiata* in the south-east of South Australia. *Australian Forestry* 30, 51-63.

Kjaer, E.D. (2004) Tropical hardwood breeding and genetic resources. In: Burley, J., Evans, J. and Younquist, J.A. (eds) *Encyclopedia of Forest Sciences.* Elsevier Academic Press, Amsterdam, pp. 1527-1532.

Libby, W.J. (2004) Propagation technology for forest trees. In: Burley, J., Evans, J. and Younquist, J.A. (eds) *Encyclopedia of Forest Sciences.* Elsevier Academic Press, Amsterdam, pp. 237-244.

Linder, S. and Axelsson, B. (1982) Changes in carbon uptake and allocation patterns as a result of irrigation and fertilization in a young *Pinus sylvestris* stand. In: Waring, R.H. (ed.) *Carbon Uptake and Allocation in Subalpine Ecosystems as a Key to Management.* Forest Research Laboratory, Oregon State University, Corvallis, USA, pp. 38-44.

Maclaren, J.P. (1993) Radiata Pine Growers' Manual. *FRI Bulletin No. 184*, New Zealand Forest Research Institute, Rotorua, New Zealand.

Maclaren, J.P. (1996) Environmental Effects of Planted Forests in New Zealand. *FRI Bulletin No. 198*, New Zealand Forest Research Institute, Rotorua, New Zealand.

Maginnis, S. and Jackson, W. (2003) The role of planted forests in forest landscape restoration. In: *UNFF Intersessional Expert Meeting on the Role of Planted Forests in Sustainable Forest Management*, 24-30 March 2003. Ministry of Agriculture and Forestry, Wellington, New Zealand, pp. 87-99.

Nair, K.S.S. (2001) *Pest Outbreaks in Tropical Forest Plantations: Is there a Greater Risk for Exotic Tree Species?* Center for International Forestry Research, Bogor, Indonesia.

Nambiar, E.K.S. (1996) Sustained productivity of forests is a continuing challenge to soil science. *Soil Science Society of America Journal* 60, 1629-1642.

Nambiar, E.K.S. and Brown, A.G. (eds) (1997) *Management of Soil, Nutrients and Water in Tropical Plantation Forests.* Australian Centre for International Agricultural Research, Canberra.

Nambiar, E.K.S and Sands, R. (1993) Competition for water and nutrients in forests. *Canadian Journal of Forest Research* 23, 1955-1968.

Powers, R.F. (1999) On the sustainability of planted forests. *New Forests* 17, 263-206.

Powers, R.F. (2001) Assessing potential sustainable wood yield. In: Evans, J. (ed.) *The Forests Handbook*, Volume 2. Blackwell Science, Oxford, pp. 105-128.

Reddy, M.V. (2002) *Management of Tropical Plantation Forests and Their Soil Litter System: Litter, Biota and Soil-Nutrient Dynamics.* Science Publishers, Enfield and Plymouth.

Sands, R. (1984) Transplanting stress in radiata pine. *Australian Forest Research* 14, 67-72.

Sweet, G.B. (1989) Keynote address - maintaining health in plantation forests. *New Zealand Journal of Forestry Science* 19(2/3), 143-154.

Taiz, L. and Zeiger, E. (1998*) Plant Physiology* (2nd Edition). Sinauer Associates, Sunderland.

Theron, K. and Bredenkamp, B.V. (2004) Stand density and stocking in plantations. In: Burley, J., Evans, J. and Younquist, J.A. (eds) *Encyclopedia of Forest Sciences*. Elsevier Academic Press, Amsterdam, pp. 829-836.

Turnbull, J.W. (2002) Tree domestication and the history of plantations. In: Squires, V.R. (ed.) *The Role of Food, Agriculture, Forestry and Fisheries and the Use of Natural Resources. Encyclopedia of Life Support Systems*. Developed under auspices of UNESCO, Eolss Publishers, Oxford. (www.eolss.net).

White, T.L., Neale, D.B. and Adams, W.T. (2003) *Forest Genetics*. CAB International, Wallingford, UK.

Will, G.M and Ballard, R. (1976) Radiata pine – soil degrader or improver. *New Zealand Journal of Forestry* 21(2), 248-252.

Zobel, B.J. and Talbert, J. (1984) *Applied Forest Tree Improvement*. The Blackburn Press, Caldwell, New Jersey, USA.

Chapter 8

Social Forestry

What is Social Forestry

There is probably no robust definition for 'social forestry'. The well-known quote 'forests are more about people than about trees' (attributed to Jack Westoby) suggests that all forestry is, or should be, social forestry and indeed this entire book is about human interactions with forests. Perhaps, in order to more closely define social forestry, the opposite question should be posed - what is 'unsocial forestry'. Unsocial forestry probably would be forestry that is remote from individuals and typified by large-scale enterprises dominated by corporations or big government and serviced by traditional forestry. Social forestry as understood today probably arose as a reaction to 'big forestry'. Probably social forestry formally commenced, or at least gained momentum and broad-scale international recognition, following the World Forestry Conference held in Jakarta in 1978 with the theme 'Forestry for People'. The original idea behind social forestry was to assist the rural poor in developing countries to use the forests in their vicinity in a sustainable manner and to their economic advantage. This probably remains the main focus but the concept has been broadened to include most 'small-scale' forestry in both rural and urban areas in both developed and developing countries. For convenience, social forestry will be sub-divided in this chapter into community forestry; farm forestry and agroforestry; and urban forestry, although the boundaries are not always distinct. Many of the issues have already been introduced in Chapter 5. Forestry as a profession developed out of scientific disciplines where it was assumed that science and technology could solve all forestry problems providing people behaved in predictable ways and did not complicate the issue. Over the last three decades, foresters and forest educators have become increasingly aware of their poor knowledge of rural sociology and extension and thankfully forestry curricula world-wide are responding to this.

Community Forestry

Discussion will be confined to the developing world. FAO (1978) defined community forestry as *'any situation that intimately involves local people in forestry activity'*. The FAO definition has been refined since 1978 to emphasize community involvement in managing forests as communal property (although not necessarily legally common property). The scope of community forestry is therefore quite wide and includes: establishing woodlots for fuel; establishing multi-purpose trees; providing fodder for livestock; hunting for bushmeat; gathering and marketing of wood, non-wood forest products and medicinal plants; and providing mulch, shade and windbreaks for agriculture. Forest activities may be critical to rural farmers in supplementing their incomes and as an emergency source of food in times of famine or crop failures. Community forestry not only applies to communities that live in the forest but also to surrounding communities that use the forest for some of their needs. Community forestry is particularly directed towards alleviating rural poverty, encouraging sustainable development, reducing deforestation, assisting the landless, empowering women, and achieving social justice for the marginalized and the disadvantaged. The overarching principle in community forestry (or community-based forest management, CBFM) is involving people at the local community level in the management of the trees and forests in their vicinity and in sharing the benefits. This includes agreed use of forested land held in common or as free access. Community forestry is fundamentally about getting people to work together towards a common goal.

In its purest form, community forestry aims for triple bottom line accountability: safeguarding the environment (arresting deforestation and conserving soil, water and biodiversity), improving economic welfare (increasing natural, physical, financial, human and social capital) and enhancing the social fabric of the community. This is all fairly idealistic and it is no wonder that there is a history of community forestry programmes going wrong if there is the expectation that all of these things will fall neatly into place. People are complex and unpredictable and communities even more so. Community forestry is a little like multiple use forestry (Chapter 6) in the sense that it brings together a range of stakeholders with different interests in an attempt to adopt and participate in a common shared approach.

Initially community forestry was a top-down approach aimed at encouraging local communities to stop destructive practices such as short fallow shifting agriculture and deforestation of upland catchments. This was followed by top-down afforestation projects to arrest the so-called fuelwood crisis (see Chapter 4). This coincided with governments taking greater interest in forest management. Some forest areas previously used as common property by local communities were appropriated by government and managed as timber reserves to support industry. This reduced access to common forest land and weakened and confused the rights and privileges local communities had over the common forested land they relied on to procure forest products and a proportion of their incomes. This was further complicated by private tenure rights being given to privileged individuals and a greater proportion of the rural poor relying on paid employment for their livelihood. The next phase was the recognition that community forestry would only work if it addressed real needs of the locals and if it was managed at the local level. The final phase was the progression from community forestry meeting the real needs of the community to community forestry being a significant income earner and a means of raising the living standard of the whole community. There has been devolution from central to decentralized management, and from national or regional government to local communities. The emphasis is now on participation of all stakeholders, but with planning and management being devolved as far as possible to the local community. It is important that national and regional governments are willing partners because community forestry still has to operate within a national or regional policy framework. Most of the same criteria that apply to community forestry apply equally to single farms and households and it is difficult and probably counterproductive to draw a line of distinction. The area of farm forestry and agroforestry will be discussed in the next section of this chapter.

Community forestry in itself may not necessarily arrest deforestation. If forest products are relatively freely available from an abundant and undervalued forest resource, farmers at the frontier are unlikely to take community forestry seriously and are unlikely to invest in forest-related activities. Community forestry can only work when there is supply pressure on forest resources valued by the community. It is unreasonable to expect the subsistence farmer in a developing country to embrace community forestry from a predominantly conservation perspective. On the other hand, some indigenous community forestry programmes, not necessarily sponsored by government, have been remarkably successful at reforestation and forest conservation. Community forestry systems in Nepal and the Ngitili system in Tanzania provide good examples (Bob Fisher, personal communication).

What Does the Community Want and What Does it Need

Community forestry only becomes a concern to the rural subsistence farmers when the forest products they rely on become scarce and substitutes are not readily available or affordable. Farmers need to recognize that a forestry solution is superior to any alternative before they will accept such a solution, which may be planting trees on farms or participating in the management of common forest land. Communities and households within communities are only likely to embrace a forestry solution if it will increase income (or reduce labour), if the products are secure, if the risks are acceptable and if the policy environment is stable and predictable (Hyde and Köhlin, 2000). Indeed farmers may well choose less risk over perceived increased income. Forestry is a long term investment and food in the short term may be preferred over fodder and fuel in the future.

A community is not homogeneous and does not respond as a cohesive group. There are hierarchies and power struggles; gender, religious and cultural considerations; and inequities in land tenure and in wealth. Not everybody will operate in the community's best interests. Sometimes the results of community forestry programmes can be quite perverse. An example of a community forestry programme that went wrong is given by Westoby (1989). India burns cow dung for fuel and does so at the expense of declining soil fertility (Chapter 4). A well-intentioned community forestry programme aimed to redress this by establishing fuelwood plantations. More than half the rural population was landless and relied on the larger richer farmers for employment. The larger farmers found that growing wood for industry was more profitable and required less employment. Consequently the rural poor became worse off and cow dung continued to be used as fuel. In recent years India has promoted joint forest management where locals have more say and acquire real benefits (Poffenberger and McGean, 1996).

The short history of community forestry is littered with top-down, aid-driven failures that were established with all the best intentions. Rural folk are suspicious of technical experts who tell them what they need for their own good. Rural subsistence farmers in the developing world have a way of life that has been handed down to them through the generations and, because their very survival is an issue, they are reluctant to take risks that they do not fully comprehend. If a community does not identify with a project and claim it as theirs, then it will probably fail. Most successful community forestry projects are those where the ideas are generated from within. This may mean that the community forester needs to live in the community to be taken seriously. Sometimes more advanced technologies are inappropriate in a community forestry project and simpler technologies proven over many decades or centuries may be better suited. Often, however, technology transfer is required but this needs to be done with care and respect. Also technology transfer requires the price of forest products to be high enough to justify it.

There are six principles in community forestry which will be discussed in turn: decentralization, devolution, participation, community-based forest management, co-management and equity. Community forestry is still evolving. There are differences in opinion about the relative importance of these principles and whether all are necessary for a community forestry programme to be successful. Arguably, the key principle is devolution of decision making and power. This, however, has been the most difficult to achieve.

Decentralization

Big forestry with a power and policy base remote from the forest has not always worked to the advantage of the subsistence farmers who need the forest as part of their livelihood. Often big forestry has had the opposite effect by alienating the very forest the rural farmers need for their livelihood and reallocating this to serve industry. Big forestry attracts big investors with little interest in the plight of the peasant farmer. On the other hand the local communities who live in close proximity to the forest and who depend on the forest for part of their livelihood are better positioned to care for the forest. They observe the forest every day. They require a range of products from the forest and it is in their immediate best interests to ensure that these products are sustainably managed. The local community is therefore likely to conserve biodiversity better than a single-interest industrial forest enterprise with its head office in a distant city. Because of their intimate involvement with the forest, it makes good common sense to include locals in its management. If the locals are not involved in management, they are likely to frustrate the efforts of central management even when central management believes they are acting in the best interests of the local community. Indeed, if the local community becomes involved in the management of their local forests, this should decrease the cost to the state in conservation and management.

Also there is an ethical issue, which is the right of an individual and a community to participate in those local matters that directly affect them.

Devolution

The idea here is that the decisions, responsibilities and rewards be devolved down to the community or sub-units in the community. All too often central authorities will devolve the responsibility but not the power. Although devolution is a key principle in community based forest management, there are relatively few examples where it has actually been achieved. Forestry departments are coming under increasing financial and political pressure to support community forestry (Poffenberger, 1996). Fisher (2003) considered that insufficient devolution and lack of tenure were the major factors limiting progress in providing livelihood benefits in community forestry programmes in Southeast and South Asia and that institutional changes and policy reforms were necessary to resolve this. Edmunds and Wollenburg (2003) evaluated case studies for local forest management in China, India and the Philippines and found that the extent of state devolution in all three countries was disappointing and inadequate.

Participation

The ideal is that all stakeholders should share the power and the responsibility and actively participate in the planning and management of forestry-related activities and share in the rewards.

Community-based Forest Management

Community-based management has its critics. Hardin (1968) introduced the concept of the 'tragedy of the commons' which claims that it is inevitable that property held in common will be overused and undervalued. The argument is that it is human nature for individuals to take more than their fair share at the expense of others if they can get away with it. This of course is true to some extent and has led to forest policies favouring either state control or privatization. However, this is unduly pessimistic. Hardin's theory is vigorously disputed in more recent literature, both on empirical and theoretical grounds. There are good examples of community-based management at the local level working very well and in the common interests of most individuals. There are certain conditions where it is obvious that some sort of shared responsibility is preferable. If a common resource is large, has multiple uses and interacts with a range of external factors, community management is likely to be more efficient. Community forestry is not a panacea to remove poverty and disadvantage and to ensure sustainable forest practices. It requires a good deal of common sense, pragmatism and compromise to work. Development may indeed be the best way of ensuring conservation at the national level and over the long term (Chapter 5). However, at the local level and over the short term, development and conservation are often contradictory and some compromise is necessary.

Co-management

Co-management (or joint management) is where the range of users and stakeholders assume joint management and control. Stakeholders may have different interests and indeed conflicting interests that need to be rationalized in a joint management model. Probably the most important partnership is between the local community and the regional or national forestry authorities. Both need each other to achieve a mutually satisfactory outcome. The central authority working alone often cannot provide a satisfactory outcome at the community level but the local community acting alone often cannot do so either. It is important that the local community acts within the policy framework of the region or nation and gets as much help and support from the centre as possible. Ideally the centre should support and encourage rather than subjugate, although the centre will still have to regulate to some extent and this may well be in the best interests of the locals. The centre may also act as a conciliator in the case of disputes and be an important agent in conflict resolution. The centre as a stakeholder may be better able to coordinate state international involvement and the provision of technical expertise. A collaborative and decentralized structure is becoming increasingly favoured by donor organizations and by international non-government organizations committed to promoting sustainable development.

National or regional government policy can apply incentives and disincentives that will impact on the success or otherwise of community forestry programmes. Incentives to promote agricultural expansion will promote deforestation rather than forest management. If the centre takes more than their reasonable share of the benefits (income) away from the locals and disperses them elsewhere, this will kill local initiative and involvement. Restrictions placed on community forestry schemes to promote sustainable forest management, conservation of soil, water and biodiversity can act as a disincentive (Arnold, 2001). These are, of course, essential outcomes but they need to be incorporated into community forestry programmes in such a way that the community still finds it a worthwhile proposition. Subsidizing the community to undertake conservation management is a blunt tool. It is better, wherever possible, to promote a commitment to sustainable management within the community. This is an important balance to achieve because, without it, the community forestry programme may not proceed at all and the region will have missed out on an opportunity to promote sustainable forest management and conservation.

Governments are rightly concerned with rehabilitating degraded lands, but sometimes expect the rural farmer to be the instrument of this even when it is of little economic benefit to the farmer. It is all very well to be concerned with afforestation of upland water catchments, with erosion control and with arresting short fallow shifting cultivation, but unless the rural farmers can see something in it for them, they are unlikely to cooperate. This is not to say that rural communities are not concerned with forest degradation and conservation issues. Rural communities are concerned because they see this as contributing to scarcity of forest resources that are important to them. Accordingly, some indigenous forest management systems have been very successful at afforestation and forest conservation. Rural communities should not, however, be expected to subsidize rehabilitation and conservation programmes.

The professional forester in a community forestry programme may have quite a different perspective on what is important than does the local community. The forester will have a different background and culture and very different skills, knowledge and institutional allegiances than the locals. A key to successful community forestry is the ability of the professional forester to understand and accommodate community perspectives on forest management (Wiersum, 1999). This is, of course, a two-way process. Professional foresters have in the past largely been trained to operate in a non-pluralistic framework, one where single and often large institutions can determine policy and practice. This is no longer the case.

Equity

There is always the risk that community management will reflect existing community power structures which may favour the wealthy over the poor, the landed over the landless, mainstream over marginalized and men over women. Women are frequently disadvantaged in rural poor communities. Women universally are the collectors of fuelwood and the gatherers of forest products. If a community forestry programme is able to materially improve the plight of women, it is more likely to be a success. However, social structures often make this difficult to achieve. The loving and sharing community is (unfortunately) an ideal and in practice certain compromises may be necessary to make a community forestry programme work. This could include, for example, some private ownership and some mixed institutional models. Power structures are usually well entrenched and hard to break down. Sometimes the most appropriate compromise is to accept them and work with them in order to remove disadvantage.

Box 8.1. Joint Forest Management in Tripura, India – by J.P. Yadav

Tripura is a hilly state, densely populated by Bengalis and tribal people. Sixty percent of the livestock population of 1.5 million graze in the forest which occupies 52% of the state and of which 67% is controlled by communities (village councils, district councils and private individuals) rather than government. Joint Forest Management (JFM) was introduced in 1991 to try to reverse deforestation due to development activities, over-exploitation, mismanagement, human-induced disasters, political instability and floods. Forest Protection and Regeneration Committees (FPRCs) are formed at the village regional level and the committees are encouraged to register under the Societies Registration Act, 1860. They work in close coordination with Panchayati Raj Institutions and government departments.

The first attempt, a Forest Department project involving rural families living in four villages surrounding a plot of 100 ha of a proposed forest reserve, was mostly successful. By the end of 1999, 165 FPRCs had been formed covering 18,566 ha of forest. The main activities of the committees are: encouraging natural regeneration of *sal* and teak; establishing bamboo plantations; under-planting of canes; establishing and maintaining medicinal plants; and soil and water conservation. The community forests also contribute to livelihood by supplying bamboo, fuelwood, small timber, thatch, sand, tubers, leaves, honey as well as employment. The timber is yet to be harvested. If the plantations and natural forests are protected by the FPRCs, the gain will be substantial.

There are some difficulties. The JFM Committees (JFMCs) are formed by government resolutions and have no legal standing like the Panchayats. This makes them vulnerable at law. The cultivation of some minor forest products (fruits, tubers, palms, fishery, oil yielding and medicinal plants, spices, stones, bajri, fibre) is crucial for motivating people to support JFM but this is not recognized as an acceptable forest activity by the Forest Conservation Act, 1980. Members of JFMC understandably object if non-member families receive their share of the common benefit, and if these families are denied it, they can gain the benefit illegally. There is no provision to deal with offenders in JFM areas except the Indian Forest Act. (Smuggling of forest products between India and Bangladesh is a concern in border areas of the state. The problem may be so acute that even small twigs and dry leaves are removed from forest areas and the local people are not in a position to prevent it.) Other reasons for poor response and low participation of local communities in JFM are: paucity and lateness of funding; bureaucratic inflexibility; inappropriate and unmotivated staff; poor monitoring and evaluation; inconsistent prescriptions; poor marketing of forest products; and better opportunities (and less manual work) offered to villagers by other agencies and departments. Sometimes forest adjacent to a village has traditionally been used by a more-distant community. This and other incompatibilities in territorial boundaries of the Forest Department and the Civil Administration sometimes create confusion. Also, lack of personal security may be a problem for staff.

Experience so far indicates that JFM is not an easy task. Most, but probably not all of these, can be resolved. The concept of JFM is still evolving and most forest personnel are not familiar with it. Planners need to proceed gradually and at first give full attention to a few selected sites. When positive results are available, more areas can be progressively added until all degraded forest areas come under participatory management regimes which would confirm JFM as a powerful tool for forest improvement and rural development. This requires staff with the right attitude and aptitude.

JFM is an integrated holistic management approach and both peoples' representatives and influential leaders (legislators, Panchayat members, local community leaders and religious heads) should be involved. Continuous flow of information from bottom-to-top and vice versa is essential to achieve joint forest management. This will not only keep all informed but will be a tool for regular monitoring and evaluation of programmes and a vehicle for promoting JFM globally.

Has Community Forestry Been a Success

The answer is both yes and no. Although mistakes have been made, community forestry has evolved, through a learning by doing process (called 'action research' by Fisher, 2001), into a qualified success story and has moved from an experimental phase to mainstream forest management. Technical issues are usually not a problem. The tricky issues are socio-political. Community-based forest management is a collaborative arrangement between different stakeholders, all with their own agenda, some of whom will manipulate and distort things to their advantage. There is no ideal solution. Consensus is almost impossible, all conflicts will not be resolved and devolution in most instances will only be partial. Providing this is understood, management can often reach a workable compromise. There is much rhetoric about community forestry but documented success stories are still relatively rare (Fisher, 2001).

Case Studies

Community-based forest management is a compromise and its success or otherwise depends on a wide range of interacting factors that are peculiar to the local community. Consequently community forestry programmes may be quite different between countries and regions. Lawrence and Gillett (2004) have reviewed the broad range of community forestry programmes world-wide. Boxes 8.1 and 8.2 give case studies of community forestry programmes in India and Nepal respectively, each written by an expert who has been intimately involved in the programme.

The role and rights of indigenous peoples in managing their ancestral forests also falls within the scope of community forestry. The most common issues for indigenous peoples are land rights and justice. A case study for the indigenous Maori people in New Zealand is given in Box 8.3.

Balance and Common Sense

Social forestry and community forestry have been a logical reaction against traditional centralized forestry towards decentralized and devolved forest management where the forester is a facilitator rather than a regulator. This certainly was a move in the right direction but it is important not to over-react. Big forestry will remain important and governments must ultimately be responsible for policy within their borders. There is room and need for both. Some foresters are bewildered by the degree of consultation that is necessary with many stakeholders in order to achieve a genuine collaborative arrangement where all parties are heard and the best devolved community-based compromise achieved. This can be very time consuming and inefficient. The various principles and criteria for sustainable management (Chapter 6) offer a bench mark against which the diverse range of opinions and values can be evaluated (Bass, 2003).

Box 8.2. Community Forestry in Nepal – by Bob Fisher

In the 1970s and 1980s Nepal was commonly presented as the prototypical example of a case where pressure from a rapidly growing population seeking firewood and timber was leading to a disastrously high rate of deforestation.

It was increasingly recognized in the 1970s that the Forest Department's real capacity to manage state forests was limited (most forests were state forests). At the same time it was recognized that some local communities had been taking successful action to protect forests and the concept of seeking community participation in forest conservation was born. Initially the emphasis was placed on seeking local contributions to plantation and protection (with considerable success), but, by the mid-1980s there was increasing recognition that community forestry was more likely to attract local support if it contributed to the needs of local people for forest products. There was a shift (in rhetoric if not so much in practice) towards forest management rather than forest conservation. This shift was motivated by the recognition that forests were a legitimate and significant source of products needed for local livelihoods. There was, at the same time, growing awareness that indigenous forest management arrangements were quite common in various parts of the country and were frequently very successful in protecting forests and restoring forest cover. These arrangements existed in parallel with the official system and had no legal status. Research showed that they were typically based on groups of people with mutually recognized (but unofficial) rights to clearly identified areas of forests.

The practice of community forestry evolved to focus on such user groups rather than on official local government boundaries. Legislation was passed in 1993 which formally allowed the Forest Department to hand over rights to use and manage forests to formally established Forest User Groups (FUGs), subject to a management plan approved by the Forest Department. The programme was immensely popular and, by 2003, over 12,000 FUGs had been registered, over a million ha of forest land were managed by these groups and more than one million households were involved. This growth was demand driven – a good indicator of the programme popularity. The programme is internationally regarded as a good (and rare) example of a large scale programme with a sound legal basis. Among the programme strengths are the recognition of local use rights and the focus on groups with shared interests in a forest ('natural users') rather than on formal local government areas.

In terms of biophysical outcomes, the programme is generally recognized as having contributed to improved forest cover in areas under community forests. (Community forestry has been largely implemented in the Middle Hill region.) Although the biodiversity of these forests is less than that in relatively untouched natural forests, it is frequently far better than the degraded areas that often existed previously. It is probably true to say that community forestry has been less successful in providing economic benefits to individual users, although limited funds generated from sale of products have been used for community development purposes such as schools and roads. In many cases forest products, especially firewood, leaf litter (for fodder and animal bedding) and some timber are distributed, but there is a strong perception that management plans in general are much more conservation oriented than they need to be and that more could be provided. The Forest Department seems unwilling to step back from maintaining a greater level of control than is apparently intended in the legislation. A related issue is the emerging recognition that the poor have not benefited particularly and may indeed have less access to forest products than previously. The FUGs seem generally to be dominated by local elites.

The programme emerged through a combination of the initiative and leadership of a number of Nepali foresters and support from the international donor community. The strength of the movement within the Forest Department during the 1980s and 1990s was exemplary. However, some elements of the Forest Department are currently attempting to wind back some of the more progressive elements of the programme. (This is being resisted by civil society, including networks of user groups). Nevertheless, the programme remains one of the few large scale international examples of community level forest supported by legislation.

Box 8.3. Justice for New Zealand Maori - by Nora Devoe

Ten separate purchases transferred New Zealand's South Island land from its native Maori inhabitants to the Queen of England, 'the Crown', or the New Zealand government. The Treaty of Waitangi (1840) established the conditions for these purchases. The Treaty gave the Queen 'the rights and powers of Sovereignty' that Maori chiefs had held. It also 'guaranteed' Maori 'full and undisturbed possession' of resources they declined to sell, and granted Maori the rights of British subjects, including property rights. Within the boundaries of the land purchases were Maori settlements, camps, food-gathering sites, and in some cases, unspecified reserves, for which continued Maori ownership was guaranteed. These guarantees were mostly immediately abrogated (Evison, 1998).

By 1891, Maori owned less than 20% of their 1840 holdings (Te Puni Kokiri, Ministry of Maori Development, 1996); 50% of South Island Maori were landless. A further 40% had insufficient land for their sustenance. Abject poverty marked their overall condition (Waitangi Tribunal, 1991). As partial compensation for the much more valuable lands promised in the deeds of purchase, the 1906 South Island Landless Natives Act (SILNA) granted 57,498 ha to 4,064 Maori who had no or inadequate land. SILNA lands were to provide for the owners' livelihoods, but they were and remain rugged, largely forested, without access, infertile and non-arable.

Between 1840 and 1993, forests throughout New Zealand were cleared to create farms and plantations. In 1993, the Forest Act 1949 was amended to require what the government deemed 'sustainable management', which precluded natural forest clearfelling except in small patches or when the wood was not milled. Export of native timber was prohibited unless it was sustainably produced. Some SILNA owners were clearfelling to export chipwood.

Because of likely Treaty of Waitangi compensation claims, SILNA lands were exempted from the Forests Amendment Act 1993 (FAA). Successive governments sought to bring SILNA lands under this law. SILNA owners argued, and independent research confirmed, that compliance with the FAA would reduce forest value by some 90% of clearfell value (Griffiths, 1998). SILNA owners viewed the FAA as an affront to their self-determination and the loss of the option to clearfell and export as a theft of many millions of dollars. Non-SILNA forest owners also protested abridgement of their property rights (Federated Farmers of New Zealand Inc., 1999; Heath 1992). The SILNA owners' argument with the Crown waxed and waned. The Crown purchased conservation settlements over the most valuable SILNA softwood forests. The owners in these settlements relinquished their Treaty of Waitangi claims.

Under a 2004 amendment to the Forest Act 1949, the Crown further reduced its exposure to compensation claims by construing compliance with its rules as voluntary. However, the SILNA exemption to export prohibitions was repealed. Other environmental legislation was cited within the 2004 Act, giving it force over SILNA land. The 2004 Act made non-sustainable management unlikely and uneconomic. If sustainable management was voluntary, refraining from non-sustainable management was not. Without payment for the foregone market values of clearfelling, the Crown effectively coerced private landowners to finance a public good, forest conservation. Treaty of Waitangi claims against the Crown are pending. The Crown offset some of the SILNA owners' economic loss by offering limited forestry services to support sustainable management. So far, hardwood management under the Crown's terms has proven unprofitable for any forest owner.

Justice is a necessary foundation for sustainable forest management. Even when laws dictating sustainable management are enacted, they are not fully effective without forest users' cooperation. Illegal logging in New Zealand has increased with rising values of selected native timbers and the sense among landowners that they have been treated unfairly.

Farm Forestry and Agroforestry

Farm forestry can be considered as any activity that incorporates forestry related activities and the use of trees in and around the farm. This includes planting of woodlots for fuel and wood products, planting trees for windbreaks, planting trees for shelter and shade for livestock, planting trees for beautification and for conservation of biodiversity, planting trees to control soil erosion and to safeguard water quality, planting trees to lower saline water tables, and planting trees to improve soil properties and to conserve soil moisture. Agroforestry has a more restrictive definition within the broader definition of farm forestry although the terms are sometimes used interchangeably.

The history of human interactions with forests has been the relentless clearing of forests to support agriculture and grazing, and to build towns and cities (Chapter 1). However, there now is a strong move among some farmers to reverse the trend and to bring trees back into the rural agricultural landscape, for both profit and environmental protection. In poor areas this is a rational response to improving livelihood and living standards. In richer areas farmers and small investment syndicates are becoming increasingly aware of the potential of forestry to make a profit. As a result, small-scale forestry is becoming more significant. Some farmers are strongly motivated towards caring for the environment and passing the farm to their children in good condition. Indeed there is a great passion among some farmers (albeit a minority) to re-establish trees in the rural landscape. Active farm forestry associations are a feature of many countries today. As was the case for community forestry, the forestry profession has had to adapt to meet the needs of this groundswell movement. As for community forestry, this has meant a change in attitude and practice where the forester has had to change from being a regulator to a facilitator. It has also meant that farmers have been increasingly seeking forestry knowledge and expertise. Forestry educators have needed to respond to this.

What is Agroforestry

Agroforestry, in its strictest sense, can be defined as a land-use system where woody species are grown intentionally on the same land and in combination with agricultural crops and/or livestock, either simultaneously or in sequence, so that there are both ecological and economical interactions between the different components (Nair, 1993). There are many possible combinations of interacting components: agrisilviculture (crops and trees); silvopastoral (pasture/animals and trees); agrosilvopastoral (crops, pasture/animals, and trees) and more, but it is simpler and less confusing to use the generic term agroforestry. The trees may be more or less equally spaced within the other components or planted as strips, or around boundaries, and/or may overlap or be separate from the other components in time. Most attention has been given to agroforestry in the tropics. However, agroforestry as well as community forestry and the broader aspects of farm forestry, are also important in the temperate developed world. The main agroforestry practices in the tropics are summarized in Table 8.1. There are other equally respectable ways of classifying (and spelling) the various agroforestry systems and practices (e.g. Young, 1997; Huxley, 1999). It is counterproductive to try to limit the definition of agroforestry. In this chapter any connection between trees and farms, no matter how loose or ill-defined, is eligible for discussion.

What are the Advantages of Agroforestry

Plants require light, water and nutrients. If plants are separated in space or time, they can access these resources independently. As they approach each other in space and time, they will compete with each other for these resources. In agricultural and forestry monocultures the individuals compete with each other for these resources in much the same zones (leaves in the air and roots in the soil) and at the same time. Agroforestry systems introduce the possibility of improving the competitive efficiency of the total system by exploiting the zone and time differences between the competing species in capturing water, nutrients and light. Consequently it is possible for the total productivity of the trees plus agricultural crops to be greater than any one of them would be alone on the same site. Also there is the possibility that one crop may improve the growth of another (such as tree legumes providing nitrogen to crops) and to impair the growth of another (such as in allelopathy). Competition in agroforestry systems is discussed in more detail by Huxley (1999) and Wojtkowski (2002). There are many examples of agroforestry systems with trees and annual crops together being more productive than annual crops alone (Ong and Huxley, 1996). Annual crops alone do not occupy the site for all of the year. However, the increased overall (tree plus crop) productivity will not necessarily please the farmer as the yield of the marketable annual crop may well be less. If the tree component is for environmental purposes only or for long-term site enhancement, the farmer may not see this as a useful gain.

Interactions between components of agroforestry systems are complex and each agroforestry system needs to be evaluated on its merits. Sometimes agroforestry systems may increase the productivity of the system as a whole, but other times it may not do so. Many interacting physiological factors need to be considered. Trees will shade ground crops and this will have the effect of reducing the amount of radiation intercepted by the ground crop. Usually reduced intercepted radiation will reduce canopy photosynthesis and productivity of the ground crop. However, some species are tolerant of reduced shade and some even prefer it. Tree shade may increase the proportion of legumes in pasture and thereby improve pasture quality under the trees (Huxley, 1999). Crops that prefer shade are well suited to agroforestry systems. Trees may increase the humidity under their canopies, causing a decrease in the vapour pressure deficit. Depending on circumstances, trees may increase or decrease the temperature under their canopies resulting in an increase or decrease in the vapour pressure deficit. Any increase in vapour pressure deficit (a measure of the dryness of the atmosphere) may decrease photosynthesis (and therefore growth) and increase the rate of water use relative to biomass production. Tree shade may reduce soil temperatures, resulting in increased or decreased growth of ground crops, depending on the circumstances. The foliage of trees may intercept a proportion of rainfall and prevent this from reaching the ground. Trees may also preferentially direct rainfall to flow down their stems or drip from their canopies. Trees may alter (usually reduce) the wind speed over ground crops which may both increase or decrease their productivity depending on the circumstances and wind speed. Wind can promote carbon dioxide entry into leaves and thereby increase photosynthesis, but too much wind can unduly increase the rate of transpiration (which can hasten the onset of water stress) at the expense of photosynthesis.

Table 8.1. The main agroforestry practices in the tropics (adapted from Nair, 1993). (w=woody; h=herbaceous; f=fodder for grazing; and a=animals).

Agroforestry practice	Arrangement of components	Groups of components
Agrisilvicultural		
Improved fallow	Woody species planted and left to grow during the fallow phase	w: fast, preferably leguminous h: common agricultural crops
Taungya	Combined woody and agricultural during early plantation establishment	w: plantation species h: common agricultural crops
Alley cropping	Agricultural species between woody hedges	w: fast coppicing leguminous h: common agricultural crops
Multilayer tree gardens	Dense multispecies, multilayered with no organized planting arrangement	w: varying in form and growth habits h: usually absent
Multipurpose trees on crop lands	Trees random, or systematic on bunds, terraces or plot/field boundaries	w: multipurpose and fruit trees h: common agricultural crops
Plantation crop combinations	(i) Integrated multistorey (mixed, dense) (ii)Mixtures in regular arrangement (iii) Shade trees for plantation crops (iv) Intercrop with agricultural crops	w: coffee, cacao, coconut, fruit esp. in (i), fuel and fodder esp. in (iii) h: present in (iv), sometimes shade tolerant species in (i)
Homegardens	Intimate multistorey combination of trees and crops around homestead	w: fruit trees, vines h: shade tolerant agricultural crops
Trees in soil conservation and reclamation	Trees on bunds, raises, terraces +/- grass strips, and for soil reclamation	w: multipurpose and/or fruit trees h: common agricultural crops
Shelterbelts, windbreaks, live hedges	Trees around farmland/plots	w: variety of tall spreading types h: local agricultural crops
Fuelwood production	Interplanting firewood species on or around agricultural land	w: firewood species h: local agricultural crops
Silvopastoral		
Trees on rangeland or pastures	Trees scattered or systematic	w: multipurpose fodder f: present, a: present
Protein banks	Protein-rich tree fodder for cut and carry fodder production	w: leguminous fodder trees h: present, f: present
Plantation crops with pastures and animals	Example: cattle under coconuts in Southeast Asia and the south Pacific	w: plantation crops f: present, a: present
Agrosilvopastoral		
Homegardens involving animals	Intimate multistorey combination of trees, crops and animals around home	w: fruit trees, other woody species a: present
Multipurpose woody hedgerows	Woody hedges for browse, mulch, green manure and soil conservation	w: fast coppicing fodder trees, shrubs h: common agricultural crops
Apiculture with trees	Trees for honey production	w: honey producing trees
Aquaforestry	Trees lining fish ponds, leaves as forage for fish	w: trees and shrubs preferred by fish
Multipurpose woodlots	For wood, fodder, soil protection, soil reclamation, etc.	Multipurpose species

Often the main competition in agroforestry systems is below ground. Roots compete for water and nutrients and there can be both spatial and temporal segregation of roots of competing species in agroforestry systems, which may enhance their competitive efficiency. There is a common assumption that the roots of agricultural crop or pasture species do not compete greatly with tree roots for water and nutrients because crops and pasture have their roots near the surface and tree species have their roots at depth. Sometimes this may be the case, for a time at least. However, sometimes competition between tree roots and herbaceous species can be intense, particularly when the trees are small and have not established root systems at depth, such as is the case for the two-year-old radiata pine shown in Fig. 7.2. The deep roots of trees, however, are important to their long-term survival during intermittent periods when the surface soil is dry. Deep rooting is an adaptation required by woody perennials but unnecessary for annual crops and pastures. Tree roots are more opportunistic than the roots of annuals because they are there all the time, and because their fine roots can quickly regenerate from an already established extensive root framework to exploit new soil resources. An annual, on the other hand, has to start from zero. Understanding and managing competition between roots of interacting species in agroforestry systems is complex and a challenge for research (Schroth, 1999).

The final competitive outcome depends on how all of the physiological variables mentioned above interact for a given climate, site, plant and tree species, and their configurations in space and time. Agroforestry systems can sometimes, but not always, optimize competition and increase the productivity of the total system.

Agroforestry systems provide a variety of products. This increases the diversity of consumables and market opportunities and reduces the risk of catastrophic single crop failure. Agroforestry spreads risk over time and may allow the rural farmer to survive bleak times. The tree component of agroforestry systems can provide shade to stock and protection to stock and crops from wind. An agroforestry system is more complex than an agricultural monoculture and is likely to be more biodiverse. Weeds can be serious competitors with tree crops for water, nutrients and light (Chapter 7) and their control can be time consuming and costly. An agroforestry option would be to replace the weeds with crops of commercial value, particularly legumes.

Trees in an agricultural setting may, over time, improve the fertility, physical properties and organic matter content of soil under their canopies. This is particularly so for *Faidherbia albida* which regularly provides higher crop yields under its canopy in agroforestry systems (Young, 1997). Nitrogen-fixing tree species can improve soil nitrogen status on a continuing basis. This is well demonstrated in alley cropping (see Table 8.1) where continual pruning of the foliage of the leguminous hedges increases the amount of light received by the crop in the alley while providing a continuous source of green nitrogen-rich manure to the soil. Agroforestry systems have the capacity to better recover, recycle and utilize nutrients than agricultural systems alone. Trees in an agroforestry setting provide more closed (less leaky) nutrient cycling. Trees can capture nutrients leached below the root zone of agricultural annuals and recycle them back into the system rather than having them leached off-site. If, however, a more productive agroforestry system replaces a less productive traditional system such as long fallow shifting agriculture on an infertile site, the soil is likely to become less fertile and nutrients may need to be added to the systems to maintain fertility (Szott *et al.*, 1991). One of the claimed benefits of agroforestry is to arrest shifting cultivation. However shifting cultivation can be considered as the first and most tested form of agroforestry and if fallow periods are long enough, it can be an environmentally acceptable option. Also, more attention is being paid to improving fallows by planting trees to improve soil fertility.

Soil erosion depends on the amount of exposed mineral soil and the time it is exposed. Annual agricultural monocultures may expose the soil for a considerable period of time.

Agroforestry systems offer the opportunity to keep soil exposure to a minimum and therefore conserve soil and improve water quality. Trees contribute to a more persistent, complete and complex protective litter layer on the surface of the soil. Contour hedgerows of woody species can be particularly effective at arresting soil erosion. Shelterbelts and windbreaks also protect the soil against erosion. Agroforestry systems may have a range of different intermeshing root types which are more effective at holding soil against wind and water than a monoculture where all the root systems are alike (Chapter 3).

Agroforestry, like community forestry, is pluralistic and integrative, sometimes appealing on an idealistic basis rather than a pragmatic one. There is a tendency to expect agroforestry to deliver more than it is capable of doing (Huxley, 1999). Agroforestry may provide an excellent option but not always. For example in New Zealand where there has been extensive research into grazing animals under widely spaced *Pinus radiata*, the consensus is now that it is more sensible for farmers to separate tree growing and animal grazing on their properties and manage them as individual and unconnected activities (Box 8.7).

Examples of Tropical Agroforestry Systems

The wide array of possible agroforestry systems in the tropics is shown in Table 8.1. Taungya, alley cropping and homegardens will be used as examples.

Taungya

Generally the focus in agroforestry is on agricultural crops, the trees having a supporting role for producing food, fuel, fodder and/or providing environmental protection. In taungya, the emphasis is turned the other way around and the primary objective is the establishment of commercial plantations for wood production. Taungya is a tree regeneration system that developed in Southeast Asia (mainly Myanmar and Thailand). Here the government promoted the establishment of plantation trees (traditionally teak, *Tectona grandis*) by recruiting peasant labour to plant and tend the plantations. In return, the peasants were allowed to grow their crops between the rows of the trees for the first few years of the rotation. The system is long-standing and has the advantage that it can re-establish plantations at minimal cost and that it can provide agricultural land to the rural peasants thereby arresting land degradation from shifting agriculture. Taungya has the capacity to provide food, labour and stability to the peasant community. However the plantations are owned by the government or by remote private enterprises. The peasants do not get any of the financial rewards of the plantation and they do not gain land tenure. The peasants can only receive an income from the land (which they do not own) for just a few years out of a rotation length of several decades (Jordan *et al.*, 1992).

From a biological perspective, taungya is an efficient agroforestry system. Plantations in the first few years do not completely occupy the site and there is the potential for ground crops to exploit water, nutrients and light that would otherwise not be utilized by the plantation crop. Weeds are a big problem in plantation establishment (which confirms that the site is not fully occupied). If an agricultural crop replaces the weeds, this may be a win-win situation. In dry areas, competition for water between weeds and newly transplanted tree seedlings can be intense. Growing crops at the early stage of plantation establishment may reduce the growth of the trees compared to a weed-free situation but the benefits of reduced establishment costs may well outweigh this disadvantage.

Taungya is a classic example of a top-down community forestry system which effectively exploits rather than empowers the local community. The system evolved in a situation where there was a shortage of agricultural land and where the peasants had little option but to accept an arrangement with the plantation owner. Neither does taungya promote responsible stewardship. If peasant farmers do a good job and promote good growth of the plantation trees, they may be moved off the site earlier thereby being effectively punished for good practice (Gajaseni, 1992). Farmers are more likely to care more for the land if they own it. It is therefore not surprising that taungya cultivators will seek other opportunities as they arise. Traditional taungya is an agroforestry success story from the biological, ecological and environmental perspective. It is probably not, however, a social success. In order for taungya to be fair and equitable, additional support needs to be given to the rural poor to encourage them to participate. Taungya is widely practised (e.g. in Thailand, Myanmar, India, China, Philippines, Sierra Leone, Ghana, Nigeria, Sri Lanka, Costa Rica, Trinidad and Tobago). There is no doubt that taungya is beneficial to the plantation owners and sometimes, but not always, it is beneficial to the local community.

Alley Cropping

Alley cropping, also called hedgerow intercropping, is where hedges of woody perennials, usually legumes, are planted in rows, and agricultural crops are planted in the alleys between the rows. If the hedges are planted on the contours in sloping country, then they can act as a barrier to soil erosion. The legumes fix atmospheric nitrogen which contributes to the nitrogen nutrition of the site. The hedges are regularly pruned which provides additional radiation to the crop in the alley. The prunings contribute to surface litter which acts as a mulch to conserve soil water and arrest soil erosion. When it decomposes, it contributes to soil organic matter and the provision of nutrients, particularly nitrogen, to the crops in the alley. The deeper rooting hedges also ensure a more closed nutrient cycling system. The prunings can also be used for high protein food and fodder, and as fuel. The hedges can rapidly regrow in the period between the annual crops.

The best tree species for alley cropping are nodulating legumes that grow fast and are amenable to frequent pruning. Historically *Leucaena leucocephala* has been important but its future is doubtful because of defoliation by the psyllid *Heteropsylla cubana*. Other species of importance are *Calliandra calothyrsus, Inga edulis, Paraserianthes falcataria, Cajanus cajan, Flemingia macrophylla, Gliricidia sepium, Sesbania sesban, Sesbania rostrata* and *Sesbania grandiflora* (Bryan, 1999).

Alley cropping has been a mixed success. Its net value to the farmer depends on the balance between the negative effects of the hedges being a competitor with the crop for water, nutrients and light and the positive benefits associated with nitrogen fixation providing nitrogen to the crop and the additional food, fodder and fuel benefits of the trees. Alley cropping has been promoted as a means of increasing crop productivity and on this basis alone it often does not succeed and farmers have been reluctant to embrace it with enthusiasm. In some situations the contribution of nitrogen from the trees has improved crop yields but in others it has decreased it. Farmers usually do not factor the value of the trees into the overall equation. In any case, removal of tree products for fodder, food and fuel means that these are no longer available as a green mulch to improve the crop. Alley cropping is of dubious merit for areas of low rainfall, highly acidic soils and high soil fertility, as well as with poorly adapted legumes and inadequate pruning (Bryan, 1999). It is promising, however, on steep slopes where soil erosion is a problem. For example, Swinkels *et al.* (2002) found, in western Kenya, that pruning hedgerows of *Leucaena leucocephala* and *Calliandra calothyrsus* considerably increased labour at the busiest time of the year and that the hedgerows had no effect on maize yield while decreasing soil erosion. They concluded that hedgerow intercropping was more likely to be useful in providing feed for an intensive dairy operation or in controlling soil erosion than in increasing the yield of the alley crop.

Homegardens

Homegardens are a traditional form of agroforestry, exemplified in Indonesia but widespread throughout the tropical world, where the area around homes is planted with a variety of trees, shrubs and vines of different sizes and configuration to provide a range of subsistence products to the household. In near to closed canopy homegardens, agricultural crops are not present unless they are shade tolerant. In more open canopy homegardens there can be vegetable and other crops. The trees occupy different zones in space and time which accords to the complementarity and diversity ideals of agroforestry systems. There are many canopy layers, and these are patchy rather than even and organized. There is a range of tree sizes, meaning that regeneration of the different tree species is more or less even and continuous. In this respect the idealized homegarden resembles a multispecies uneven-aged forest managed under the single tree selection system (Kelty, 1999). However, in practice and as for single tree selection forestry, actual homegardens are not so exact. Homegardens have gaps and edges caused by the homestead, bordering roads and other agricultural systems. These gaps and edges can enhance diversity and provide additional temporal and spatial growing zones compared with a continuous canopy. Homegardens might also contain animals (pigs, goats) and fishponds (Wojtkowski, 1998).

The trees can provide shade, fodder, fuel, fruit, spices, vegetables, edible seeds, flowers, honey, medicines, timber for construction, and materials for craft. They can protect the soil from erosion, maintain high soil fertility and soil organic matter, and improve the microclimate for the householders and grazing animals. They can provide a variety of food that is available at different times of the year. Homegardens are sustainable to the extent that they can conserve soil and hold their nutrients within a closed cycle. If the nutrients removed from the site by harvest and/or leaching are substantial, nutrients may need to be imported from off-site in the form of organic or inorganic fertilizers. Household waste can be an important contribution to the nutrient budget of homegardens. If possible, organic residues should not be removed from the site and burning should be avoided.

Temperate Agroforestry Systems

Agroforestry has received most attention in the developing tropics in the context of land shortage, overpopulation, poverty, land degradation and deforestation. Agroforestry though is alive and well in temperate regions and for much the same reasons as in the tropics. Farmers consider agroforestry options if they can increase and/or diversify their incomes and improve resource conservation. Some (a minority) of farmers are prepared to accept a

Box 8.4. Traditional Agroforestry in Europe - by Etienne Saur

Europe has a large diversity of climates (from Boreal to Mediterranean), cultures and land-use practices. Traditional agroforestry systems are numerous and have evolved since the very early age of agriculture in response to environmental, technical and social constraints. The first agroforests were farming activities introduced into native forest after controlled thinning. One example is sylvopastoralism, which once was common where grazing capability was limited by poor soils or populated areas. Forest grazing is an ancient practice still used in many parts of Europe, especially in Mediterranean and mountain forests. Edible species like *Castanea sativa* and *Quercus* spp. were favoured when part of the local vegetation. The most successful and extended farming in native forests in Europe is an agrosylvopastoral system call '*Dehesa*' in Western Spain and '*Montado*' in Portugal. This covers an area of 4-5 million ha and was developed in a semi-arid climate under randomly scattered oaks (*Quercus ilex, Q. suber, Q. faginea*) with a typical density of 30-80 trees/ha. The undercrop is generally a cereal followed by 5-10 years of pasture management. Pigs, sheep and cattle benefit from the grass production, sweet acorns and microclimatic improvement. The whole system is considerably more productive than the separate cultures alone and in addition supports wood, cork and fruit production as well as hunting, but at the cost of intense manual labour and difficulty in using large machinery.

Later, farmers started breeding fruit-trees and developed intercropped orchards to optimize available water and light. Many systems were developed before the Roman Empire and the Mediterranean climate favoured multi-strata associations like (a) cereal cultivation between olive trees (650,000 ha in Greece, 79,000 ha in Sicily and 3000 ha in France remaining today), (b) tree-grapevine-crop associations such as '*hautain*' (where the tree acts as a stake for the grapevine - 2 million ha in Italy before 1950), '*joualle*' (a mixed plantation of fruit trees-and grapevines), (c) complex systems comparable to tropical home gardens (*Coltura promiscua*-grapevine-wheat-tomato-peas on terraces in Tuscany), and (d) '*Huerta*' (irrigated market-garden under peach, apricot, orange, and cherry, e.g. in Valencia in Spain and Roussillon in France). Intercropped walnut orchards were very popular until the 19th century in the southern part of France (e.g. '*noyeraie du Dauphiné et de Dordogne*') mostly based on rotational agroforestry practices. Grass orchards became very common in colder, more humid oceanic areas (Herefordshire in England, Normandy in France), and in mountain landscapes (France, Switzerland, Germany, Austria). Grass orchards cover 3 million ha over Europe mainly with apple-pear-plum-cherry trees planted at 40-80 stems/ha. The income from the fruit, particularly when processed on the farm (juice, cider, alcohol), could be 3-4 times that of the dairy. Grass orchards also conserve insects, bats and birds.

Forest trees on farmland developed later, after the Middle Ages, for fuelwood and to support agriculture under marginal conditions like steep slopes and excess rain or wind. During the 17th century agronomists developed 'bocage' from selected forest species planted in hedgerows around the fields. There were 4-5 million km of hedgerows in Europe at the beginning of the 19th century (United Kingdom, France, South Scandinavia, Switzerland, Austria, Slovenia, Romania). Modern hedgerows were developed during the 19th century as windbreaks (Southern France, Austria, Poland).

Traditional European agroforests confirmed the superiority of tree associations with farm production in most European conditions due to soil erosion control, improved soil fertility, microclimatic changes and biodiversity enhancement. The exception was fertile soils free of floods, high winds and drought where open-field cultivation was more successful (eastern England, Central France, North Germany, Czech Republic, Hungary, Slovakia, Poland).The green revolution (since 1950) led to most trees on farms being cleared and replaced by monocultures with high inputs of fertilizer and chemicals. Most agroforestry systems have disappeared and the larger ones like '*Dehesa*', '*bocage*' and grass orchards are fragile and endangered ecosystems. The green revolution increased food production but Europe now faces challenges like flood risk, mineral pollution, droughts, biodiversity loss, increased pest risk, and, ironically, over-production associated with decreased quality of food.

Box 8.5. The Future of Agroforestry in Europe – by Etienne Saur

Traditional agroforests have no chance of survival in Europe today because of the cost of labour, unless publicly subsidized and/or associated with high value products. As an example, the future of the remaining Normandy grass orchards is to sell the image of cattle grazing under flowering apple trees as an extra-cost on the best quality cider or Camembert cheese in conjunction with agro-tourism development. In the same way, the pig industry in the '*Dehesa*' is healthy thanks to a very high quality ham sold on the international market as the luxury product '*pata negra*' with a gastronomic status close to caviar or *foie-gras*. In this case, the appellation '*Real Ibérico de bellota*' certifies the breed of pig and the feeding regime from acorns in the 'wild' environment of the *Dehesa* agroforest. South European countries give high priority to policies promoting quality food production, much of which goes to the agro-food industry. Specifically, France has long been characterized by the diversity of its local quality agricultural production, from *Appellation d'Origine Contrôlée* wines and cheeses to local meat, vegetable and fruit *produits de terroir* that form a critical part of both the image and the economic strength of French farming. A raft of policy support mechanisms exists to develop and reinforce high added value products which are often rooted in individual localities and local savoir faire, dependent upon relatively small and concentrated production chains, often accompanied by some form of specific labelling and usually including some on-farm product transformation.

Apart from these niche products rooted in history, the future of agroforestry is to invent new systems based on agronomical knowledge from the past and from ongoing scientific research but fully compatible with mechanized agriculture and most of all, with European economy and policy. Research, mainly from the SAFE UE programme (Silvoarable Agroforestry for Europe) indicates that modern agroforestry production systems can be efficient, innovative, environmentally friendly and profitable. The two main systems retained for European situations are high quality timber trees in crop and pasture. Fodder trees have been widely explored but are not suited to the current market.

Tree configurations are determined by cultivation equipment with 10-14m between rows of trees (50-80 stems/hectare) and enlarged parcels. Wide spacing and lack of thinning necessitates the use of genetically improved trees with individual tree shelters and appropriate pruning in order to provide high-value timber. European broadleaved species are chosen for the demanding cabinetwork and marquetry wood chain (wild cherry - *Prunus avium*, service tree - *Sorbus domestica*, red oak - *Quercus rubra*, walnut - *Juglans* spp., sycamore - *Acer pseudoplatanus*, ash - *Fraxinus excelsior*). Generally, agroforestry is not eligible for either agricultural or forestry subsidies because it is not considered to be either, and this is an enormous competitive handicap. However, since very recently, French farmers and landowners can be subsidized for agroforestry: crops planted between the trees are eligible for Common Agricultural Policy (CAP) payments and tree rows are eligible for an annual payment to compensate for the loss of income due to afforestation of agricultural land. In some cases the agroforestry system is eligible for specific agri-environmental aid. At the European level, current regulations do not recognize agroforestry but the working group of the SAFE consortium is supporting new propositions to be added to the Common Agricultural Policy to be enforced in 2005 in the European Union.

In conclusion, growing high quality trees in association with arable crops/pasture in European fields may improve the sustainability of farming systems (compulsory fallow use, better quality of the agricultural products), provide new temperate wood products competitive with tropical wood imports, create novel landscapes of high value, help to control fire, diversify farmers' incomes, sustain country employment, increase carbon sequestration, and 're-educate' the farmer on tree management and environmental consciousness. New agroforestry systems are still embryonic in Europe but well positioned for a very quick start in the new CAP heading to innovative production methods that support environmentally friendly quality products that the public wants.

direct or opportunity cost to establish trees on their farms for environmental conservation, but the majority are only likely to do so if there is some form of incentive to plant or disincentive to not plant. Government policy may be required to achieve this. Also resident full-time farmers are not the sole occupiers of the rural landscape and their numbers are falling in many temperate areas. Non-resident owners and part-time 'hobby' farmers are significant. Hobby farmers can be both poor and good custodians of the land. The profit component is not so essential to hobby farmers and they may embrace conservation measures due to 'eco-guilt' (Williams *et al.,* 1997) or a genuine desire to improve the land because they can afford to do so. Incorporating trees in the rural landscape ideally should be coordinated at the landscape level rather than at the individual farm level because of off-site effects. For example, in southern Australia, tree removal over the last century has resulted in saline water tables rising and breaching the soil surface at lower parts of the catchment. This has taken large areas of agricultural land out of productivity. Replanting with trees is seen as one way (among others) of lowering water tables and rehabilitating the land, and planting trees on farms on middle slopes of catchments will benefit farmers on lower slopes. Farm forestry groups supported by government policy can assist in achieving broader landscape level planning.

Trees were originally the enemy of the farmer who saw them as an obstacle to be cleared. Availability of agricultural land is no longer an issue in much of the temperate world and farmers and rural planning authorities are increasingly seeing the benefits of returning trees to the rural landscape. Trees can be part of the temperate rural landscape for a variety of reasons: provision of wood and non-wood products, providing shelterbelts and windbreaks to protect stock and crops; establishing or repairing riparian strips to safeguard water quality; controlling soil erosion on slopes; ameliorating saline sites; enhancing biodiversity; providing hunting opportunities; rehabilitating degraded land; disposing of nutrient rich sewage effluent and animal wastes; reducing the use of agrochemicals; retiring marginal land no longer required for crop production; and for beauty and recreation. Trees can increase rural property values considerably, but this represents a capital gain rather than a profit. Agroforestry can improve animal welfare, increase carbon sequestration and reduce fire risk. Agroforestry practices include trees widely spaced over pasture or crops, as rows in alley cropping, in woodlots, in windbreaks, along property boundaries and road edges, in riparian strips, on contours, and scattered singly or in groups. Alley cropping in the temperate world is more likely to be crops grown in the alley between rows of horticultural trees rather than leguminous hedges. Agroforestry in temperate climates is very suitable for growing high value tree crops such as walnut in North America and *Sorbus domestica* in Europe. Some find agroforestry landscapes to be more attractive than either forest or open land. This is because the scattered trees provide shade and interest but long-distance visibility is not impeded unduly.

Examples of Temperate Agroforestry Practices

Examples of agroforestry systems and practices in Europe, North America, temperate Australia, Argentina, New Zealand and temperate China are given in Gordon and Newman (1997). Agroforestry practices in the United Kingdom are reviewed by Hislop and Claridge (2000). Selected examples are given here for North America (USA and Canada) and China. Europe has a long tradition of agroforestry practices, particularly in Mediterranean areas,

Box 8.6. Farm Forestry/Agroforestry in New Zealand - by Nick Ledgard

Around half of New Zealand is farm land. The majority was converted from forest/shrubland to pasture during the last 150 years. Over that time the interest in planting more trees on farms has generally been low, despite there being excellent growing sites, and good commercial reasons for doing so.

In New Zealand the term 'agroforestry' used to create mental images of wide-spaced (80-100 stems/ha), pruned (to 6 m) radiata pine, underneath which contented sheep and cows are grazing. Such a two-tier 'silvopastoral' farming system was widely promoted. Today, good examples can only be found on a few farms. Why is this? As one keen farm forester stated 'Trying to teach farmers to grow trees and graze animals on the same patch of ground is like trying to teach a child to ride a bicycle and juggle at the same time – only the most gifted will succeed.' Forestry problems quickly arise as wide-spaced pines on fertile land grow too fast, become unstable and sinuous, while growing large branches, which make pruning for a small diameter knotty core very difficult. In addition, a close watch has to be kept for browse damage to trees, and as they mature, animal problems manifest themselves in the form of poorer performance (declining pasture quality and quantity) and the likes of increasing tree debris in sheep wool. Forestry advisors quickly realized that if they wanted more trees on farms, they were much more likely to succeed if they recommended growing 300 trees on one fenced-off hectare, with animals grazing alone on another two, rather than to recommend 300 trees widely spaced over three hectares.

These days it is well recognized that farm forestry is much more than just woodlots and plantations on farm land. Trees have their traditional roles to play in providing shelter and shade, protecting soils from erosion, and making landscapes more attractive. Recently, new roles have arisen in the form of forage feed banks and treatment areas for the disposal of dairy farm effluent. In addition, the consumers of New Zealand's high value farm products are demanding that the goods they purchase come from sustainably managed and animal-friendly farm environments. Trees are an integral part of such production systems, and it is not surprising that virtually all the major winners of present-day sustainable farming awards feature trees in their farm management systems.

Ironically, despite the above benefits, and New Zealand having an active Farm Forestry Association with 3000 members in over 30 branches nationwide, a widespread farmer indifference towards trees and forestry continues to persist. The reasons are various. The history of woody vegetation clearance continues to create suspicions about forestry. Long-term investments in timber growing are not encouraged by the facts that average farm tenure is only 11 years, and that there are no financial incentives (subsidies) to bridge the considerable gap between investment and return. Many farmers have had bad experiences with trees - falling on fences, providing good habitat for pests such as rabbits and possums, and creating localized shade which encourages concentration of grazing animals. Others complain about low commercial returns, although this is often due to poor siting and silviculture and/or weak marketing.

Such experiences are frequently the consequence of the major reason for low farmer appreciation of trees and forestry – a lack of basic silvicultural knowledge, due to few forestry learning opportunities in secondary and tertiary education and training institutes. For example, despite the country having world-leading agricultural centres at Lincoln and Massey universities, forestry courses are only recent introductions and are not widely promoted as an integral part of efficient land use. This reflects not only in most farmers being ill-informed about trees and forestry, but also in the fact that the majority of those employed in land-use policy making, advice and administration at local government and national level are equally naïve on the topic.

It is for these reasons, that trees and forestry struggle for acceptance on most New Zealand farms, despite their having excellent environments for tree growth, and there being good commercial and environmental reasons supporting their increased use.

dating back to the beginning of the age of agriculture. Etienne Saur discusses traditional agroforestry systems in Europe in Box 8.4 and of the future of agroforestry in Europe in Box 8.5. Nick Ledgard discusses the development of agroforestry in New Zealand in Box 8.6 and Rowan Reid discusses farm forestry in Australia in Box 8.7.

Alley Cropping with Black Walnut in North America

Black walnut (*Juglans nigra*) is an excellent agroforestry species for alley cropping. It has very high value as a premium furniture timber, it provides valuable edible nuts, it is an attractive tree, it can grow quickly and it is easy to manage. Black walnut is a deciduous species with a longer leafless period than most other deciduous angiosperms in the region and consequently more annual radiation can be intercepted by the crop. Even when the tree is in full foliage, it still lets through more light than many other species. For nut and wood production the trees need to be thinned to the best 75 trees per hectare within the first 25 to 30 years. Crops can be grown in the rows until shading precludes it. This will occur earlier for light demanding crops such as maize and later for shade tolerating crops such as ginseng. When crops can no longer be grown, the space between the trees can be used for forage grasses which, for some reason, appear to have greater yields and higher forage quality accompanied by low density plantings of walnut than when trees are absent (Williams *et al.*, 1997).

Winter wheat is a preferred intercrop because it grows during the walnut's dormancy and leafless period. Soybean and milo compete more directly with the walnut for water and nutrients, although the walnut has roots below the rooting zone of the row crops. Soybean in summer and wheat in winter are very efficient during the early stages when there is still plenty of light. This can then be followed by pasture production and grazing. The various combinations are evaluated in Table 8.2. Managing for timber alone delivered a real rate of return of 6.7% whereas the most complex agroforestry combination delivered 10.9%. Other specialty crops such as Christmas trees, red clover, vegetables, and landscaping plants have also been successfully planted in walnut agroforestry systems (Garrett *et al.*, 1991).

Table 8.2. Present net worth (PNW) @7.5% and internal rate of return (IRR) of alternative black walnut agroforestry regimes on a good quality site and rotation length of 60 years (abstracted from Garrett *et al.*, 1991).

Management regime	PNW ($ per acre)	IRR (%)
Timber	-575	6.7
Timber and nuts	134	7.7
Timber, nuts and wheat	822	9.4
Timber, nuts, wheat, soybeans, fescue hay and grazing	1290	10.9

Crops Interplanted with Paulownia in China

Paulownia elongata (Lankao Pawtung) is a fast growing deciduous native tree of China. It has a strong light wood suitable for a wide variety of uses, and it provides fodder for animals and nectar for honey production. The leaf arrangement allows a relatively large amount of light to penetrate through to ground crops. The leaves fall late which protects the crops from winter frost damage, and the new leaves also emerge late which improves light acquisition by the crop in spring. The tree is deep rooted which minimizes root competition. Common crops are winter wheat, soybean, millet, peanuts, oilseed rape, garlic, cotton, and sweet potato. The trees are either planted in rows or as 'four sides plantations.' Four sides plantations are where trees are planted along roads, rivers and canals and around houses and villages. When *Paulownia* is planted in rows, the trees are 5 metres apart in north-south rows and 10 metres between rows if the wood is more profitable than the crop, 15-20 metres between rows where the wood and crop are about equivalent in value and 30-50 metres between rows where the crop is considered to be more important. It is claimed that eight-year-old *Paulownia* at a spacing of 5 x 20 metres can receive almost 80% of available light in the middle of the alleys and over 60% under the trees, that wind velocity can be reduced by 20-50%, that evaporation can be decreased by 10% during the day and by 4% during the night, that soil water is conserved, and that soil temperatures are increased in summer and reduced in winter. Even so, competition for water has been shown to be significant in drier areas and yields near to the trees decline as the trees get older and produce more shade (Wu and Zhu, 1997).

Box 8.7. Australian Agroforestry is about Farmers Growing Trees for Conservation and Profit - by Rowan Reid

Most definitions of agroforestry focus on the role the trees play and their location or arrangement. The USDA, for example, adopts this approach identifying five basic types of agroforestry practices: alley cropping, windbreaks, riparian buffer strips, silvopastoral and forest farming. In Australia, the term Farm Forestry is more widely used. Again the emphasis is on the role of the trees and what the practice looks like. Because most proponents are foresters, farm forestry has come to mean commercial timber production on farmland occupying part of a continuum between small scale plantings on farms and large industrial plantations.

Adopting and promoting definitions based on predefined land use practices or distinguishing systems on the basis of scale or intention is problematic. If someone rejects the model, such as pines on pasture, they also tend to reject the approach. This is critical in a field where locally appropriate practices are yet to be fully developed for most areas let alone for individual farmers.

Growing trees on Australian farms is seen as a means of tackling land degradation, improving water quality, enhancing biodiversity and lifting rural incomes. The fact is that the vast majority of the land targeted for revegetation is controlled by farmers. So, it is the farmers who will decide if trees are planted and for what purpose. For this reason I argue that farm forestry and agroforestry definitions should relate to the process by which these forests are established and managed. My definition is simple: 'Agroforestry is the commitment of resources by farmers, alone or in partnerships, towards the establishment or management of forests on their land.'

If a farmer plants a forest, whatever the purpose or configuration, it is agroforestry. Having accepted this, the argument shifts from what the practice looks like to what the farmer wants and what type of tree growing might suit their situation. It also changes our focus from the paddock level to the farm level. It is at the farm level that decisions about land use are made and opportunity for forestry to contribute to the economic, social and environmental wellbeing of rural communities is best realized.

When looking at the whole farm enterprise the problems of competition for land that plague conventional agroforestry options are often irrelevant. Trees can be grown on land unsuitable for agriculture, such as along water courses, or in arrangements that enhance agricultural production. Research shows that many Australian farmers do want to grow trees. Their primary interest is shelter (75%) and land protection (50%). Interestingly, around 30% of those who plant trees consider nature conservation and wildlife benefits to be an important driver and 10% even mention improved aesthetics. However, few farmers (about 1%) see commercial timber production as a primary purpose for growing trees.

Rather than trying to convince farmers to dedicate more of their productive land to elaborate timber production options the key is to link timber production to their other interests. Generating income from trees grown initially for conservation is seen as a bonus or 'icing on the cake'. This multipurpose approach is very different to conventional plantation forestry because it requires that each planting is carefully designed to suit the particular situation.

But how can this be a viable way to grow commercial tree products like timber? In fact, farmers may actually have a comparative advantage over industrial forest growers because of the multiple benefits. We're finding that farmers actually prefer to grow specialty timber species and are comfortable with long rotations because they receive real rewards while the trees are growing. This goes contrary to conventional economic wisdom that argues for uniform, large scale, short-rotation, monocultural plantations.

The economics of multipurpose agroforestry is different. The need for trees justifies the cost of establishment. The environmental and agricultural services trees provide justify their presence on the farm. The only question is whether or not the timber is viable to harvest considering the going market price, harvesting costs and the possible loss of valued services. To keep this option alive the farmers need to manage their trees with a focus on high log quality and reduced harvesting cost.

Research into agroforestry and farm forestry represents a different type for forestry because it involves farmers. It is critical that we acknowledge the varied interests and opportunities facing the millions of farmers who make the decisions that ultimately affect the way much of the world's land is managed. Agroforestry research in Australia is rightly focused on multipurpose design principles, silvicultural management, harvesting technology and processing. Extension initiatives, like our own Australian Master TreeGrower Program, help farmers and their supporters devise unique and elegant forestry options that they are proud of.

Australian agroforestry has moved on from the early days of wide spaced pines on pasture. Fortunately, well managed forests on farms, whatever the species or arrangement, generally also provide real environmental and social benefits to the wider community. That's why most of us are so passionate about trees, isn't it?

Urban Forestry

Towns and cities may have many trees. Indeed, many cities contain more trees than the surrounding rural landscape. The trees are found in the streets, municipal parks, gardens and reserves, golf courses, cemeteries, around streams, on private property, on catchments, in greenbelts, and indeed most everywhere. Together this forms the urban forest. There are several definitions of urban forestry, all similar in most respects to that of Nilsson *et al.* (2001a) who define urban forestry as 'the establishment, management, planning and design of trees and forest stands with amenity values, situated in or near urban areas.'

The variety of trees and other plant species in the urban forest can be very large. Urban trees are a very important resource for beautification, conservation of biodiversity, shade, recreation and amenity. Trees can trap air pollutants, conserve energy by reducing temperatures in and around buildings, screen unsightly structures, and reduce noise. Trees in the urban environment can reduce storm runoff and improve water quality. Green areas in and around cities are known to contribute to the good health and general well-being of their citizens. Trees can also be an important source of food, wood and non-wood forest products, particularly in the developing world (Konijnendijk, 2004). There is ample evidence to show that suburbs or streets with abundant trees have higher property values than suburbs and streets with sparse trees. The same follows for suburbs and streets that are close to parks and other wooded areas. The most elite suburbs are often those that have many large trees. Trees can also be a nuisance in the urban environment. They are probably the main cause of argument with neighbours. They can blow down in the wind causing damage to both person and property. They can compete with power transmission lines. Leaf fall, particularly of deciduous species, can block drains, cripple rail networks and be a general nuisance. Tree roots can invade water pipes and disturb and break building foundations. Falling limbs can be a hazard. Old, large and over-mature trees are one of the biggest problems. They are both the most desirable but also the biggest nuisance and the most dangerous of the trees in the city. The cost of removing large trees in an urban environment can be substantial. The urban forest resource is very large. The United States Department of Agriculture calculated that one third of all area in USA under broad metropolitan control was covered in trees and that this equated to eight percent of the entire country and one quarter of all US trees (Konijnendijk, 2004). It follows that the urban forest resource is an important carbon sink that needs to be included in strategies to mitigate global warming (see Chapter 6).

Many people visit metropolitan parks and gardens for recreation and to many this is their main experience of 'nature'. Most people choose to recreate close to their homes and consequently many more people visit urban forests than forests in the countryside. Most urbanites learn about 'nature' from green areas in the city rather than in the countryside.

Several cultures meet in maintaining trees in the urban landscape. First there are the arborists who often have a horticultural background and are employed by municipal authorities to look after the tree components of parks and gardens. They also consult with and provide a service to private property holders about the care and maintenance of their trees. Next there are the landscape architects and designers who see trees and other plants as an integral part of the overall urban landscape. They are motivated by landscape planning and design. Finally, there are the foresters who see trees and other plants as part of a complex and interacting community, like a forest. Foresters bring an ecological and a sustainable management background. They see the potential of urban forests to provide sustained production of environmental, social and economic benefits. Urban foresters also have a culture of recycling, arising from their ecosystem management background which can contribute to waste management in urban environments.

There is no suggestion that urban forestry should take over the other well-established cultures. Rather, urban foresters can bring a different way of thinking and a different set of skills into the urban landscape that can complement the culture and skill base of other authorities concerned with trees in the city (arborists, landscape architects, engineers, legislators, utility managers). Foresters are skilled at managing trees as a community (a forest) for a range of uses for a range of stakeholders. On the other hand, arborists are very skilled at the care and maintenance of trees, but see them as individuals rather than part of a community. There is a very good case that arboriculture should be part of urban forestry subjects taught in university forestry courses. Indeed, there is a good case that the multiple-use sustained ecosystem management culture of forestry courses will provide a more appropriate background to city tree/forest management than the traditional horticulture background which is based on food and agriculture.

The urban forest is the ecosystem containing all of the trees, plants and associated animals in the urban environment, both in and around the city. The management of the urban forest is a complex of biological, social, political and environmental factors where management is usually carried out by a range of different authorities, community organizations and private landowners. Urban forestry advocates coordinated management across the city estate with sustainable objectives well into the future. Clearly this is more difficult to achieve than managing a forest in the countryside with a single owner. Also, management of trees in the city is surrounded by many more rules and regulations than forestry practised in the rural environment. Managers of the urban forests need to be aware of interactions with buildings, utilities, pavements, drainage and sewers and of tree hazards to person and property. The great variety of trees and plants in the urban forests bring a different and more complex perspective to integrated pest management. The use of agrichemicals in the urban environment may be more restrictive than in the rural environment. Having said this, however, the most wasteful and irresponsible user of agrichemicals is probably the urban home gardener.

The urban environment can be very hostile to trees, particularly in some of the city canyons created by high buildings. Temperatures can be extreme, heavy traffic can cause considerable air pollution, trees may be shaded for most or all of the day, root space may be restricted by buildings and pavements, soils may be compacted, de-icing salts may be used in cold environments, trees may be damaged (accidentally and intentionally), city lighting may be continuous throughout the night and dogs can modify the nutrient balance. There are many hazards and the failure rate of new plantings in streets can be very high. Not all tree species are suited to streetscapes and parks. Successful species have not only to withstand the hazards of the urban environment but they need (among other things) to be easy to raise and grow, require little maintenance, not cause damage to building foundations, not have roots that invade sewers and drains, not have suckers that break through pavements or invade private gardens, not have a propensity to weediness, and be attractive and of size and growth rate appropriate for the purpose. Practical considerations about establishing trees in the urban environment are given by Nilsson *et al.* (2001b). For this reason exotic species are sometimes preferred although exotics also have often been preferred by colonizing immigrants to remind them of their homeland. Arguments about the appropriateness or otherwise of exotics in metropolitan streets and parks can be intense.

There are many stakeholders in urban forestry, not necessarily with compatible interests. To this extent urban forestry is like community forestry and to many is just another form of community forestry that just happens to occur in the urban environment. Urban forestry may be represented within municipal government by a city forestry department that works within an overall planning authority. Alternatively, there may be one city forester working alongside others in a parks and gardens group. There are many possible administrative structures. Most likely, urban foresters will be part of a municipal team where they need to sell the ideals of sustainable ecosystem management to a broader group who may well see trees in the city from a different perspective. Urban forestry is multi-disciplinary.

The standard of tree care and maintenance in cities is very variable. Sometimes it is very obvious that street trees have been pruned under the guidance of city engineers who are more interested in the protection of structures and have no understanding of the growth habits of the trees. There certainly is a pressing need for well-trained professional tree managers in the city environment, including professional urban foresters. Urban forestry, among other things, seeks to manage the trees in a city as a renewable resource that can produce a range of benefits, which may include timber production. Clearly, timber production is not an overriding concern of the urban forester. Management for amenity is the major objective. Perhaps the greatest contribution that urban foresters can bring to a city is their skills in planning and in ecosystem management.

The concept of urban forestry commenced at the University of Toronto in 1965 (Koch, 2000) and is probably most advanced in the USA (see Kuser, 2000). However, urban forestry is now a world wide phenomenon. The beneficiaries of urban forestry are the local urban community, and various sections of the community may be keen to be involved in the management of public forested land in the city. The challenge is to mobilize and focus community participation and where appropriate to devolve management and power to community groups.

Conclusion

This chapter, and indeed the whole book, has highlighted the human dimensions of forest management, mis-management or lack of management. Much has happened in the last few decades. There has been a trend from institutionalized authoritarian forestry to collaborative community-based forestry. Forest management for environmental services is now recognized, in principle at least, as of paramount importance. It is generally acknowledged that, in developing countries, sustainable forest management should go hand in hand with sustainable development. Forest management remains a controversial issue capable of arousing great passion and considerable conflict. The role of the modern forester has become that of a facilitator and a conflict manager. The forester will always base her or his advice on the best evidence-based research and the best predictive models. However, this will be only one component of a solution that considers social, political, economic and ecological consequences.

References and Further Reading

Arnold, J.E.M. (2001) *25 years of Community Forestry*. Food and Agriculture Organization of the United Nations, Rome.

Ashton, M.S. and Montagnini, F. (eds) (1999) *The Silvicultural Basis for Agroforestry Systems*. CRC Press, Boca Raton.

Bass, S. (2003) The importance of social values. In: Evans, J. (ed) *The Forests Handbook, Volume 1*. Blackwell Science, Oxford, pp. 362-371.

Brown, D. (1999) *Principles and Practice of Forest Co-management: Evidence from West-Central Africa*. European Union Tropical Forestry Paper 2, ODI, London.

Bryan, J.A. (1999) Nitrogen fixation of leguminous trees in traditional and modern agroforestry systems. In: Ashton, M.S. and Montagnini, F. (eds) *The Silvicultural Basis for Agroforestry Systems*. CRC Press, Boca Raton, pp. 161-182.

Buck, L.E., Lassoie, J P. and Fernandes, E.C.M. (eds) (1998*) Agroforestry in Sustainable Agricultural Systems*. Lewis Publishers, Boca Raton.

Buck, L.E., Geisler, G.C., Schelhas, J. and Wollenberg, E. (eds) (2001*) Biological Diversity: Balancing Interests Through Adaptive Collaborative Management*. CRC Press: Boca Raton, Florida,

Edmunds, D. and Wollenberg, E. (eds) (2003) *Local Forest Management; The Impacts of Devolution Policies*. Earthscan, London.

Evison, H. (1998) Waitutu's century of deceit and poverty. *Indigena* March, pp 4-5.

FAO (1978) *Forestry for local community development*. Forestry Paper 7, Rome.

Federated Farmers of New Zealand Inc. (1999) *Submission to the Transport and Environment Select Committee*, Parliament of New Zealand, Clerk of the Select Committee, Bowen House, Parliament Buildings, Wellington, 2 pp.

Fisher, R.J. (1999) Devolution and decentralization of forest management in Asia and the Pacific. *Unasylva* 50(1999), 3-5.

Fisher, R.J. (2001) Experiences, challenges, and prospects for collaborative management of protected areas: an international perspective. In: Buck, L.E., Geisler, G.C., Schelhas, J. and Wollenberg, E. (eds) *Biological Diversity: Balancing Interests Through Adaptive Collaborative Management.* CRC Press: Boca Raton, Florida, pp 81-96.

Fisher, R.J. (2003) Innovations, persistence and change: reflections on the state of community forestry. In: RECOFTC and FAO Community Forestry: *Current Innovations and Experiences.* Regional Community Forestry Training Center (RECOFTC) and Food and Agricultural Organization of the United Nations. Bangkok, Thailand.

Gajaseni, A. (1992) Socioeconomic aspects of taungya. In: Jordan, C.F., Gajaseni, J. and Watanabe, H. (eds) *Taungya – Forest Plantations with Agriculture in Southeast Asia.* CAB International, Wallingford, pp. 18-31.

Garrett, H.E., Jones, J.E., Kurtz, W.B and Slusher, J.P. (1991) Black walnut (*Juglans nigra* L.) agroforestry – its design and potential as a land-use alternative. *The Forestry Chronicle* 67(3), 213-218.

Gordon, A.M. and Newman, S.M. (eds) (1997) *Temperate Agroforestry Systems.* CAB International, Wallingford.

Grey, G.W. (1996) *The Urban Forest – Comprehensive Management.* Wiley, New York.

Griffiths, A. (1998) Sustainable management of private native forests in New Zealand: what's in it for the landowner? *New Zealand Forestry* 42(4): 13-16.

Hardin, G. (1968) The tragedy of the commons. *Science* 162, 1243-1248.

Heath, H. 1992. Forests Amendment Bill- (un)sustainable management? *New Zealand Forestry* 37(3): 4-6.

Hislop, M. and Claridge, J. (2000) *Agroforestry in the UK.* Forestry Commission Bulletin 122, Forestry Commission, Edinburgh.

Hodge, S.J. (1995) *Creating and Managing Woodlands around Towns.* Forestry Commission Handbook 11, HMSO, London.

Huxley, P. (1999) *Tropical Agroforestry.* Blackwell Science, Oxford.

Hyde, W.F. and Köhlin G. (2000) Social forestry reconsidered. *Silva Fennica* 34(3), 285-314.

Jarvis, P.G. (ed.) (1991) *Agroforestry: Principles and Practice*, Elsevier, Amsterdam.

Jordan, C.F., Gajaseni, J. and Watanabe, H. (eds) (1992) *Taungya – Forest Plantations with Agriculture in Southeast Asia.* CAB International, Wallingford.

Kelty, M.J. (1999) Species interactions, stand structure, and productivity in agroforestry systems. In: Ashton, M.S. and Montagnini, F. (eds) *The Silvicultural Basis for Agroforestry Systems.* CRC Press, Boca Raton, pp 183-205.

Koch, J. (2000) The origins of urban forestry. In: Kuser, J.E. (ed.) *Handbook of Urban and Community Forestry in the Northeast.* Kluwer Academic Publishers, New York, pp. 1-10.

Konijnendijk, C.C. (2004) Urban forestry. In: Burley, J., Evans, J. and Younquist, J.A. (eds) *Encyclopedia of Forest Sciences.* Elsevier Academic Press, Amsterdam, pp. 471-478.

Kuser, J.E. (ed.) (2000) *Handbook of Urban and Community Forestry in the Northeast.* Kluwer Academic Publishers, New York.

Lawrence, A. and Gillett, S. (2004) Joint and collaborative forest management. In: Burley, J., Evans, J. and Younquist, J.A. (eds) *Encyclopedia of Forest Sciences.* Elsevier Academic Press, Amsterdam, pp. 1143-1157.

Nair, P.K.R. (1993) *Agroforestry Systems in the Tropics.* Kluwer Academic Publishers, Dordrecht.

Nilsson, K., Randrup, T.B. and Wandall, B.M. (2001a) Trees in the urban environment. . In: Evans, J. (ed.) *The Forests Handbook, Volume 1.* Blackwell Science, Oxford, pp.347-361.

Nilsson, K., Randrup, T.B. and Wandall, B.L.M. (2001b) Trees in the urban environment: some practical considerations. In: Evans, J. (ed.) *The Forests Handbook, Volume 2.* Blackwell Science, Oxford, pp.260-287.

Ong, C.K. and Huxley, P. (eds) (1996) *Tree-Crop Interactions.* CAB International, Wallingford.

Ostrom, E. (1990) *Governing the Commons: the Evolution of Institutions for Collective Action.* Cambridge University Press, Cambridge.

Palo, M and Uusivuori, J. (eds) (1999) *World Forests, Society and Environment.* Kluwer Academic Publishers. Dordrecht.

Poffenberger, M. (1996) *Communities and Forest Management.* A report by the IUCN Working Group on Community Involvement in Forest Management with Recommendations to the Intergovernmental Panel on Forests, IUCN, Washington, DC.

Poffenberger, M. and McGean, B. (1996) *Village Voices, Forest Choices – Joint Forest Management in India.* Oxford University Press, Delhi.

Saxena, N.C and Ballabh, V. (eds) (1995) *Farm Forestry in South Asia.* Sage Publications, New Delhi.

Schroff, G. (1999) A review of belowground interactions in agroforestry, focussing on mechanisms and management options. *Agroforestry Systems* 43, 5-34.

Schroff, G., da Fonseca, G.A.B., Harvey, C.A., Gascon, C., Vasconcelos, H.L. and Izac, A-M.N. (2004) *Agroforestry and Biodiversity Conservation in Tropical Landscapes.* Island Press, Washington.

Swinkels, R.A., Shepherd, K.D., Franzel, S., Ndufa, J.K., Ohlsson, E. and Sjogren, H. (2002) Assessing the adoption potential of hedgerow intercropping for improving soil fertility in western Kenya. In: Franzel, S. and Scherr, S.J. (eds) *Trees on the Farm – Assessing the Adoption Potential of Agroforestry Practices in Africa.* CAB International, Wallingford, pp. 89-110.

Szott, L.T., Fernandes, E.C.M. and Sanchez, P.A. (1991) Soil-plant interactions in agroforestry systems. In: Jarvis, P.G. (ed.) *Agroforestry: Principles and Practice.* Elsevier, Amsterdam, pp. 127-152.

TPK (Te Puni Kokiri, Ministry of Maori Development) (1996) *Maori Land Information Database,* Te Puni Kokiri, Wellington, New Zealand.

Waitangi Tribunal (1991) *Ngai Tahu Report Volumes One, Two, and Three.* Brooker and Friends, Ltd., Wellington, New Zealand.

Westoby, J. (1989) *Introduction to World Forestry: People and their Trees.* Blackwell, Oxford and New York.

Wiersum, K.F. (1999) *Social Forestry: Changing Perspectives in Forestry Science or Practice.* Thesis from Wageningen Agricultural University, the Netherlands.

Wiersum, K.F. (2004) Social and community forestry. In: Burley, J., Evans, J. and Younquist, J.A. (eds) *Encyclopedia of Forest Sciences.* Elsevier Academic Press, Amsterdam, pp. 1136-1143.

Williams, P.A., Gordon, A.M., Garrett, H.E. and Buck, L. (1997) Agroforestry in North America and its role in farming systems. In: Gordon, A.M. and Newman, S.M. (eds) *Temperate Agroforestry Systems.* CAB International, Wallingford, pp. 9-84.

Wojtkowski, P.A. (1998*) The Theory and Practice of Agroforestry Design.* Science Publishers Inc., Enfield.

Wojtkowski, P.A. (2002) *Agroecological Perspectives in Agronomy, Forestry and Agroforestry.* Science Publishers, Inc., Enfield.

Wu, Y and Zhu, Z. (1997) Temperate agroforestry in China. In: Gordon, A.M. and Newman, S.M. (eds) *Temperate Agroforestry Systems.* CAB International, Wallingford, pp. 149-179.

Young, A. (1989) *Agroforestry for Soil Conservation.* CAB International, Wallingford.

Young, A. (1997) *Agroforestry for Soil Management.* CAB International, Wallingford.

General Index

Index of Places (Historical, Geographical and Political)

Index of Trees and Other Plants

257